# British Naval Gun Mountings

# HYDRAULIC MOUNTING FOR 2-13·5 INCH 45 CAL: GUNS.

SECTIONAL ELEVATION.

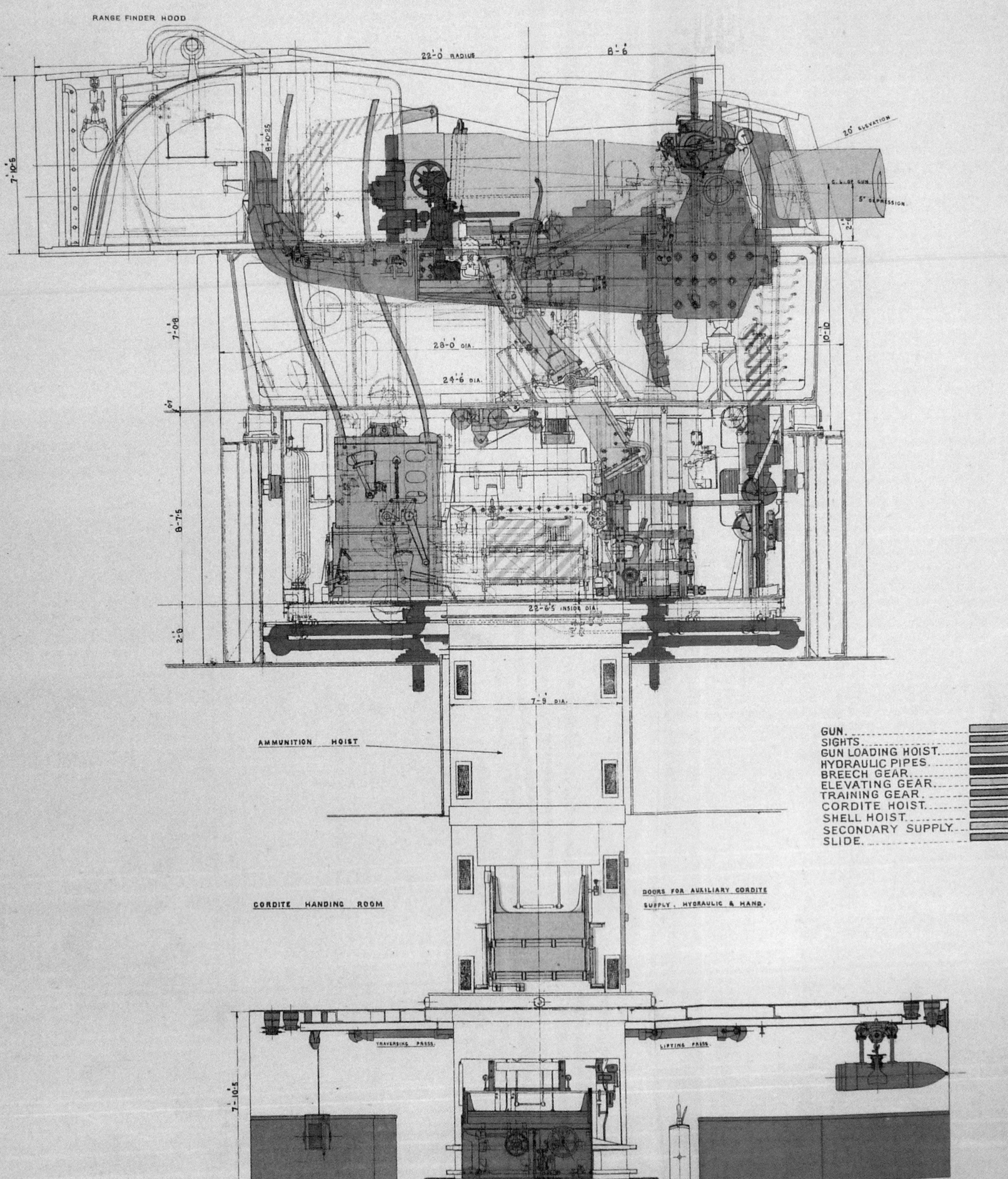

# British Naval Gun Mountings

## From 1890: 18-inch to 4.5-inch Mark 8

IAN BUXTON

Seaforth
PUBLISHING

**Frontispiece:** The profile of Coventry Ordnance Works first heavy gun mounting the 13.5-inch Mark III twin is usefully colour-coded (see jacket) and dimensioned. The roller path diameter was 24ft 6in, ammunition hoist diameter 7ft 6in and height from working chamber floor to gunhouse roof 24ft 6in. With two 76-ton guns, the rotating weight was about 590 tons. *(ADM 186/137)*



First published in Great Britain in 2024 by
Seaforth Publishing
An imprint of Pen & Sword Books Ltd
George House, Beevor Street,
Barnsley S71 1HN

www.seaforthpublishing.com
Email info@seaforthpublishing.com

*British Library Cataloguing in Publication Data*
A CIP data record for this book is available from the British Library

ISBN 978-1-3990-5996-1 (Hardback)

ISBN 978-1-3990-3099-1 (ePub)

ISBN 978-1-3990-3101-1 (PDF)



Pen & Sword Books Limited incorporates the imprints of Atlas, Archaeology, Aviation, Discovery, Family History, Fiction, History, Maritime, Military, Military Classics, Politics, Select, Transport, True Crime, Air World, Frontline Publishing, Leo Cooper, Remember When, Seaforth Publishing, The Praetorian Press, Wharncliffe Local History, Wharncliffe Transport, Wharncliffe True Crime and White Owl.

Typeset and designed by Stephen Dent
Printed and bound in China by 1010 Printing International Ltd

MIX
Paper | Supporting responsible forestry
FSC® C016973

# Contents

# Abbreviations

| | |
|---|---|
| AC | Alternating current |
| APC | Armour piercing, capped (projectile) |
| AW | Armstrong, Whitworth |
| B | Barbette mounting, *eg* B.VII = Mark VII |
| BAE | BAE Systems (Barrow archives) |
| BD | Between Decks mounting recessed into deck *eg* QE modernised twin 4.5in |
| BL | Breech Loading gun (separate projectile and bagged propellant) |
| BM | Breech mechanism |
| BR | Book of Reference (official publication) |
| C | Cemented (armour) |
| cal | Calibre: length of gun/bore |
| CB | Confidential Book (official publication) |
| CD | Coast defence |
| CG | Centre of gravity |
| CL | Cammell Laird |
| COW | Coventry Ordnance Works |
| CP | Central Pivot mounting (bolted to deck) |
| CPC | Common, pointed cap (shell) |
| crh | Calibres radius of head |
| CUL | Cambridge University Library |
| DC | Direct current |
| DNC | Director of Naval Construction |
| DNO | Director of Naval Ordnance |
| DOD | Director of Dockyards |
| efc | Equivalent full charge |
| EOC | Elswick Ordnance Company |
| ESC | English Steel Corporation |
| grt | gross registered tons (measure of merchant ship size) |
| HA | High angle mounting |
| HACS | High angle control system |
| HA/LA | High Angle/Low Angle mounting, *ie* dual purpose |
| HE | High explosive or High effect |
| H&W | Harland & Wolff |
| hp | horsepower |
| HV | High velocity |
| INA | Institution of Naval Architects |
| IWM | Imperial War Museum |
| KC | Krupp cemented (armour) |
| KNC | Krupp non cemented (armour) |
| kW | kilowatt |
| LA | Low angle mounting |
| LCG | Landing Craft, Gun |
| lb | pounds (weight) |
| MD | Modified cordite (propellant) |
| Mk | Mark |
| MTSC | Marine Technology Special Collection, Newcastle University |
| MV | Muzzle velocity |
| NC | Non cemented (armour) |
| NCA | Naval Construction & Armaments |
| NMM | National Maritime Museum |
| NOD | Naval Ordnance Department |
| NRS | National Records of Scotland |
| P | Pedestal mounting (bolted to deck) |
| pdr | pounder (gun) |
| QF | Quick Firing gun (fixed or separate projectile and cartridge case) |
| rpg | rounds per gun (per minute) |
| RGF | Royal Gun Factory, Woolwich |
| RN | Royal Navy |
| RP | Remote Power (Control) |
| SAP | Semi armour piercing (projectile); SAPC = capped |
| SC | Solventless carbamite (propellant) |
| TNA | The National Archives |
| TS | Transmitting Station |
| UCS | Upper Clyde Shipbuilders |
| UD | Upper Deck mounting, *eg* depot ship twin 4.5in |
| V-A | Vickers-Armstrongs |
| VSG | Variable Speed Gear (a hydraulic motor) |
| VT | Variable time (proximity fuze) |
| WW1 | First World War |
| WW2 | Second World War |

# CHAPTER ONE

# Introduction

HEAVY GUN MOUNTINGS dominated warship design for over half a century. They were among the heaviest, most complex and expensive engineering artefacts of their day, with very few companies capable of building them. The main armament of a battleship made up about a quarter of its cost and took as long to build as the ship itself, so was on the critical path of the building process. This is the story of how these remarkable products were designed, built and worked. The Admiralty called them 'non-transferable', as they were 'made to order' for a specific ship and built into its hull structure rather than being simply bolted to the deck. Beneath the visible turret with the actual guns were the complex operating machinery and ammunition hoists. Below deck, the equipment and the magazines were protected by armour, while the gunhouse itself was heavily armoured to protect the crew and prevent a hit setting off an ammunition explosion.

Over a century and a half, the Royal Navy had installed a large number of designs of smaller calibre weapons, mostly hand operated. Even today in an era of guided missiles, most significant ships retain one medium calibre gun, albeit mainly for shore bombardment rather than anti-ship or anti-aircraft roles. But the most interesting period followed the Naval Defence Act of 1889, whose large new construction programme spurred the development of naval ordnance and encouraged more engineering companies to produce the ships and their armament. Until the end of the Second World War, the focus was on battleships, where main armament calibres ranged from 10in up to 18in. After the First World War, all cruisers and some destroyers had larger power operated mountings, from 8in to 4.5in calibre. Only brief mention is made of the smaller 'transferable' mountings which ranged from 9.2in down to 3in or 12pdr, usually hand-worked with a separate ammunition supply, while close range anti-aircraft weapons are excluded. The term 'mounting' is generally used to describe the complete installation from ammunition supply right up to the guns themselves. The word 'turret' tends to be used more loosely although should really only apply to the visible rotating portion above the deck.

While some aspects of the design, manufacture and operation have been included in books on warship design including battleships, this is the first to draw together all aspects of these weapon systems, with a wide range of drawings and photographs, many never previously published. There is a modest degree of overlap with the author's previous books *The Battleship Builders*, *Big Gun Monitors* and *Battleship Duke of York* but also much new material, especially on production, costs and the specialist companies. There were only three such British companies: Armstrong, Whitworth at Elswick and Scotswood near Newcastle (although its Openshaw works near Manchester did make a few), Vickers at Barrow in Furness and the short-lived Coventry Ordnance Works, whose Scotstoun works near Glasgow was later taken over by Harland & Wolff. The Royal Gun Factory at Woolwich made a large number of transferable gun mountings and 'carriages' for the Army but never produced a satisfactory heavy gun mounting, although it did make guns for them.

Company financial records are now often the only source of information on contracts, many not previously accessed, which have produced much new information on their production and profitability. A Dreadnought typically had five twin 12in mountings, priced at about £75,000 each including guns, weighed about 500 tons and took two years to build, or about £150 per ton, albeit usually including a healthy profit margin. For the price of one mounting, two cargo ships of about 7000 tons carrying capacity could be built in nine months, at a cost of about £15 per ton empty weight. Twenty mainline railway steam locomotives could be built for the same price at a cost of about £40 per ton. To give an idea of 2020s values, multiply by about 120 for pre-WW1 and about 60 for WW2 era.

Some reports of the operation of the guns and mountings in action are included to give a general flavour rather than a comprehensive account of each. There never seems to have been an overall Admiralty review of the effectiveness of different gun mountings, so assessment has to be based on some action reports, passing references and anecdotal accounts.

Much of the data is presented in tables, but sources often conflict on the numbers, so I have selected what seems most appropriate and most official for the service and period. For example, several different figures exist for muzzle velocity (and corresponding range) but can often be explained as applying to a new or part worn gun, or different weight and temperature of the cordite charge, or different projectiles, so are indicative rather than definitive.

In order to include more new material, I have limited the amount of information and drawings of gun mountings already described elsewhere, for example in *Big Gun Monitors*. Adding a rare selection of photographs of mountings under demolition has added a new perspective, revealing parts of the interior not seen before. A few such mountings can still be seen in preserved ships or mounted ashore, notably the 6in cruiser *Belfast* in London, and 4.5in destroyer *Vampire* in Sydney while for the largest calibres there are eight battleships in the US.

For mountings of the era prior to this book, see the 1893 *Artillery: Its Progress and Present Position* by E W Lloyd and A G Hadcock, a detailed account by two industry insiders.

CHAPTER TWO

# The Heavy Gun Mounting

THE FUNCTION OF A HEAVY GUN MOUNTING was to allow the guns to be loaded, elevated, trained and fired in accordance with a fire control system to hit an enemy target.

Battleship gun mountings were even more impressive pieces of heavy engineering than their guns. By about 1910, they had evolved into efficient and reliable mechanisms. A twin 15in Mark I mounting had a revolving weight of about 750–800 tons depending on gunhouse protection thickness (of which only 200 tons was the guns themselves) and took 18–27 months to build and cost in 1914 about £114,000 excluding guns. An armoured gunhouse protected the guns, loading and sighting mechanisms. Beneath that 'turret' was the turntable, the top of which formed the gunhouse floor supporting the trunnions holding the gun slides and cradles. It revolved on a machined track on top of a fixed ring bulkhead built into the hull. Below that was the working chamber with machinery to train and elevate the guns and transfer shells and charges to the loading cages taking them up into the gunhouse. Ammunition hoists inside a trunk rotated with the mounting, reaching down to the magazines and shell rooms at the bottom of the ship. The whole mounting could have a total height of up to 60ft in superfiring mountings. The profile and section show a typical mounting and its components. *(See colour plates on page 145.)*

**Right:** Two stages of gunhouse erection are illustrated by *Barham* at the John Brown shipyard at Clydebank on 12 April 1915. Mounting Y has the armoured sides fitted with two gun cradles ready to receive the 15in guns to be followed by the roof plates with their square centre support. The trapezoidal shape at the rear is the local control position. X mounting has its 3in thick gunhouse floor fitted, whose edge has been slotted and drilled to secure the armour plates while the trunnion supports await the two gun slides and cradles. The hole in the rear is an access manhole into the turret. *(NRS)*

## THE WORKINGS

By the start of WW1, the features and configuration of heavy gun mountings had stabilised. They had to be strong enough to withstand the stresses of firing and transmit them to the supporting structure. Ammunition had to be supplied quickly and safely to the guns, which had to be aimed to hit the predicted position of the target. And it had to be as light as possible and compact but well protected, with parts readily accessible for maintenance. To this end, the main components of a battleship mounting usually comprised:

*Gunhouse*: the rotating turret containing the guns, loading mechanisms, sights and sometimes a rangefinder. It provided protection to the guns and crew from weather, blast from nearby guns and from enemy shells or bombs, so were heavily armoured.

*Turntable*: a substantial structure which supported the gunhouse above and ammunition supply mechanisms beneath, the elevating and training mechanisms, all rotating on a roller path.

*Working chamber*: directly below the turntable, it transferred ammunition from the main hoist from the magazines to the gun loading cage which lifted shell and cordite charges up rails into the gunhouse. It could also contain ready-use shells and auxiliary equipment.

*Ammunition trunk*: an enclosed vertical shaft reaching down to magazine level, which rotated with the turret above, containing cages which hoisted one shell and its cordite charges to serve each gun.

*Magazines and shell rooms*: compartments low down in the ship where shells and cordite charges were stowed when not in action. They contained mechanisms for bringing them from stowage areas to the bottom of the main hoist through flashtight doors.

*Power supply system*: with the great weight of shells, power assistance was essential for operating the many systems and to achieve an acceptable rate of fire. Usually hydraulically powered, with high pressure fluid distributed via pumps and pipes. Hand operation could be used in emergency but was much slower.

General descriptions were given in the following confidential books which illustrated the key features and working principles:

*Power Worked Mountings* 1921.[1]
*Ordnance Engineering Notes* 1956.[2]

For each major mounting there was a handbook with a detailed description and drawings, many of which have been drawn on for this book.

[1] ADM 186/255 & 256
[2] The National Archives. ADM 234/782

The ammunition trunk reached from magazine levels up to the working chamber. On 28 January 1943 the monitor *Abercrombie*'s 15in trunk was moved out of Elswick's Nº 24 Shop for shipping at Vickers-Armstrongs' Naval Yard downstream. The pair of doors for two half cordite charges are seen above the (closed) shell door. A shell loading bogie (not seen) traversed around the toothed rack to bring the projectile from the shell room into line with the revolving hoist as it trained. The turntable and working chamber behind will be the next components erected. *(Vickers)*

Most of the components for *Colossus'* twin 12in A mounting lie on the dockside at Greenock on 18 August 1910. The 25.5-ton gun slide for the right gun holds the recoil cylinders and contains the chain rammer in the loading arm. Behind is the turntable and working chamber marked TURRET A 112G (Vickers contract number). To the right, Scotts' 100-ton derrick crane is about to lift the cylindrical ammunition trunk. *(NRS)*

Looking inside *Barham's* B barbette on 7 June 1915 shows the 7 to 10in thick armour plates surrounding the ring bulkhead which supports all the rotating parts of the mounting. The training rack is fixed inside the bulkhead to whose upper surface is attached the machined lower roller path, on which 48 rollers are held in place by a live ring. In A mounting the turntable has already been installed with trunnion supports now awaiting the gunhouse floor. There seems to be a lot of men watching and waiting, some perched on precarious wooden planks. *(NRS)*

Each gun was held in a cast steel *cradle* which was supported by and recoiled along the machined surface of a forged steel *slide* which could be elevated. The slide was supported by trunnions and brackets attached to the gunhouse floor, to which was attached the elevating mechanism. Slides were interchangeable for a particular Mark of mounting and were a main component of spare mountings. A *loading arm* extended from the rear of the slide with trays on to which the *gun loading cage* delivered first the shell then the charges (usually two quarter charges at a time) to be rammed into the gun *breech* by a retractable hydraulically operated chain rammer. *(See colour plates on pages 146 and 147)*

The usual *elevating mechanism* was a hydraulic ram hinged at the turntable end and at the slide. It was fitted with stops to prevent elevation or depression beyond the designed limits. By arranging the centre of gravity (CG) of all the elevating parts to be in line with the trunnions (if necessary with a balance weight over the breech), less force was needed. Medium calibre guns usually had a toothed rack attached to the slide driven by a pinion – see 8in (page 164). The main supporting structure of the turntable was built up of steel girders and was deep enough for the gun well to accommodate the recoil of the breech at maximum elevation. Fixed to its underside was the upper *roller path*, machined at a slight angle to take the tapered rollers which allowed the whole mounting to rotate. The lower roller path was fixed to the *ring bulkhead* which was part of the ship's structure, taking the whole weight of the mounting. The rollers, typically 48, were flanged at each end which prevented the turntable from being displaced either by recoil forces or by the ship rolling. They were held in place by a live ring, which was fitted in segments, allowing replacement. *(See colour plate on page 150)*

The *training mechanism* consisted of two hydraulic drives to overcome not only the friction of the revolving mounting but also the inertia to move it from rest, at angles of heel up to 10°. By arranging the CG close to the centre of rotation, less power was required. Only one was clutched in at a time, the other being a back-up. Although hand training was an alternative in earlier or in smaller mountings, it was hardly practical for a mass that could be around a thousand tons. The training engine in earlier mountings was usually a six-cylinder reciprocating machine but from the 13.5in mountings onwards this was replaced by an ingenious *swashplate* motor. Usually fitted on the turntable, the motor drove a vertical shaft with a pinion at its bottom end which engaged with a toothed rack fitted inside the ring bulkhead. Buffers and stops were fitted to prevent the mounting training beyond safe limits. A swashplate machine took a steady input such as from an electric motor and by adjusting internal wedges and a number of cylinders could regulate the speed and direction of its output, either fluid flow (pump) or mechanical drive (motor). In its VSG (Variable Speed Gear) form it was widely fitted in mountings large and small. *(See colour plate on page 147)*

The *working chamber* was directly under the turntable with the two lifted as a complete unit during erection at the ship, topped by the gunhouse floor, typically 3in thick in heavy mountings. The main hoist delivered into a flashtight enclosure, with an overhead radial crane taking shells to waiting trays from which telescopic rammers could push them (and charges) into the *gun loading cage*, already lowered from the gunhouse. The cage had three levels: one for shell and two for charges. An ordnance artificer was usually stationed in the working chamber, in addition to the operating crew. *(See colour plate on page 150)*

The *main hoist* cages (one for each gun) were in two parts, with the charges (each of a weight that could usually be manhandled) in the upper, the shell at the lower, all usually horizontal. Wires from hydraulic ram presses with sheaves at each end were fitted in the working chamber to lift the two cages as one. At the bottom of the hoist were *shell bogies* (one per gun) which could first be rotated round the

The turntable and working chamber of *Barham*'s Y mounting made by Vickers (145G) being lifted aboard on 9 March 1915. It is being lifted by the 150-ton cantilever crane, probably near its limit. The older 150-ton derrick crane is seen across the basin at Clydebank. The underside of the turntable has been machined to form the upper roller path. The 6in casemate is empty and will be plated over before completion, as it was decided that the gun would be too close to the waterline to be fired in a seaway. *(NRS)*

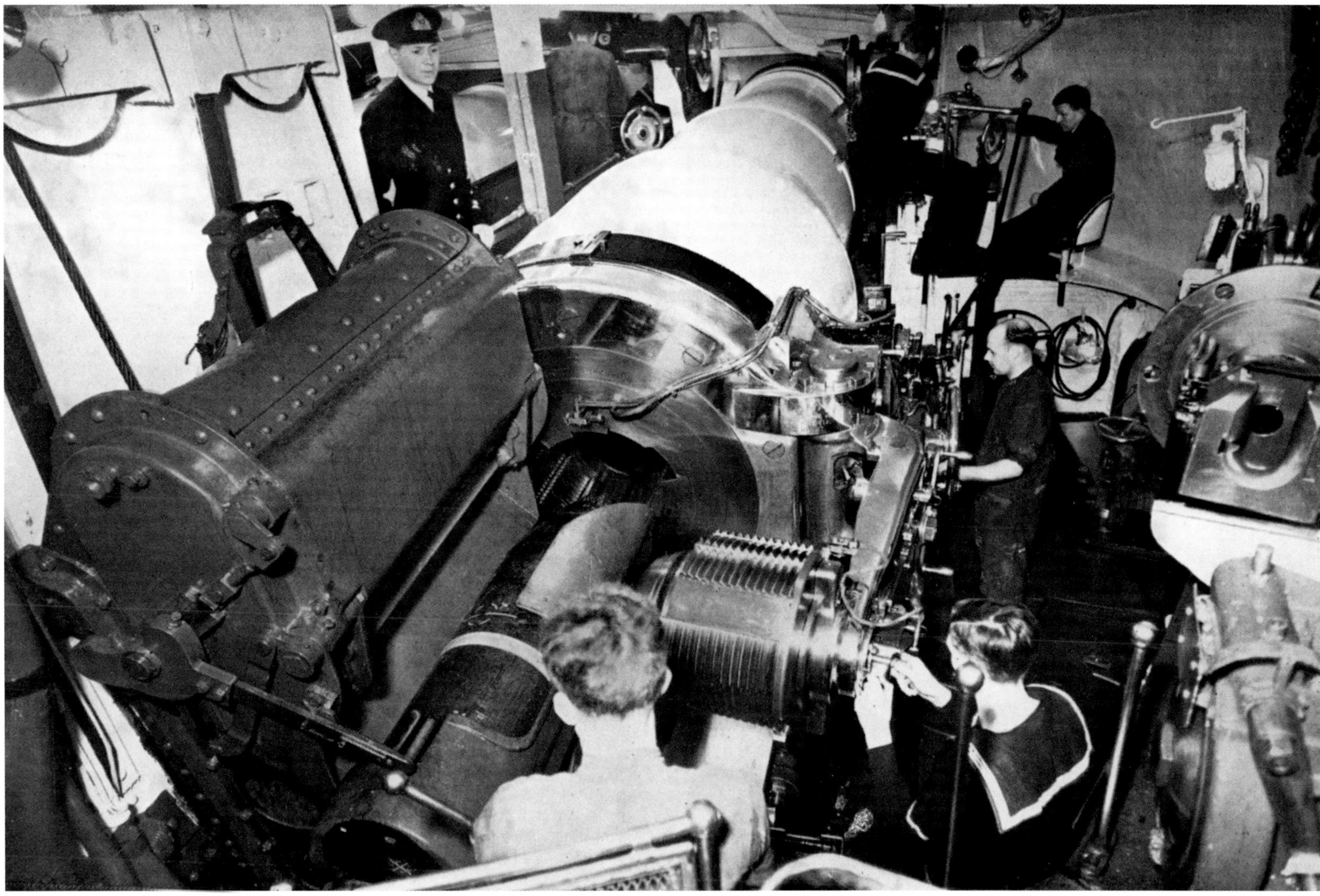

**Above:** A view inside one of *Iron Duke*'s 13.5in gunhouses while a gunnery training ship. The 1400lb shell is about to be rammed into the right breech, to be followed by the four quarter charges still in the upper part of the gun loading cage when it is lowered. A 6pdr sub-calibre gun is seen at right. *(Author's collection)*

**Right:** A 14in Mark VII gun, cradle and slide awaits trial installation in a mounting on one of the pits at Elswick. This whole 125-ton unit will elevate about the trunnion pins at the forward end of the cradle. One of the two recoil cylinders is underneath with run-out cylinder on top. The rectangular block is a 11.5-ton balance weight which surrounds the breech whose slim operating cylinder is seen below the breech open/close handwheel. *(Tyne & Wear Archives)*

hoist to take a shell from the fixed structure, then clutched to the rotating trunk and traversed round a rack by hand drive to line up with the flashtight opening to the shell cage. For flashtightness the trunk had no other holes in it. It usually also contained an access ladder and shafts through which shells could be lifted vertically in emergency.

The *recoil* of the gun after firing was absorbed by one or two recoil cylinders fitted below the breech, their forward end fixed to the slide, their rear end to the gun breech ring. Inside the cylinder hydraulic fluid was forced past a ram with small clearances. The *run-out* mechanism then forced the gun back to its loading position with a hydraulic ram. But during WW1 it was found that when directors fired half the guns together in salvoes or all in broadsides, it put such a load on the hydraulic system in addition to all the other loads that the whole cycle was slowed down. Furthermore, using all-angle loading at high elevations placed a further load by pushing the shells 'uphill'. Therefore, mountings were gradually converted to pneumatic recoil, or recuperation as it was sometimes called, which also absorbed some of the recoil. That required a supply of high-pressure air from flasks (usually four per gun) stored in the trunk at 4000lb/sq in, reduced to 1250 for working. That air was also required for *air blast* which was used to clear a fired gun of noxious fumes which would otherwise flow into the gunhouse.

Water was also piped to the breech to *wash out* any smouldering residues from the cordite. It was also used to cool the *obturator* which sealed the breech block of breech loading guns. (In quick firing guns the brass cartridge case formed the seal.) Water was also required to drench any cordite fires with spray or cool charges about to burn, supplied if on the mounting from the hydraulic main or if in the magazine from the ship's fire main. Cordite would burn in open air producing only noxious gases, but in a confined space could build up pressure and cause an explosion. That pressure could be relieved by blow-off vent holes in the gunhouse rear plate or by loosely secured hatches used to embark ammunition to the magazines, allowing the gases to escape. If the fire proved a serious risk, the whole magazine could be flooded with sea water taken through valves in the ship's bottom. Gunhouses also had regular ventilation fans and ducts supplied through holes in the rear gunhouse floor to keep temperatures comfortable; heaters were supplied for cold weather.

*Flashtightness* became even more important after several losses from magazine explosions after Jutland. It was essential that there was no direct path from a burning charge or flames to ignite others in the ammunition chain from the gunhouse down to the magazines. Zones were arranged with flashtight doors so that each only contained one charge at a time. As well as flashtight enclosures at the top and bottom of the main hoist, a further set of flashtight scuttles was fitted between magazine and cordite handing room which fed the main hoist. *(See colour plate on page 150)*

Numerous *interlocks* were fitted to ensure that only one flashtight opening was open at a time. They were also extensively fitted to prevent any potentially dangerous operation, such as only closing the breech after the gun loading cage had been lowered. Other examples include: the main cage could not be raised until both shell and cordite loading doors were closed; and transfer of ammunition operations could not be worked out of sequence. *(See colour plate on page 147)*

**Above:** A view inside one of *Malaya's* 15in shell rooms, showing a 1920lb projectile being lifted from a stowage bin by a grab which will move it on the overhead rail to the main hoist. *(Author's collection)*

*Shell room machinery* had to be substantial to handle shells which could weigh up to a ton. Shells were stored in bins, usually stowed fore and aft three to eight high, separated by rope grommets which protected the copper driving bands. Overhead rails with travelling grabs plumbed each bin and transferred the shells to the hoist area. Shells were embarked by lowering vertically through a series of hatches in each deck, as illustrated in *Battleship Duke of York*, pp46-7.

*Hydraulic power* was supplied by steam driven pumps, usually one per turret. Reciprocating engines drove the pumps until steam turbines came in for the *King George V* class battleships. The fluid was usually distilled water with one part in 60 of vegetable oil and soft soap added for lubrication. Pressure pipes at typically 1650lb/sq in were usually duplicated port and starboard, with exhaust pipes returning the fluid to make-up feed tanks near each pump. The system was the engineering department's responsibility until the pipes reached each mounting. *Walking pipes* carried hydraulic fluid from the fixed to rotating structure, both supply and exhaust. Its joints like many other hydraulic systems under pressure often used leather seals, cheap and effective if well maintained. *(See colour plates on pages 146, 150)*

Other components included:

*Safety depression gear* to prevent guns pointing at a dangerous angle, for example at angles of depression when pointing directly ahead. *Locking bolts* to secure the whole mounting from rotating or the slide from elevating when not in operation. Later worm drive training gear achieved the same purpose. *Training clips* fitted under the roller path to prevent a mounting lifting off the rollers in a seaway or from the upward thrust at the forward end as the gun fired. At recoil there was a much heavier downward thrust at the rear end, taken by the rollers; it could amount to over 300 tons per gun. *Voice pipes* were fitted to allow communication between different parts of the mounting. *Lateral guide rollers* at the bottom of the trunk kept it centred. *Weather protection* was required around openings in the gunhouse or between the rotating and fixed barbette structure. Gun ports had to be large enough to allow the gun to reach its maximum elevation, but both gun blast and seas could enter. For guns with modest elevation canvas blast bags might be sufficient to seal the hole, but later curved mantlet plates were fitted – see *Nelson* profile (page 161). *Sub-calibre* guns, typically 6pdr, were kept in the gunhouse which could be inserted in the bore of the main gun to provide training for the crew without the expense and bore wear of firing full charge rounds,

*The Big Gun* by Peter Hodges, a former ordnance engineer, includes more detailed descriptions and drawings of many of the above components.

## OPERATING MECHANISMS

British practice favoured hydraulic actuation of gun mountings over electric, steam or compressed air. Dawson outlined the considerations in both 1901 and 1909.[3] Precise control of motions, reliability and a high rate of aimed fire were paramount. The motions included elevating the gun and slide, training the mounting, recoil and run-out of each gun after firing, and all the stages of hoisting and loading shell and charge from magazine to breech.

Hydraulic mechanisms required a pump, usually steam powered, to provide fresh water at a pressure of 1100lb/sq in or more through a system of pipes and valves. Creep control was easy with direct action from the cylinders (sometimes called presses), it was smooth, near silent and with no wild heat and easily repaired. Disadvantages included leaky joints, large diameter pipes and risk of freezing. Indeed, Armstrong had perfected such machinery from its earliest days in the 1850s with dockside cranes and equipment, so naturally preferred such power for its ordnance; Vickers followed suit. Later hydraulic fluid added soluble Argolene oil for lubrication or switched to mineral oil completely for lighter calibre mountings.

In the case of steam, long runs of hot piping for pressure and exhaust were required running through many compartments, with many joints which could leak and scalding was a risk. Condensed water had to be drained, while precise creep control was difficult. Compressed air required very high pressures but at least could be exhausted to atmosphere. Electric power had the advantage of using cables rather than pipes but required extra generating capacity. The then high-speed motors required gearing, while starting and creep control was difficult, and circuits could easily fail with diagnostics difficult. Interlocks were more difficult to arrange. Although the US Navy used electrics successfully (and the RN inherited the US design of twin 14in mountings in the 1915 monitors), RN experience was not encouraging. Battlecruiser *Invincible* had two electric Armstrong turrets and two Vickers when completed in 1909 but these were replaced by hydraulics in 1914. But what was found suitable for power-worked medium calibre mountings post WW1 was the fitting below the turntable of an electrically powered VSG swashplate hydraulic oil pump, which actuated training and elevating gear and rammers. It was not until the 1950s that fully electrical mountings were used, with half of the *Tiger* class cruiser 6in being so powered, the other half being electro-hydraulic.

## MANUFACTURING THE MOUNTINGS

By 1911 the Admiralty had wrested control of procurement not only of gun mountings but also guns from the War Office. It could thus determine design, demand and funding for both, bringing in both Coventry Ordnance Works and Beardmore for heavy calibre guns. It was able to pursue first the 13.5in then 15in within its own budgets. The Naval Ordnance Department was then able to build up better relationships with contractors.

All heavy gun mounting designs were made by private industry to meet NOD specifications. Gun mounting contracts for a particular ship covered the construction and testing of the mountings at the maker's works, transport to the shipyard, erection on board ship, placing the guns into the mountings plus construction of any spare mountings. Spare mountings were ordered usually one per ship, but for the heaviest mountings comprised only one gun cradle, slide, trunnions, elevating gear and breech operating gear, rather than the entire revolving unit. Gun bodies and breech mechanisms were ordered and manufactured separately, kept in stock at dockyards and Woolwich and only issued to ships when required.

Each manufacturer had a number of workshops where the various components could be made before final erection. Such a complex mechanism had to be assembled in the erecting shop to check that all components and connections fitted and worked, prior to installing in the ship. Pits up to 50ft deep and 30ft in diameter were sunk below the workshop floor lined with cast iron segments like a tunnel, simulating the arrangement in a ship, served by 60- or 100-ton electric overhead travelling cranes. Bays alongside were used to assemble major components such as turntables and working chambers before erection in the pit. Trials of all the hydraulic systems, hoists and interlocks would be carried out in the presence of Admiralty inspectors, before dismantling for shipment to the shipyard. Since most shipyards before 1930 could only lift up to about 150 tons at their fitting out quays, the mountings had to be designed so that no component exceeded that weight – usually the turntable/working chamber was the heaviest item, as the gunhouse could have its heavy armour plates and guns fitted one at a time.

The manufacture of such gun mountings involved a wide range of engineering skills. Draughtsmen produced hundreds of drawings (773 for the first 15in) defining all the parts, to be issued to workshops and sub-contractors. Wooden mock-ups were built to check clearances and avoid interference between moving parts, with interlocks required to prevent dangerous simultaneous movements. Rough forgings received from steelworks, such as for gun slides, needed careful machining, requiring turners and most mechanical engineering trades in both heavy and light machine shops. Key components were machined using jigs to ensure dimensional accuracy and interchangeability of parts. Foundries cast less highly stressed components such as bearing housings and valve bodies. Hydraulic machinery was made for gun elevating, training and

[3] Dawson, A Trevor. *The Engineering of Ordnance*. Junior Institution of Engineers, 1909

recoil mechanisms and ammunition hoists. The associated pipework and valves needed plumbers and coppersmiths. Boilermakers, platers, caulkers and riveters made the supporting steel structure for ammunition trunks, the cylindrical working chamber, the turntable floor, girders and side plates, and various supports. Gun shield plates were received from the armour makers to be erected on the gunhouse floor.

All the components were brought together in the erecting shop for testing, assembled by fitters from the bottom up in a pit, complete apart from the gunhouse roof and minor fittings. Supporting trades included cranemen and riggers to move the components, sheet metal workers,

**Above:** The new gun mounting pit built in Nº 24 Shop at Elswick in 1938, 48ft deep and 46/30ft in diameter. *(Tyne & Wear Archives)*

**Right:** The 15in Mark I/N mounting for *Abercrombie* fully erected in Elswick's Nº 24 Shop on 23 December 1943. It had originally been made as a possible replacement for *Furious'* 18in but had lain at Chatham Dockyard for 24 years. The guns have their 12-ton balance weights fitted to achieve 30° elevation. The circular platform is for a 20mm Oerlikon to be fitted on the turret roof. What looks like a 13.5in gun lies alongside, probably one of those planned for the Turkish coast defences but to be converted to a 13.5/8in hyper velocity gun for Dover. *(Vickers)*

painters, electricians and labourers. After nine to twelve months in the pit, the mounting was disassembled for transport to the shipyard and re-erected in the ship. Erection on board involved both shipyard and gun mounting contractor staff. A key feature of the shipyard work was the preparation of the ring bulkhead with its roller path upon which the whole weight of the mounting revolved. A special planing machine was placed in each barbette, carefully centred and levelled to machine the top surface of the roller path. This had to be absolutely parallel with all the other turrets and the director, as measured against a datum plane on the hull, so that when each rotated, they did so in exactly the same plane to point correctly at the target.

## MANUFACTURER'S OUTPUT CLAIMS

The manufacturers were always keen to extol their products. Brassey noted their claims.[4] Sir Andrew Noble, Armstrong's chairman, wrote in April 1909:

> From our Elswick yard alone we could, if need be, deliver three Dreadnoughts complete in two and a half years from the date of order, and could further continue delivering at the rate of three vessels each fifteen subsequent months while, with the assistance of our Openshaw works, we should have no difficulty in supplying and completing their armaments without interfering with such armaments as we may be making for other Powers.

Colonel T E Vickers had made a similar optimistic claim the previous month:

> The firm could lay down and complete, arm, and fit ready for service in every respect three battleships of the Dreadnought type in three years from the date of approval of drawings without going outside its own factories. Thereafter their firm could deliver Dreadnoughts complete for service at the rate of one in every six months, in addition to existing contracts … The firm could supply the artillery equipment complete for seven Dreadnoughts in three years.

That implied that they could make a dozen heavy gun mountings a year, a rate nearly double anything they had achieved hitherto.

Even Coventry Ordnance Works claimed that they were now in a position to undertake the supply of the largest calibre of naval guns, in addition to the mountings of two and a half battleships per annum, after thirteen months from the receipt of order. That equated to about twelve mountings a year, when all they actually achieved was an average of three a year over the next five years.

## GUN MOUNTING CONTRACTS

The Admiralty had standard gun mounting contracts. As well as details of the particular mountings to be supplied, they issued general conditions relating to the performance of the work. A selection of clauses from the 1930s is summarised, containing a number of archaic requirements dating from earlier days:[5]

> The gun-mountings, machinery, articles or things are to be constructed to the satisfaction of the Commissioners for executing the office of Lord High Admiral of the United Kingdom … whose officers must be afforded full facilities for inspecting the gun mountings … while they are being constructed.
>
> No Member of Parliament shall be admitted to any share of the contract.
>
> The Contractor shall pay rates of wages and observe hours of labour not less favourable than those commonly recognised by employers … in the trade in the district where the work is carried out.
>
> No persons other than British subjects shall be employed on the work or have access to the drawings and specifications …
>
> The dimensions of the several parts must be in strict accordance with the drawings and each part must be interchangeable …
>
> All the materials used in the construction of the gun-mountings shall be of the qualities and strengths approved by the Admiralty and must pass the tests required by the Admiralty …
>
> If the gun-mountings are delivered by water, they must be shipped in British vessels, manned entirely by British subjects.
>
> The storage of any parts of the machinery in the Dockyard will be provided by the Commissioners till the vessel is ready to receive it.
>
> The Contractor shall deliver the whole of the gun-mountings into the charge of the Officer at the Receiving Establishment on or before the 'prescribed date' which expression shall mean … weeks from date of acceptance of tender …
>
> Notwithstanding anything contained in the Contract, the Commissioners reserve to themselves the right to limit the total sum to be paid … during any financial year to such amount as they may decide … [It was difficult to diverge much from the sums approved under each heading in the Navy Estimates voted by Parliament each financial year.]

## SCHEDULE OF PAYMENTS

Various schemes were available, typical of which was:

> One-fifth when 50% of the material has been delivered at the Contractor's works, satisfactorily tested and stamped by the Overseer.
>
> One fifth when 95% of the material has been delivered …
>
> One fifth when 50% of the machining has been completed.
>
> One fifth when 95% of the machinery has been properly fitted, assembled and completed at the Contractor's works and is ready for delivery.
>
> One fifth when the whole of the machinery required to be supplied has been manufactured and delivered at the contract

[4] *The Naval Annual 1910*, p302

[5] NMM: Contract for HMS *Anson* (*Duke of York*) 234/584

destination; all trials carried out and finally accepted by the Admiralty. [In practice for heavy gun mountings, the contract was broken down into about fifteen 'items' such as 'turntable', 'gun slide', 'training gear' etc and the payment schedule applied to each item.]

All defects shall be made good at the Contractor's expense on the understanding that this undertaking shall be operative for a period of twelve months from the date of complete delivery of the gun-mounting.

## DEMARCATION

The hull specification outlined the demarcation between the gun mounting manufacturer and the shipbuilder. The 20 pages for the *King George V* class 14in mountings are summarised:

Shipbuilder to provide, fit and machine: circular supports under lower roller path (ring bulkhead), training rack, structure for lateral guide rollers. Machine surfaces on the barbette armour for upper apron (for watertightness). Cut access holes and passages for pipes. Fresh water supply for gun washout; drainage. Provide and fit voice pipes. Provide and fit seatings for training and depression cut-offs. Provide tanks for hydraulic fluid; steam heating; spare part stowage.

Seatings on fixed structure to accommodate gun mountings: machine surfaces of lower roller paths with ship waterborne, parallel to waterline. Max tilt 2½ minutes of arc in any direction. Supports strong enough to stand stresses when all guns fired.

Shipbuilder to allow erector [of gun mountings] all necessary facilities for erecting on board at a suitable berth. Erector to load and unload materials [from transport vessel] and lift on board. Shipbuilder to provide lifting power, steam, compressed air and electrical power. Gun mounting contractor to provide and fit gun mountings, shield, turntable, roller path, rollers, training rack, pipes and valves for power transmission outside revolving structure.

Guns and breech mechanisms will be supplied by Admiralty to contractor. Shipbuilder to transport guns to ship, hoist on board and secure in cradle. Provide sub-calibre guns and fit in stowage.

## GUN MOUNTING REVIEW 1932

In March 1932, the Principal Supply Officers Committee reviewed the provision of gun mountings.[6] Their report took as a basis a notional War Emergency building programme prepared in 1929, *ie* the warships required on top of the normal annual newbuilding programmes. This comprised building over a 36-month period four new battleships, ten 8in cruisers, sixteen 6in cruisers, three aircraft carriers, 36 destroyers, 21 submarines plus minelayers, minesweepers, sloops and depot ships, with their associated gun mountings.

Before 1914 heavy power-worked gun mountings were only fitted in capital ships and armoured cruisers, but by the 1920s, all cruisers required such mountings. At the same time transferable mountings became more complex requiring more precision engineering, so were less suitable for emergency manufacture by general engineering firms. The Committee reviewed the industrial capacity for manufacturing naval gun mountings (but not guns), assuming that in an emergency the companies would have full staffs (presumably no military call up of their skilled manpower), would work a night shift ensuring that their specialist machine tools were continuously worked and that material requirements would be met. On that basis they concluded that the mountings for the capital ships could be produced within the construction time of the ships (3 years), but that for cruisers the limited manufacturing capacity would determine the number of ships that could be built. But that would require the whole of Vickers-Armstrongs' Elswick capacity, whereas at present a large proportion of its gun mounting capacity was being used for armoured fighting vehicles and commercial work; furthermore its steel foundry and forge shop had been closed. It also commented that during the war Elswick had produced a thousand torpedo tubes and while such were now manufactured in the dockyards, Elswick's facilities might be needed again. The situation at Barrow was complicated as the same shops were used both for gun mountings and marine engines, but should be able to produce both if the demand for marine engines was limited.

For transferable mountings, there existed a large stock for naval auxiliaries and defensively equipped merchant ships, so only about 600 would be required, albeit more complex. But as well as engineering companies, Vickers-Armstrongs' facilities would still be needed for the more specialised mountings, for example, those power operated. Director firing gear was seen as an adjunct to gun mountings, now concentrated at V-A's Crayford works since nearby Erith had been closed. Barrow still made the large director control towers for big ships. Armour for gun turrets made up only a small proportion of the total required for a ship, so demand could be assessed together.

The Committee compared the situation in 1914 with now. In 1914 the general engineering industry was thriving, with works fully equipped and manned, so could readily be adapted for less complex gun mounting manufacture. The armament firms were fully occupied with the large annual naval programmes, Vickers' Erith and Armstrong's Openshaw works near Manchester were operating, but now both were no longer available. Coventry Ordnance Works' Scotstoun heavy gun mounting plant had been sold off to Harland & Wolff, and was now only used for general engineering. The pre-war armament firms had large technical staffs and experienced gun mounting erectors, both at their works and in the shipyards, so when war broke out, it was only necessary to speed up production and take over ships and gun mountings being made for foreign navies. As the pre-war heavy naval mounting orders were completed, more of the armaments companies' production could be switched to the massive artillery demands of the Army.

The situation at Woolwich was that as the modest number of naval transferable gun mountings were completed, its output could be concentrated on the Army; the committee considered that that would remain the situation (for mountings, not guns). At Barrow output from

[6] The National Archives SUPP 3/43/7. Naval Gun Mountings, Director Firing Gear and Components.

1916 was concentrated on the Army (*eg* shells, howitzers and railway gun mountings) and on marine engines, including the large demand for submarine diesels. Elswick (and nearby Scotswood) were able to focus on Army artillery and munitions. Openshaw made heavy steel castings and forgings and machined parts for Elswick. Scotstoun had made mountings up to 15in (1916 *Royal Sovereign* and *Renown*) then focussed on Army work. Post-war, Erith's transferable gun mounting work had been moved to Barrow and its director gear to Crayford.

By the 1930s, output of power-worked non-transferable mountings depended on the number of erection pits available. A triple 16in required seven months in a pit and a twin 8in four months (both less than actual). Now all mountings from 6in upwards required pits, although short trunk mountings as in the *Leander*s could be built on the shop floor on gantries if the overhead crane clearance allowed, but long trunk mountings needed pits. The four pits at Openshaw suitable for 8in mountings had now been dismantled and the five at Scotstoun for 15in had been filled in with loose rubble, but its workshop machinery was still much the same as 1916, so could be brought back into use in about six months.

In an Emergency, they recommended using all eight pits at Elswick, four for battleships (for which it was better equipped), and four for 8in cruisers, with gantries used for short trunk 6in. The twelve largest at Barrow would be used, four for secondary 6in for battleships, four for 8in cruisers, four for 6in. In addition, hydraulic pumping machinery, shell room machinery and hoists would be required. The Committee concluded '... it will be vital to allocate probably the whole of Elswick works (except the gun and munition shops) and a large portion of the machine shops at Barrow to Heavy Turret Mountings for the Navy.'

For transferable and high angle mountings, the Committee noted that during the war Woolwich and the armament firms could no longer cope with the increased demand for gun carriages for the Army and the enormous number to arm naval auxiliaries and merchant ships. Some 11,500 naval gun mountings from 3pdr to 18in had been produced during the war, hence requiring the resources of ordinary engineering firms. They assumed that suitable gun mounting designs would be available in an Emergency for production to start immediately, although recognising that with the additional complication of modern mountings, about 25% extra effort would be required.

The Committee detailed the demand in a number of tables, excluding 0.5in machine guns and minor vessels, summarised below.

**Gun mountings to be manufactured in an emergency**

| Mounting | Number per ship | Ships | Total mountings | Made by |
|---|---|---|---|---|
| Battleships | | | | |
| Triple 16in | 3 | 4 | 12 | Elswick |
| Or twin 12–16in | 4 | 4 | 16 | Elswick |
| Twin 6in | 6 | 4 | 24 | Barrow |
| Twin 4in Mk XVII BD | 4 | 4 | 16 | Barrow |
| 8-barrel pom-pom | 4 | 4 | 16 | Barrow |
| 10,000-ton Cruisers | | | | |
| Twin 8in | 4 | 10 | 40 | Elswick & Barrow |
| Twin 4in Mk XVIII UD | 2 | 10 | 20 | Barrow, Woolwich & Industry 1 |
| Twin 4in HA Mk IV | 2 | 10 | 20 | Woolwich & Industry 2 |
| 4-barrel pom-pom | 2 | 10 | 20 | Barrow |
| 6000-ton Cruisers | | | | |
| Twin 6in | 4 | 16 | 64 | Elswick & Barrow |
| Twin 4in HA Mk IV | 4 | 16 | 64 | Woolwich & Industry 2 |
| 15,000-ton Aircraft Carriers | | | | |
| Single 4.7in HA Mk XII | 16 | 3 | 48 | Barrow |
| 8-barrel pom-pom | 2 | 3 | 6 | Barrow |
| Destroyers | | | | |
| Single 4.7in CP XIV | 4 | 36 | 144 | Barrow, Woolwich & Industry 3 |
| Submarines | | | | |
| Single 4in S.1 | 1 | 21 | 21 | Barrow, Industry 4 |

Industry 1 = Vulcan Foundry, Newton-le-Willows; W Marshall, Sons & Co, Gainsborough.
Industry 2 = W Marshall, Sons & Co, Gainsborough; Beyer, Peacock & Co, Manchester; Hunslet Engineering Co, Leeds; Greenwood & Batley, Leeds.
Industry 3 = Ruston & Hornsby, Lincoln; Peter Brotherhood, Peterborough; W Beardmore & Co, Glasgow; R Hoe & Co, London SE.
Industry 4 = R Stephenson, Darlington; North British Locomotive, Glasgow.

CHAPTER THREE

# Sir W G Armstrong, Whitworth & Co Ltd

WILLIAM GEORGE ARMSTRONG abandoned a career in law at the relatively late age of 37 to follow his interest in hydraulics, applied to the working of machinery and dock cranes. In 1847 he established a works at Elswick on the north bank of the Tyne just less than two miles west of the centre of Newcastle. Here, Armstrong designed

**Below:** Whitworth was the only competitor to Armstrong making heavy mountings in the 1890s but was taken over by the latter in 1897. This view of Openshaw's main gun mounting shop was probably taken at that time when 12in B.III mountings were being made, with working chambers, turntables and a roller path live ring in evidence. *(Cambridge University Library)*

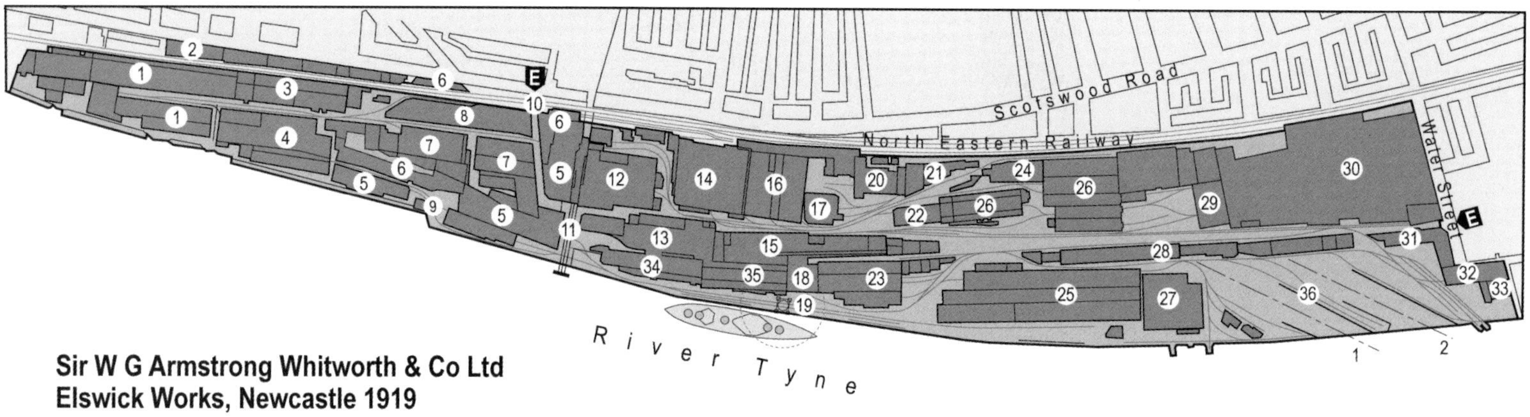


**Sir W G Armstrong Whitworth & Co Ltd**
**Elswick Works, Newcastle 1919**

1. Engine works erecting shops (1860)
2. Pattern shop
3. Crane and bridge shop (1847)
4. Bridge yard
5. Gun carriage & mounting fabrication shops (No 17 Shop) (1887)
6. Offices
7. Gun machining shops (No 11 Shop)(1883)
8. Forge shop (No 4 Shop) (1919)
9. Torpedo department
10. Main entrance
11. Waggon-way to coal staith
12. Foundry (No 33 Shop) (1909)
13. Machine & cartridge case shop
14. Gun carriage shop (No 29 shop) (1896)
15. Smith's shop (No 35 Shop) (1881)
16. Main Store (1901)
17. Steam hammer & forge shop (No 31 Shop)
18. 18-inch mounting shop (1916)
19. 150-ton crane
20. Coiling shop (No 32 Shop)
21. Shell shop (No 8 Shop) (1885)
22. Gun shrinking pits (No 13 Shop) (1910)
23. Turret & machine shop with gun pits (No 24 Shop) (1910)
24. 18-inch gun shop (1915)
25. Ordnance structural shops (No 130 and 135 Shops) (1919)
26. Gun wiring and inspection shop (No 18 Shop) (1887 and 1896)
27. Shipyard plate shop
28. Tool & machine shops
29. 5,000-ton press shop
30. Steel works and foundry (1883 and 1897)
31. Offices & laboratory
32. Timber store
33. Sawmill
34. Gun machining shop (No 6 Shop) (1901)
35. Gun mounting and erection shop (No 7 Shop) (1901)
36. Shipbuilding berths

area outside works
area of works
buildings
E entrance

Feet 250 500 1000
Metres 150 300

**Elswick plans.** Armstrong's Elswick works west of Newcastle was continually extended from its start in 1847 to WW1. Its rambling site included shops for building guns, gun mountings, dock machinery and ships, plus a steel works and foundry. The shipyard closed in 1919, the area being used in WW2 for new tank building and shell shops. *(Ian Johnston)*

**Armstrong, Whitworth & Co Ltd, Elswick Works 1921**
**No 24 Gun Mounting Shop**

Feet 125 250
Metres 37.5 75

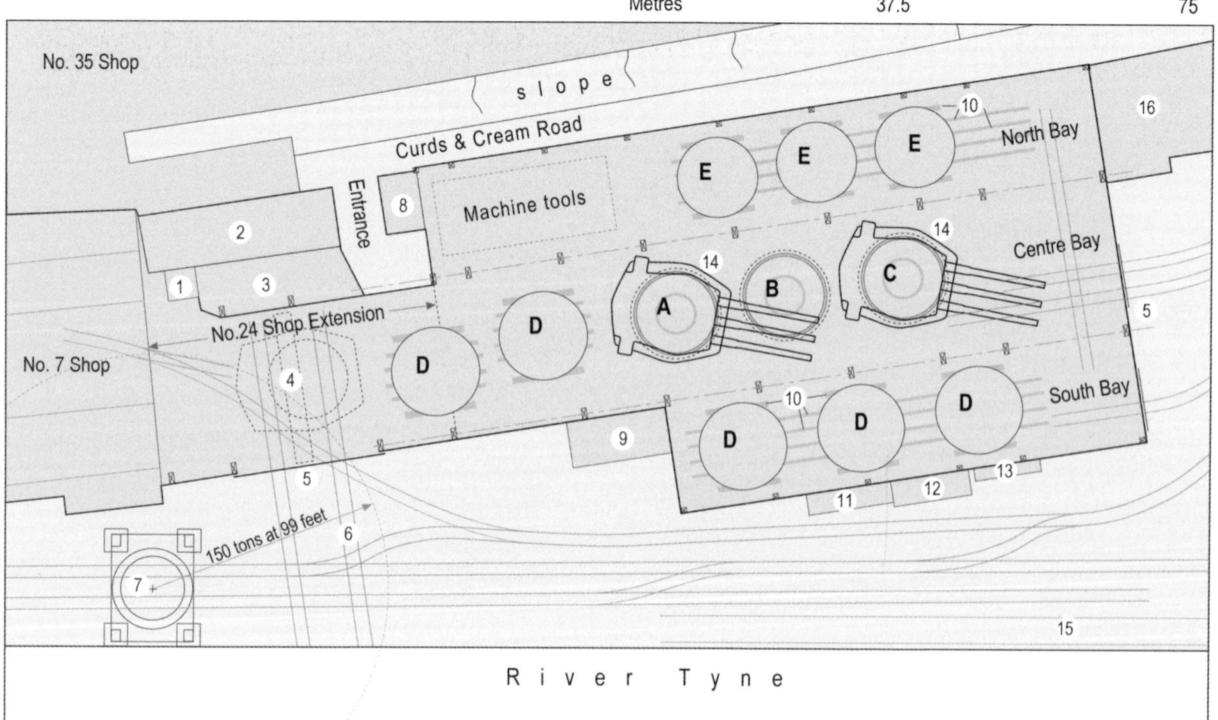


1. Accumulator
2. Pump house for hydraulic crane
3. General store
4. 100-ton overhead crane
5. Sliding doors
6. Bogie tracks
7. 150-ton hydraulic jib crane
8. Time office
9. General store
10. Concrete beds
11. Blacksmiths shop
12. Tool room
13. Lavatory
14. 16-in triple mounting
15. Travelling crane
16. Pumping plant

**A** West pit
**B** Centre pit
**C** East pit
**D** Turntable berths
**E** Working chamber berths

**Above and pages 25–30, 32:** Armstrong Whitworth produced a booklet *Guns & Mountings: Naval Artillery Section* about 1913. It had 21 photographs illustrating the various workshops, taken by Elswick photographer James Goold. This selection gives a good impression of an ordnance works at its busiest. *(Tyne & Wear Archives and Museums DS.VA/9/1/2)*

**Above:** Nº 17 Shop with a turntable being turned over by a 50-ton crane and a gunshield being erected which looks like a twin 12in for what became later *Agincourt.* The nearer crane is 50 tons, the further 40 tons.

and manufactured hydraulically powered dock equipment for the new and expanding ports around the British Isles and abroad. As the hydraulics business grew, Elswick became a major engineering centre for other mechanical and structural products such as winding engines, bridges and pumping engines.

Although Armstrong's contribution to the field of hydraulics was significant enough on its own, his entry into the business for which his company became best known – ordnance – seemed to be unplanned and quite by chance. Soon after the Crimean War (1853–56) broke out, he and James Rendel railed against what they considered to be the archaic nature of military ordnance, in particular field guns. The outcome of this interest was that Armstrong resolved to forward ideas for a new type of field gun to the Secretary of State for War. Armstrong's design was an engineered product far in advance of anything then in existence, employing a breech-loading gun with a rifled barrel firing an elongated steel shell. This produced a more rapid and accurate fire than was possible with existing weaponry. Over the next two years, Armstrong provided a variety of guns to be evaluated by the Ordnance Committee to consider the type of gun most suitable for future use in the Army. Armstrong's proposals did not go unchallenged however, and were contested by Joseph Whitworth, a talented engineer renowned for precision measuring methods in engineering and the screw thread named after him, who had established a large works at Openshaw near Manchester, where machinery, tools, steel forgings and guns were manufactured. The Committee found in favour of Armstrong and he was appointed in 1859 as Engineer of Rifled Ordnance at the Royal

Nº 7 Shop erected complete mountings in pits with three being erected plus an ammunition trunk. The line shafting at left appears to be driving machines in the gallery making smaller components.

Gun Factory, Woolwich, receiving a knighthood in recognition of his contribution to British arms. To avoid a conflict of interest, Armstrong set up the Elswick Ordnance Company, separate from his hydraulic machinery company, but with manufacturing capacity on the Elswick site. In 1863 Armstrong resigned from his government post and in the following year he merged the ordnance works with the rest of his business at Elswick to form Sir W G Armstrong & Co which, by the mid-1860s, employed about 4000 persons, most of whom were in the Ordnance Department. More detail on the company history is included in *The Battleship Builders*.[1]

With an uncertain future for ordnance manufacture at Elswick, largely shorn of British government contracts, Armstrong successfully developed his business with overseas governments. He arranged with Charles Mitchell for warship hulls to be built at the latter's Low Walker yard on the Tyne. This arrangement with Mitchell was formalised in 1882 when the business became Sir W G Armstrong, Mitchell & Co Ltd. The following year the decision was taken to add shipbuilding berths for warship construction at the eastern end of the Elswick works and it was from here that many innovative protected cruisers were launched from 1885. At the same time a steel works covering ten acres at the eastern end of the Elswick works was commissioned to produce steel by the Siemens-Martin process for forgings and ordnance components.

Naval architect George Rendel turned his attention from gunboats to the machinery required to work naval guns – the hydraulic mechanisms necessary to repeatedly supply projectile and propellant to guns which in turn had to be trained and elevated (*ie* aimed at a target), as well as fired. A gun mounting department was consequently established where the design and manufacture of these elaborate and large mechanisms could be undertaken. Another key individual adding to the success of

[1] Also, Warren, Kenneth. *Armstrong of Elswick.* Macmillan Press, 1989

Nº 7 Shop with what looks like a twin 13.5in mounting with one gun in place, with nearer a smaller mounting, probably a 12in. The great height enabled completed mountings to be lifted over the others to reach the door leading out to the 150-ton crane on the quayside.

the company was Andrew Noble, a captain in the Royal Artillery and ballistics expert, who joined the company in 1860, eventually to become Armstrong's successor as managing director and Chairman. Josiah Vavasseur's London Ordnance Works had developed a hydraulic gun mounting which recoiled up an inclined slide. Unable to meet the demands for it, Vavasseur merged his company with Armstrong's in 1883, becoming a director until his death in 1908.

By the 1880s, the sprawling but congested works at Elswick were capable of turning out every type of modern warship and its armament, the only one of its kind in the UK. There were however two major exceptions in this otherwise seamless array of warship components, which was the capacity to manufacture propelling machinery and armour plate. In the former, Armstrong was well served by several marine engineering works on Tyneside. They did however enter the marine engineering market after WW1, but this was to provide civilian work in the absence of warship and armaments orders. Some marine steam reciprocating engines were built at Elswick, while a Sulzer diesel engine licence was taken out in 1921 with some two dozen such engines built at Elswick and Scotswood, together with steam locomotives, before the latter was taken over for ordnance work in 1936.

In 1897, Armstrong amalgamated with rival Sir Joseph Whitworth & Co Ltd. The new company, Sir W G Armstrong, Whitworth & Co Ltd, set about further extending its capability by laying down a new armour plate mill at Openshaw. In 1899, the company opened a new works at Scotswood to the west of Elswick for the manufacture of munitions. Extended ordnance shops were built at Elswick including Nº 7 Shop for erecting naval gun mountings in 1901, followed by Nº 24 Shop in 1910. Alongside these shops Armstrong built a 150-ton crane in 1907

The new Nº 24 Shop with a twin 13.5in nearly complete in its pit in the centre bay. On the floor are trunnion supports (left), gunhouse armour plates (centre) and the second gun (right).

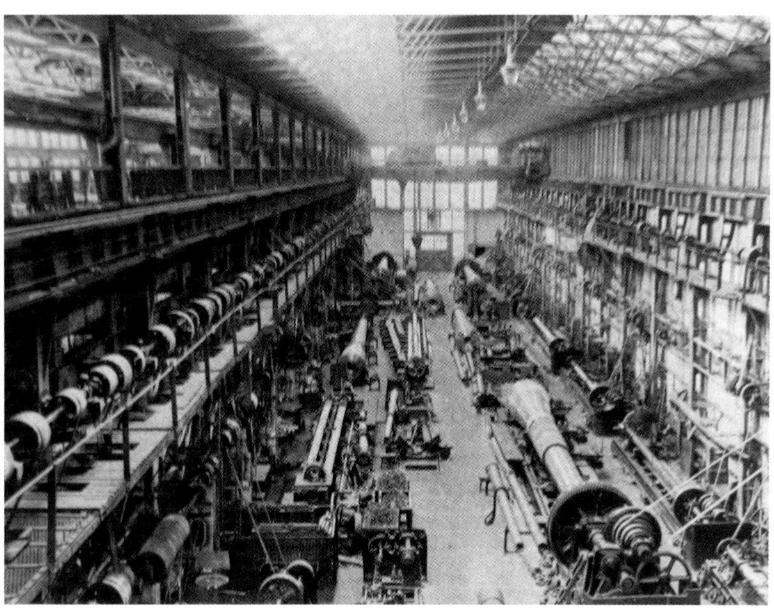

Part of Nº 6 Shop with 12in, 13.5in and 14in guns being turned on the outside and rifled.

for installing the heaviest mountings in ships, replacing the previous sheerlegs. Armstrong's company had come a long way over the preceding fifty years and now stood as a large manufacturing concern producing all manner of engineering products but principally ordnance, armour and warships. By the end of the nineteenth century the company was employing about 20,000 across its three main sites.

The company produced extensive publicity material. In 1905 a lavish well illustrated 240-page compendium entitled *Warships and War Material* went into great detail about the various works and products and the warships built.[2] A more widely circulated 1912 booklet *The Elswick Ordnance Company* chronicled the company history before going on to describe the facilities. It noted 40 large workshops at Elswick making naval and land service guns and mountings, gun sights,

[2] Cambridge University Library, Manuscripts. Vickers Archive 2000.

The north bay in Nº 24 Shop with three turntables being machined upside down prior to erection in the centre bay. At the far end is a working chamber and turntable topped by their gunhouse floor with trunnion supports.

torpedo tubes, shells and cartridge cases plus supporting facilities like drawing offices, laboratories, pattern making and inspection. It recorded that from 1893 to 1911 they had made 5266 guns of 3in to 14in calibre and 5476 gun mountings of 3in to 14in calibre, a goodly proportion for the War Office and for export. It included its Openshaw works making gun mountings up to 12in, guns up to 13.5in, gun shields and armour plate. Nearby Scotswood made shells, fuzes and motor cars. It mentioned a devastating fire in July 1899 which gutted vital Nºs 6 and 7 Shops, making parts for and erecting heavy mountings. By 1901 the shops had been rebuilt and extended, but deliveries of 12in mountings were delayed. Parliamentary paper 'Arrears of Shipbuilding'[3] noted that *Formidable*'s mountings were delivered 13 months after their contract date, *London* 12 months (both further delayed when machinery contractor Earle went bankrupt), *Implacable* 10 months and *Bulwark* 8 months.

The first battleship to be built at Elswick was the turret ship *Victoria*, completed in 1890, with one twin 16.25in mounting. Its shipbuilding facilities continued to be expanded with more and larger berths. But eventually these too were overtaken by the increasing size of dreadnoughts. The arrangement of Tyne bridges, particularly the Newcastle high level and swing bridges, imposed width and height restrictions on ships passing down river from Elswick which effectively limited continuing battleship construction at Elswick. Instead, the Board agreed to purchase ground at High Walker, just up river from the original Armstrong Mitchell yard, where a new shipyard – subsequently known as the Naval Yard – was laid out. The new yard was finally completed in 1914, the first ship to be laid down there in October 1913 being battleship *Malaya*. The battleship *Nelson* was laid down in 1923 but it was not until 1937 that Naval Yard's last battleship, *King George V*, was laid down, albeit under the Vickers-Armstrongs name.

The Armstrong Whitworth annual accounts between 1902 and 1914

[3] Cd 1055, 1902

The centre bay of Nº 11 Shop with two gun jackets being machined internally to slot over the wire winding applied to the A tube.

include figures for the Elswick divisions.[4] The Ordnance division was the largest, employing between 8500 and 12,500, averaging 10,057 or 62.4% of the works average of 16,109. Next came the Elswick shipyard with 3368, Engine works 1554 and Steel works 1130. Annual Ordnance orders averaged £2.834M from 1903 to 1913 but fluctuated widely from year to year; in 1914 that figure nearly trebled. The Elswick accounts gave a breakdown of Ordnance division profits. The average for the years 1907–1912 is shown in the table, when fifty twin naval gun mountings were made. Gun mountings contributed about 64%. No breakdown was given between naval and land ordnance, although the later accounts when the company had become Vickers-Armstrongs did so (see Appendix 2). The net profit is slightly overstated as no depreciation was charged at that stage.

| **Average Profit 1907–12** | £ | % |
|---|---|---|
| Gun mountings | 232,381 | 63.96 |
| Guns | 75,567 | 20.80 |
| Shell & fuzes | 12,472 | 3.43 |
| Steel forgings | 6,160 | 1.70 |
| Royalties | 6,650 | 1.83 |
| Miscellaneous | 30,097 | 8.28 |
| Manufacturing profit | 363,326 | 100.00 |
| Charges (overheads) | 172,453 | |
| Net profit | 190,873 | |

Over that period, Ordnance profits made up 23% of group net profits totalling £825,067. Openshaw contributed 47% but that was mostly due to its making armour which benefited from the very high prices set by the cartel of five armour makers, rather than great efficiency. Ordnance's return on capital of £1.492M was 13.6%, half that of Openshaw.[5]

[4] Tyne & Wear Archives DS.VA/3/2/1 3

[5] *The Battleship Builders*, p245

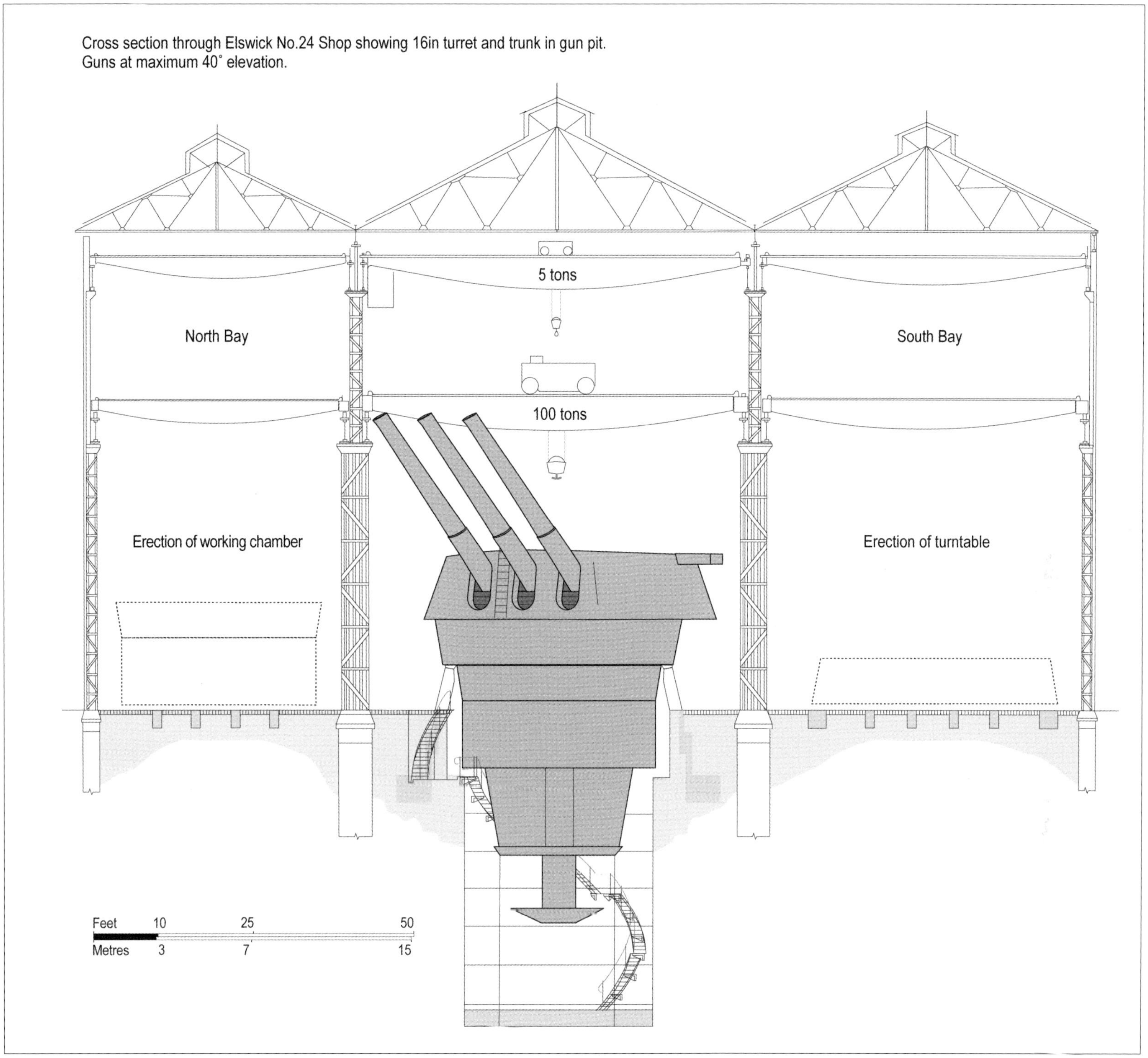


Although existing Armstrong records do not give details of each major gun mounting order (unlike Vickers), it can be calculated that between 1901 and 1912, it made 79 twin mountings of 12in and above at Elswick. It did publish some statistics of its impressive wartime record, noting that in November 1918 the company had 78,000 employees, 21,000 being women. That was up from 47,500 in July 1915.[6]

[6] Tyne & Wear Archives DS.VA/9/20/1-5

**New Guns Made at Elswick and Openshaw**

| *Calibre* | *Admiralty* | *Ministry of Munitions* |
|---|---|---|
| 5.5in | 2 | |
| 6in | 85 | 42 |
| 7.5in Mark VI | 9 | |
| 9.2in | 4 | 13 |
| 12in | 3 | 56 |
| 14in | 12 | 4 |
| 15in Mark I | 27 | |
| 18in | 3 | |

N° 18C Shop with finished guns being examined before being taken for proof trials at Ridsdale. The furthest guns are 13.5in, those to the right 4.7in and 6in. A series of projectiles and smaller guns are lined up at the left. The floor is brick paved as was common with engineering workshops.

**Output of Armstrong works Aug 1914–Nov 1918**

| | |
|---|---|
| Guns 3pdr to 18in | 13,003 |
| Carriages (*ie* mountings) | 12,093 |
| Shells, million | 14.5 |
| Cartridges, million | 21.1 |

**Approx average monthly output**

| | *1915* | *1918* |
|---|---|---|
| Guns | 120 | 540 |
| Mountings | 110 | 560 |

Post-war, Armstrong Whitworth suffered from the dearth of gun mounting orders. It did however build the six triple 16in mountings for battleships *Nelson* and *Rodney* and some 8in mountings for 'County' class cruisers at both Elswick and Openshaw – pages 110 and 118. But from 1925 it was making losses of over £500k a year while its finances became increasingly strained after its unsuccessful venture into forestry and newsprint in Newfoundland. In 1927 it was forced into a merger with Vickers to become Vickers-Armstrongs. That included the Elswick works which continued to make gun mountings with the new company concentrating gun making at Elswick.

CHAPTER FOUR

# Vickers at Barrow

### VICKERS ENTERS THE MARKET

The Sheffield specialist steelmaker Vickers Sons & Co Ltd had been making armour plate for the Admiralty since 1888. Encouraged by the British government to make guns, it started making them also in 1888. Orders resulted for larger calibre guns for warships procured following the Naval Defence Act. The Vickers brothers, Albert and Tom, had been watching Armstrong dominate the production of naval ordnance and warships for export from its extensive works at Elswick. In 1896 they headhunted 30-year-old Lieutenant Trevor Dawson, a naval officer who was an experimental officer at Woolwich, previously having been gunnery officer in cruiser *Iris* and ironclad *Iron Duke* and then on the staff of the gunnery school *Excellent* at Portsmouth. Trebilcock noted his 'fresh intelligence, tried scientific capability and his expert knowledge of the armament users' requirements plus his practised ease in liaison with service officials'.[1] Dawson saw that as well as making naval guns (at Sheffield), it was necessary to acquire the capability of manufacturing heavy gun mountings. At that time, only Armstrong and Whitworth at Openshaw were able to make such – the Royal Gun Factory at Woolwich only made lighter mountings and guns. When they merged in 1897 to become Armstrong Whitworth, an effective monopoly was created. Vickers then sought to take over a suitable shipyard but found that many like Palmers on the Tyne or Fairfield on the Clyde were on restricted riverside sites. But the Naval Construction & Armaments yard at Barrow in Furness, a modest sized company which could build warships and their machinery, had space for expansion. Their engine works could be developed to include gun mounting manufacture, whereas Armstrong's Elswick site was constricted between the River Tyne and a railway. Vickers bought N C & A in 1897 for £425,000 with help from Rothschild. They also bought out Maxim at Crayford and Nordenfelt at Erith making lighter ordnance and machine guns. The company was then renamed Vickers, Sons & Maxim Ltd, with a capital of £2.5M.

In the meantime, observing orders for 12in gun mountings going to both Armstrong and Whitworth, Vickers was determined to secure an order for battleship gun mountings from the Admiralty, which was also keen to have an alternative supplier. Although with no experience as yet in manufacturing such or having associated costing records, they knew they had to put together a cost estimate and tender a price comparable with what the Admiralty were paying Armstrong. With

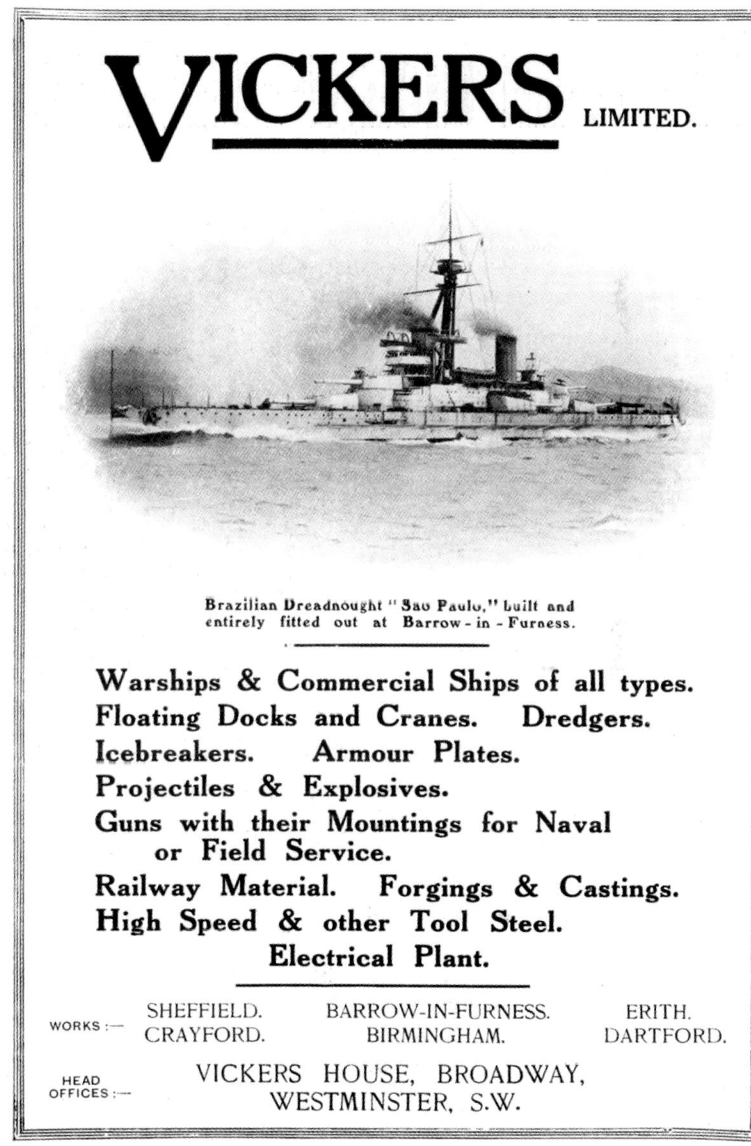


This advertisement from *Cassier's Magazine* December 1911 shows the range of Vickers products.

[1] Trebilcock, Clive. *The Vickers Brothers*. Europa, 1977, p46.

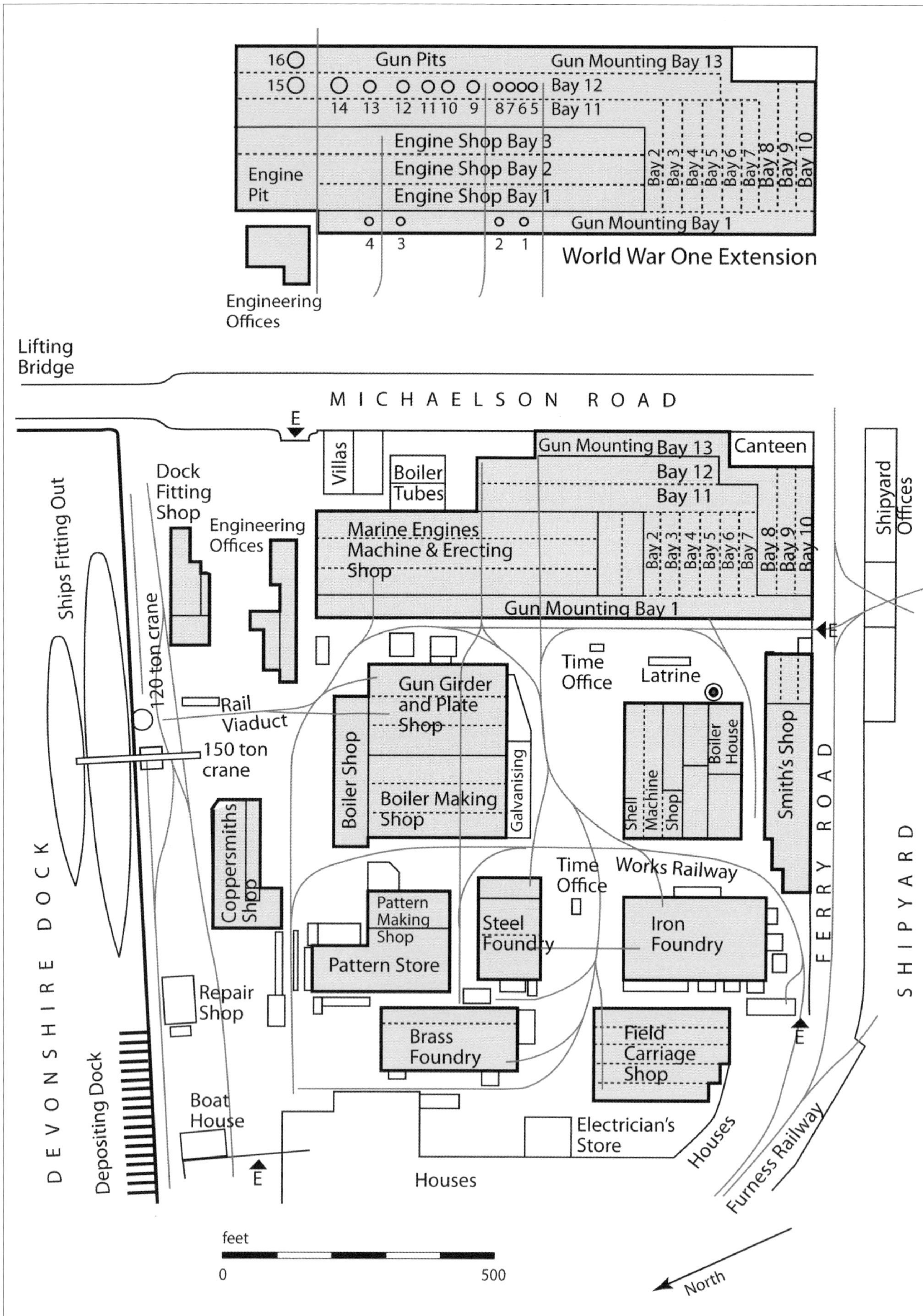


Barrow plan, showing the part of the Vickers works used for ordnance work, north of the shipyard. The original marine engine works was extended from 1898 to include gun mounting shops. By WW1 there were sixteen pits which could be used to erect mountings up to 15in. In 1937 a new shop was built north of the boiler shop with three pits for 14in quadruple mountings, with a new 250-ton crane replacing the 150-ton one at Devonshire Dock. *(Ian Johnston)*

Dawson's Navy contacts and with Vickers supplying guns and armour to the Admiralty, it is likely that they got indications of the sort of figure that was being paid. They also produced an improved design which could load at all angles of elevation – guns on previous mountings had to be brought to a fixed angle to load, a slow process. They were conscious of the need to cut their price below cost and did not even charge for erection of the mountings on board.[2] They were awarded the contract on 19 March 1898 to build the two twin 12in mountings for *Vengeance* for delivery in June 1899. They were also awarded the contract for her hull and machinery to be built at Barrow. Their tender price for the mountings was only £52,128, although that excluded the cost of the gunshield protection and barbette armour which were priced on a weight basis at a cost per ton agreed each year by the Admiralty and the armour manufacturers. The guns and armour for the hull were ordered separately, with Vickers getting a large share. Only five days later Vickers were awarded a contract for £53,250 for the two 12in mountings for *Irresistible* to be built at Chatham dockyard; they had tendered for no fewer than three battleships to be built in naval dockyards.[3]

Vickers was concerned to get a fair return on its investment, Dawson writing to the Admiralty on 20 July 1899:

> ... We have laid out on works at Sheffield, Barrow and Erith upwards of £1 million. Our designs should not be handed to competitors, but we are prepared to grant licences. Our only important competition for foreign orders is from Armstrong Whitworth, with whom at the present time we are carrying on a serious patent law case in Chancery Court. Notwithstanding, we are prepared to allow them to manufacture our naval mountings for Great Britain and the Colonies at a royalty of 5% of selling price, provided we are given a fair and equitable proportion of the orders to be placed. ... If Their Lordships desire to manufacture mountings embodying our patents and designs in Government factories we would ask only a nominal royalty of £1 per mounting.

Unfortunately no follow up correspondence was found.[4]

Vickers, ever conscious of the need for publicity, supplied detailed information on its activities at its new works at Barrow in a well-illustrated article in the prestigious weekly journal *Engineering* of 26 July 1901. For *Vengeance* they pointed out that for the first time, one company had built for a battleship the hull, machinery, gun mountings and armour – as well as some of the guns, a claim Armstrong would never be able to make, having no marine engine building capacity. Although her mountings were made at a loss, their learning curve was steep, with costs being steadily reduced so that by 1902 the contract for *Exmouth*'s two 12in mountings produced a profit. In 1900 they tendered successfully for 9.2in single mountings for coast defence for the War Office. Thereafter gun mountings from Barrow, and their associated gunhouse armour and guns from Sheffield, produced substantial profits. At the same time Vickers built up their expertise in the design of guns and breech mechanisms, as well as smaller guns at their Erith works.

In 1902 they commissioned Alex Richardson of *Engineering* to produce a lavish 200-page publicity book on the Vickers company, weighing in at 9lb. The company also took care to recruit very capable senior staff, such as Vincent Caillard for finance and engineer James McKechnie, who together with Dawson all became directors. This was quite different from Armstrong Whitworth, which after the retirement of Lord Armstrong became the fiefdom of Sir Andrew Noble who recruited his sons and other ineffectual names with little relevant expertise to the board, with the result that this geriatric group could not compete with the more thrusting Vickers. For example, the latter pioneered submarine construction from 1900 and then diesel engine development, whilst senior Armstrong naval architects Philip Watts and later Tennyson d'Eyncourt could be headhunted to become the prestigious Directors of Naval Construction at the Admiralty. In 1902 Vickers was paying James Horne its chief draughtsman in the gun mounting department £1050 a year, as much as a Rear Admiral in the Navy. Dawson received bonuses in 1906–1908 totalling £42,382, the same as Basil Zaharoff, Vickers salesman extraordinaire, say £5M in today's money.[5]

## THE BARROW GUN MOUNTING WORKS

Vickers wasted no time in developing its newly acquired Barrow works to build the full range of ever larger naval vessels. Over the six years 1898 to 1903 capital expenditure at Barrow amounted to £933,462, building new gun mounting shops and modernising the shipyard.[6] In the shipyard the fourteen narrow building berths were changed to ten larger ones, served by new travelling cranes on gantries, and its workshops were extended. Across Ferry Road, there was plenty of space to expand the engine works. The existing three-bay marine engine building shop of 520ft x 150ft was now surrounded on three sides by new gun mounting workshops. To the west Bay N$^{o}$ 1 was 1000ft x 45ft with large circular planing machines for turntables with gun mounting pits N$^{os}$ 1–4, served by cranes up to 60 tons. To the south, shorter 230ft long Bays N$^{os}$ 2 to 10 produced smaller components, N$^{o}$ 2 with milling machines, N$^{o}$ 7 planing machines, N$^{o}$ 9 for gun slides. Galleries over N$^{os}$ 8 and 10 were used to make hydraulic machinery components. To the east, N$^{os}$ 11 to 13 were added, each 50ft wide, the latter including erection of smaller mountings. N$^{o}$ 11 was for erecting garrison mountings (*ie* coast defence), N$^{o}$ 12 had eight pits, N$^{os}$ 5 to 12 for mountings up to 12in, soon to be extended northwards with three more large pits, N$^{os}$ 13 to 15. Bay 13 was two storeys at its southern end, with finished parts stored below and brass finishing above. This bay too was later extended at its northern end sweeping away two managers' houses and housing Pit N$^{o}$ 16; the area of Bays 11

[2] Cambridge University Library, Manuscripts, Vickers 1116 Contract Book No 1

[3] *Ibid*, Vickers 1146

[4] *Ibid*, Vickers 1146

[5] Sheffield City Archives, Vickers accounts Group 1988

[6] Vickers, Sons & Maxim Annual accounts 1903. BAE Systems, Barrow shipyard archives

Vickers extended the existing marine engine building shops at Barrow to include gun mounting shops. An extensive write-up in *Engineering* of 26 July 1901 included this view of Bays N^os^ 9 and 10 making gun slides and in the foreground the roller path of a twin 6in mounting.

to 13 was doubled.[7] The whole complex of about 60 acres was managed as one unit under the engineering director James McKechnie, sharing a labour force who had similar skills; about 1500 were working on gun mountings alone by 1900.

Part of the existing boiler making shop was converted to fabricating steel plates (up to 1¾in thick) and girders for the riveted structures of the mountings. Those parts were assembled into up to five turntables at a time in an added 50ft wide bay, served by cranes up to 100 tons. New foundries for iron, steel and brass castings were built on open ground to the west. A new shop was built for making shells up to 12in and another for field gun carriages. All the shops were served by standard gauge rail tracks which connected up with the Furness Railway with a line down Ferry Road, with narrow gauge rails in each bay to move items around. In all 780 machine tools, British, American and German, were installed across the shipyard and engine works. By 1901 the Barrow works was employing 10,300 workers, up 90% from N C & A days, with a weekly wage bill of £16,500 (average £1.60).[8] The influx caused a housing shortage at Barrow such that Vickers bought the redundant passenger liner *Alaska* as an accommodation vessel and started building over a thousand new houses at Vickerstown on Walney Island opposite the shipyard, then only connected by ferry, replaced in 1908 by an opening bridge, when Ferry Road became Bridge Road.

Vickers Erith works, started by Nordenfelt, manufactured transferable mountings up to 6in calibre and machine guns, while Crayford, started by Maxim, made guns up to 3in. The heavier guns and armour continued to be made at Sheffield.

While still N C & A, *Engineering* had published a detailed description of the works over five issues from 7 August to 18 September 1896. Quite possibly this encouraged the Vickers directors to take a closer look at the company which they took over the following year. Vickers itself was ever aware of the need for publicity and invited visits from staff of technical publications. Their extensive Sheffield works were featured in well-illustrated issues of *Engineering* between 1 October and 31 December 1897. Once the extensions at Barrow were completed, Vickers invited members of the Institution of Mechanical Engineers to visit in July 1901. On that occasion shipyard director James Dunn gave a very detailed paper with many diagrams on 'Shipbuilding Works'.

[7] *Engineering*, 26 July 1901, pp110–120

[8] *Ibid*, p110

Vickers Bay Nº 1 gun mounting shop in full swing in 1911, which had been built alongside the old engineering shops behind the wall to the left. The turret nearing completion is *Orion*'s Y (Contract 115G). The turntable next to it is a 12in for *Australia* (121G). The line shafting to the right powered tools too small to justify their own electric drive. The driver of the 75-ton crane seems to have got dressed up for the photographer. *(Dock Museum 6484)*

Dawson then published an extensive paper on 'Naval Ordnance' (*ie* guns and mountings), reproduced over three issues of *Engineering* in August 1901.

Ever ready to invest around 10% of profits in research and development, Vickers were now keen to bid for Admiralty contracts. Only days after taking over N C & A, on 6 July 1897 they were successful with the order for battleship *Vengeance* of the 1897–98 Programme. That was for hull and machinery, with planned delivery 12 July 1900. Most of her armour came from Sheffield under a separate contract. Her two twin 12in B.V gun mountings were Vickers' first heavy mountings from Barrow. Although work could not start until the new shops were well under way, they had quoted an optimistic delivery date of June 1899. Their tender price was £52,128, but Vickers recognised that this was a loss leading price. Tender details are given on page 60. In the event the ship was two years late and made a loss on the mountings of £35,912, 36% on cost.

Up to the merger in 1927, Vickers and Armstrong notionally

*Vengeance* was the first warship ordered with Vickers gun mountings in March 1898, although the 12,950-ton pre-dreadnought was not completed until April 1902. Her two twin 12in B.V mountings were built at a loss, as a low price had been tendered to break into the market. Vickers also built the hull and machinery (also at a loss) and her armour and guns – the only company able to build an entire warship. *(World Ship Society)*

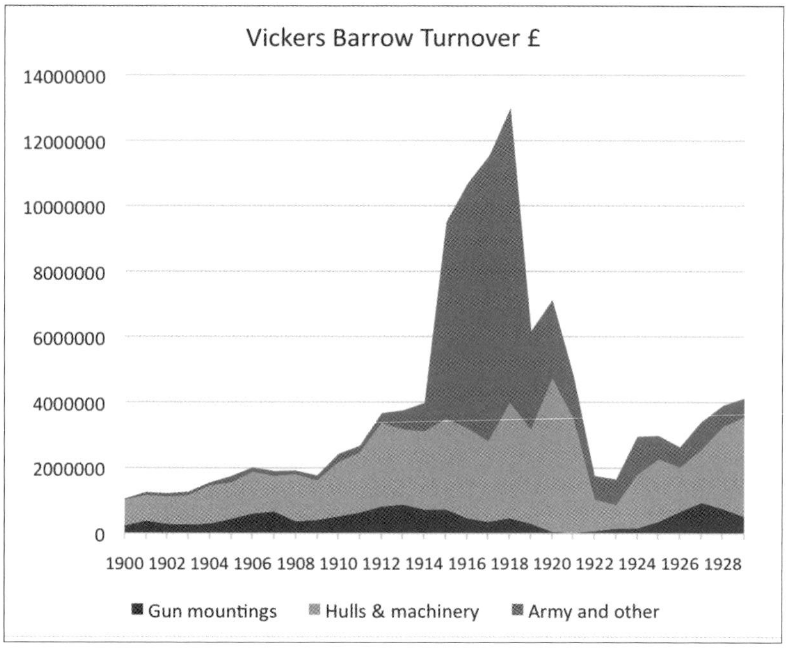


competed for naval armament contracts. In practice they colluded on prices, keeping them on average at highly profitable levels before WW1, with examples on page 64. To an extent this was a defence mechanism providing a cushion against highly variable government annual ordnance orders. In a few cases, mountings for a ship were supplied by both companies – *eg* to *Dreadnought, Thunderer* and *Erin* – to speed delivery in the former's case. For export orders, there was private agreement that the 'winning' company would pay an agreed percentage of the contract price to the 'losing' company.[9] There was often significant international competition for ordnance export contracts, but Vickers was able to keep a large share of these based as much on technical excellence as on keen pricing. This was a win-win situation for the two companies, but it did maintain a substantial level of national production capacity, which was badly needed in WW1. For a similar reason, the Admiralty awarded contracts roughly equally between the two companies, but were pleased when a third company, Coventry Ordnance Works, started making heavy gun mountings from 1909, putting a downward pressure on prices, and reducing profit margins.

[9] *The Battleship Builders*, p229

Non-transferable mountings were ordered for a specific ship, but transferable mountings were ordered in batches, and allocated by the Admiralty to ships (or depots as spares) as necessary. While initially most of these latter had been built at Erith, by 1930 most were being made at Barrow. Transferable mountings were small enough to be transported by rail to dockyards who until about 1904 installed the entire armament of ships built by private shipyards. Non-transferable mountings were erected in pits at the works for testing, then dismantled for transport to the shipyard, usually in a coaster. In 1903 Barrow had installed a German 150-ton crane for such work, building a second British 150-ton crane in 1908.[10]

For the years up to WW1, the turnover of the gun mounting department was between about £250k and £850k, constituting 20–35% of the total, the rest being ship hulls and machinery, submarines, ammunition, Army artillery equipment and general engineering. These latter dominated during WW1 and the early 1920s until gun mounting orders picked up again in the late 1920s. During WW1 huge orders for ammunition were received from the Ministry of Munitions with Shells being the largest department in terms of turnover in 1916–1918. But in April 1922, only 1457 were working in Ordnance out of a Barrow total of 8252 (18%) [11]

Appendix 1 lists major gun mounting contracts.

[10] *Ibid*, p149
[11] Vickers Ltd Annual accounts 1916–22. BAE Systems, Barrow shipyard archives

## CHAPTER FIVE

# Coventry Ordnance Works

CHARLES CAMMELL had noted that fellow Sheffield steel and armour makers Vickers and John Brown had successfully taken over shipyards enabling them to take profitable orders for export warships, including the Japanese battleships *Mikasa* at Barrow and *Asahi* at Clydebank. So Cammell acquired Birkenhead shipbuilder Laird Brothers in 1903, forming the Cammell Laird company, which was already in the process of expanding its works through the Tranmere Bay Development Company. Vickers having started gun mounting manufacture at Barrow, Cammell acquired the Mulliner-Wigley company in January 1903 for £142,000. Its 60-acre works at Coventry, acquired in 1902, had been producing field guns and army vehicles for the Boer War. John Brown took a half share in the business in 1905. Soon after, Cammell Laird sold half their share to Fairfield, half of whose ordinary share capital CL had just purchased. A new company was then formed: the Coventry Ordnance Works Ltd with 50% John Brown and 25% each Cammell Laird and Fairfield.

By 1905 the Cammell Laird Board was able to report that the Coventry works were actively employed. However, this was only for smaller calibre ordnance such as transferable naval mountings and field guns, as the 16-acre Scotstoun works for erecting heavy mountings was not completed until 1907. It also had a gun testing range at Freiston on the Wash and a shell filling plant at Cliffe on the Thames. The Coventry works was featured in an extensive write up in *The Engineer* of 31 May 1907. Some machine tools were made by the company; others were bought from America. Both British and German overhead travelling cranes of up to 75 tons were installed. Investment totalled about £1 million.[1]

There was plenty of capacity for transferable mountings at Armstrong Whitworth and Vickers and Woolwich. The latter was rarely fully employed in peacetime, being seen partly as reserve wartime capacity. But from 1907 Government policy was to allocate about one-third of guns to the Royal Gun Factory and two-thirds to industry which enabled COW to get a modest share of smaller calibre gun orders. [2]

COW struggled to break into the duopoly of Elswick and Barrow for the heavier naval gun mountings. The latter objected to the newcomer, claiming the need to standardise; each was manufacturing to the other's design if that had been chosen by the Admiralty. COW's managing director Herbert Mulliner persistently lobbied the Admiralty for orders. He claimed to have inside information that Krupp was greatly expanding its ordnance manufacturing capacity at Essen, such that the Germans could produce as many battleships a year as Britain. During 1908 this information was taken up by the Conservative opposition in Parliament, pressing for more battleships, while the Liberal government was trying to reduce naval expenditure. Ultimately the government was forced by a public campaign to agree to not only an order for four capital ships (three battleships and one battlecruiser) in the 1909 Navy Estimates but also four more 'contingent' on events in Germany. Mulliner's extravagant claims made COW unpopular with the Admiralty, and the COW Board forced Mulliner to resign in June 1909. At the same time, Cammell Laird had only just got back on to the Admiralty list of suppliers, having been taken off when it was found to have falsified tests for steel and armour plate in 1906. Francis Elgar, formerly Director of Dockyards at the Admiralty, had been brought in as Chairman. But he died in 1909 after cleaning up the company and weeding out senior personnel, so not only was Cammell Laird looking for a new Chairman but also a new managing director for its subsidiary COW. Perhaps recalling Elgar's antecedents, the Chairman of COW, Charles Ellis of parent company John Brown, approached Reginald Bacon, who had made a name for himself at the Admiralty introducing new technologies, first as Inspecting Captain of Submarines in 1900, then as the first commanding officer of *Dreadnought* in 1906, then as Director Naval Ordnance from August 1907. As DNO Bacon had succeeded Jellicoe who in 1908 became Third Sea Lord and Controller in charge of all materiel including ships, machinery, ordnance and armour. According to Bacon's autobiography, Jellicoe advised Bacon that the latter was lined up to become the next Controller in 1910. The 46-year old Bacon, newly promoted to Rear Admiral, coveted this position, which he also saw as a possible stepping stone to command of a major fleet, as Jellicoe subsequently achieved. But he recognised that as he was short of time in command of seagoing squadrons, that last step was unlikely to happen. So when Ellis approached him, offering him a reported £7000 a year as managing director of COW – the same as

[1] Warren, Kenneth. *Steel, Ships and Men*. Liverpool University Press, 1998, p144

[2] Breer, Andrew. 'British Industrial Policy Concerning the Heavy Ordnance Industry, 1900–1917', p82. King's College London thesis, 2015

Elgar – he accepted; this was over six times his £1095 naval salary.[3] From COW's point of view this was a coup. Bacon would bring not only Admiralty contacts but also inside information on ordnance policy and contractual information from competitors. At that time there was no bar on Government officials taking up lucrative positions in industry, indeed Armstrong also did the same to strengthen its somewhat geriatric Board (in 1912 Rear Admiral Sir Charles Ottley, formerly Director of Naval Intelligence, Sir George Murray, ex-Treasury and Sir Percy Girouard, ex-Colonial Office). Bacon immediately resigned as DNO, taking up his new post at COW in November 1909, initially using its London office at 30 Broadway. He then moved to the Manor House, Wolston, 5 miles south east of the Coventry works, where the 1911 Census shows him living with his wife and daughter, with a butler, governess and chauffeur plus six other servants.

The situation at COW was not encouraging. It was operating at under half its capacity, had no lucrative foreign orders and was employing fewer than half of Elswick's ordnance works numbers. It made a loss of £50,000 in 1910 and paid no dividends.[4] The only bright spot was an order for its first heavy gun mountings. In August 1909 COW had submitted a tender for twin 13.5in mountings for the 1909 Programme battleships. The Admiralty wished to have a third supplier of such, rightly suspecting that Armstrong and Vickers were colluding to keep prices high both for Admiralty and foreign orders – indeed their profit margin was then around 50% on cost.[5] Just as Bacon was appointed, COW were formally awarded the contract for *Conqueror*'s mountings on 10 December 1909, worth over £300k. Excluding gunhouse armour which had the same fixed price per ton for all mounting suppliers, this was about 16% less than their rivals. A proviso was that no arrangements were to be made with Armstrong or Vickers, so that COW had to develop its own detailed design, designated 13.5in Mark III. The inexperienced company also had to get the three vital hydraulic pumps from elsewhere. Even if Bacon had probably brought with him ordnance drawings from the Admiralty, plus technical and financial information from recent tenders, COW was still faced with the problem of designing the myriad of components that made up each mounting to meet the delivery date of 20 October 1911. Inevitably COW encountered the steep learning curve that Vickers had ten years earlier trying to break into the market. Delays occurred even with placing simple orders for the gunhouse armour. Most components including hoists were manufactured at Coventry and then railed to Scotstoun for erection. Scotstoun made the larger components including turntable and working chamber, which were erected and tested in its five pits, 36ft in diameter and 45ft deep. *Conqueror*'s builders were Beardmore, also struggling to complete on time their first battleship contract at their new Dalmuir shipyard. The mountings were not ready for trials until mid-1912; fortunately COW's contract did not include the usual £50 per day late penalty. Indeed *King George V* of the 1910 Programme completed a day ahead of *Conqueror*. Embarrassingly and with the First Lord, Churchill, on board, the first gun trials were a failure, the complex hoist systems differing from previous ships; see Chapter 10 and Appendix 3 for Elswick's report. Modifications had to be made, with COW making a large loss on this contract.

Bacon was keen to get overseas orders, but was handicapped by COW's lack of a track record with the Admiralty. He appointed fellow RN officer, Captain T E Crease, essentially as overseas salesman, and appointed agents in possible customer countries. (Crease went on to become the warship building representative at the trade association the Shipbuilding Conference.) Bacon himself made sales trips abroad, spending some time in Japan in 1913 with senior Japanese naval officers whom he had helped in the past, probably with submarines. But the Japanese were keen to build up their own warship industries, only buying British warships, machinery and ordnance in order to develop their own expertise – a form of industrial espionage, especially as they sent a large number of staff to 'oversee' each contract. They had already set up a joint venture with Armstrong and Vickers to form the Japan Steel Works (Nihon Seiko-Sho 25% share) to manufacture ordnance at Muroran. Side trips by Bacon to Canada, China and Russia brought no results, and COW had no success with Turkey or South America where their competitors were well established.

At the company, Bacon's arrogant quarterdeck manner and lack of commercial experience did not endear him to his managers, often overruling their advice and promising unattainable delivery dates. In 1912 the managers submitted a range of complaints about Bacon to the Board, detailed in Warren.[6] Despite misgivings, the Board decided to keep Bacon in post, despite the bad atmosphere in the company. Meantime losses were building up and the parent companies were reluctant to keep bailing COW out, especially Fairfield whose finances were not robust. Fresh funds had been raised in 1910 with the issue of £600,000 of debentures, but more loans were required from the parent companies.

Until 1911, the Director of Army Contracts had been responsible for the ordering of guns for the Admiralty, ostensibly to avoid competition for manufacturing resources between the two services. But from about 1909 the Admiralty had had the freedom to place their own orders for gun mountings where they felt it advisable, and wanted the same for guns, which were the subject of a separate Vote in the Navy Estimates. Unlike Armstrong and Vickers, COW could only build one set of battleship mountings at a time but the Admiralty deemed it prudent to keep them afloat, first with an order for five twin 13.5in for *Ajax* in 1911 and another five for *Benbow* in 1912. CL benefited with an order from COW in July 1911 for 951 tons of *Ajax*'s gunshield armour worth £117,000. In January 1912 they also got an order for 32 6in mountings, probably P.VII for cruisers. COW also benefited from a pre-war Admiralty decision to place all gun mounting orders with industry, allowing Woolwich to concentrate on guns and carriages for the Army. Modest numbers of 12in, 13.5in and later 15in guns had been made at Coventry from about 1909, with forgings supplied from Cammell Laird and John Brown at Sheffield. In 1914 it got an order for 26 4in BL.VII guns, typically battleship secondary armament. Pre-war they had

[3] Warren, *Steel, Ships and Men*, p148

[4] Warren, *Steel, Ships and Men*, pp146, 149

[5] *The Battleship Builders*, p190

[6] Warren, *Steel, Ships and Men*, pp151–3

designed a successful 4.5in howitzer for the War Office, so that by the end of WW1 2325 of these guns and 2799 carriages had been made.[7]

So by 1914 COW was in a healthier situation, employing about 4000 at Coventry and 1000 at Scotstoun, making a profit that year of £58,000. The Coventry Syndicate (*ie* COW and the three shipbuilders) had at last been successful with a large foreign order, namely two cruisers (Cammell Laird) and four destroyers (two John Brown, two Fairfield) for Greece, financed by a large loan. Incomplete when war broke out, all were taken over by the Admiralty early in 1915. The cruisers would each have ten 5.5in 50-calibre guns designed by COW, the destroyers three 4in each. From 1915 production ramped up, more for the War Office than the Admiralty, with the works expanded. Four 15in mountings had been ordered for *Royal Sovereign* in 1913 and four for *Renown* in 1914, but when the latter was redesigned as a battlecruiser early in 1915, the fourth mounting was switched to *Courageous*; all were completed by 1916. Bacon gives a list of COW's impressive output during WW1.[8] In addition to the eight 15in and fifteen 13.5in twin turrets, he notes 174 6in, 65 5.5in (Nos 1–58 for the Admiralty), 385 4in single mountings plus conversions, and new smaller guns, repaired guns and ammunition; also much for land service especially howitzers. Post Jutland COW also converted 33 13.5in mountings from hydraulic to pneumatic run-out.

COW also rapidly designed a 15in howitzer for the Admiralty, making twelve by early 1915. Bacon offered his services to take command of these guns in France; Churchill accepted, so he resigned from COW, Colonel John Mansell succeeding him as managing director. Bacon went on to become Admiral in command at Dover. Four twin 15in Mark II for *Hood* class *Rodney* to be built by Fairfield were ordered at COW in 1916. When the mountings were cancelled, COW bought them as scrap but modified one turntable in 1920 to take three 15in guns to be tested in monitor *Lord Clive*.

After the Armistice, COW like the other ordnance manufacturers recognised that they faced a dearth of orders. A proposal to the government to support it as a nucleus armament factory, requiring a subsidy of £250,000 a year, was turned down. The parent companies decided to dispose of the company's works, Coventry going to the newly formed English Electric company in December 1918, now an industrial estate.[9] Scotstoun was sold to Harland & Wolff in 1920 for £475,000, a good price.[10] It was to be used for expanding diesel engine production; H&W saw a great future for making diesel engines, mostly to the Danish Burmeister & Wain design, at both its Belfast and Glasgow works. Four of the five gun mounting pits were filled in during 1921. The Coventry Ordnance Works company went into voluntary liquidation in May 1925.[11] Due to lack of orders, the works were put on care and maintenance between 1927 and 1936.

However, with rearmament starting in 1936, with the limited capacity of Vickers-Armstrongs and the risk from air attack, the Admiralty recognised the need for a third independent producer of the heavier gun mountings. Reinstating Scotstoun's gun mounting capacity would also minimise transport of mountings to Clyde shipyards. It became H&W's Scotstoun Ordnance Works, initially with only 9 staff and 87 workers. The first orders in 1936 were for sixteen single 4.7in Mark XVIII mountings for destroyers. This enabled H&W to order new machine tools and plan extensions to their shops. The old gun mounting pits were cleared of rubble and the loading dock dredged. About £311,000 was spent initially.[12] A second expansion saw additional shops built west of the original ones, nearly doubling their area,

[7] Bacon, Reginald. *From 1900 Onward*. Hutchinson, 1940, p385
[8] *Ibid*, p384
[9] Warren, *Steel, Ships and Men*, p180
[10] Moss, Michael & Hume, John. *Shipbuilders to the World*. Blackstaff Press, 1986, p222
[11] Warren, *Steel, Ships and Men*, p179

This advertisement for Coventry Ordnance Works from the 1919 edition of *Jane's Fighting Ships* was an attempt to attract new business as wartime orders dried up.

[12] Directorate of Naval Ordnance, War Production. CAB 102/526

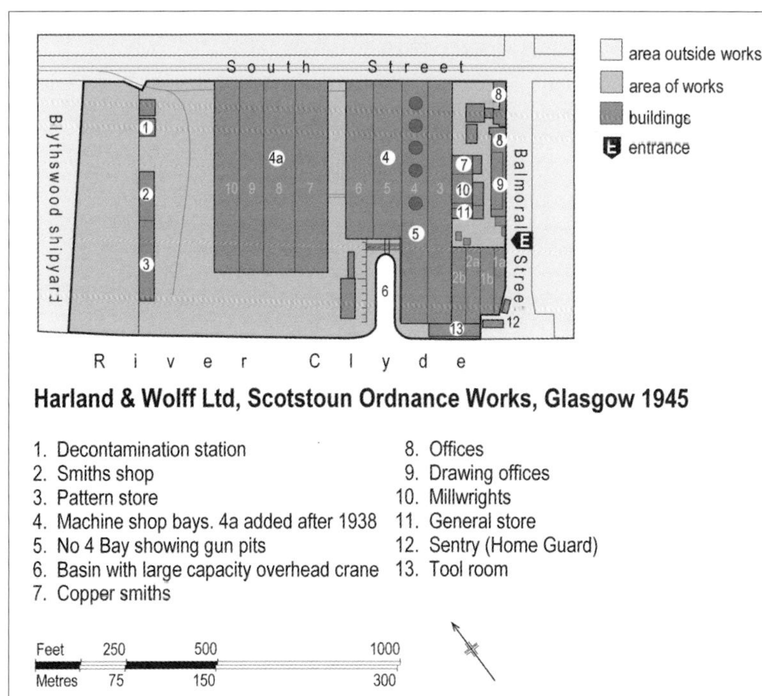


**Scotstoun plan.** Coventry Ordnance Works had built its Scotstoun works on the north bank of the Clyde in 1907 for erecting heavy gun mountings. Closed between the wars it was brought back into ordnance work in 1936 and operated by Harland & Wolff, with the Admiralty paying for new shops built to the west of the loading dock. *(Ian Johnston)*

An erection shop at Scotstoun after Harland & Wolff had reopened the ordnance works, showing a twin 4.7in CP XIX, probably for 'J' or 'K' class destroyers, about 1939. *(Harland & Wolff)*

financed largely by the Admiralty. A second order was for fifty twin 4.7in Mark XIX for the 'J', 'K' and 'N' class destroyers, with output planned for four per month. The three 25-ton mountings for *Kelly* were shipped to the Tyne in a coaster in July 1939, to be lifted aboard by the floating crane *Titan II*. It was followed by 24 twin 5.25in Mark I mountings for three of the new battleships (40 originally ordered but some were spares and others were later cancelled) – see page 192.

By 1939 employment had built up to 1000, although, as ever, skilled men were in short supply. Four of the original five pits were used from 1942 for modernising the four twin 15in removed from *Courageous* and *Glorious* for *Vanguard*. Up to and during WW2, 143 single 4.7in Mark XVIII were completed, some arming the gun armed landing craft (LCG) of 1943, 91 twin 4.7in Mark XIX, 64 single 4.7in Mark XXII for destroyers, 60 single 4.5in Mark V plus 50 of the latter with RPC also for destroyers; 250 single 20mm Oerlikon Mark IIIA mountings were also made. They also designed the single 2pdr Mark VIII, although they were manufactured elsewhere, including at H&W's Finnieston engine works. The works were seriously damaged by land mines in the Clydebank air raid of 14 March 1941, with production not resumed until 4 April.[13] Part of the Coventry works was taken over as an Admiralty gun mounting store.

About 1944, it received orders for the new twin 4.5in Mark 6 mounting for destroyers, but most were cancelled; only six were in hand in 1946. Unlike after WW1, the Admiralty continued to use Scotstoun after WW2. The Mark 6s continued to be produced in modest numbers, perhaps no more than one or two a year, the last contract being in 1966. A mothballed 4.5in mounting had been stored at the head of the loading dock in the 1960s, and the last Reg Nº 231 was shipped to Vickers at Barrow in 1969, presumably for storage. Although the works had also been making compressors and car body presses, it had been losing money from 1966. H&W closed the works in December 1969, having sold them to Albion Motors of Glasgow for £800,000 to develop their truck manufacture.[14] Many of the buildings were still in use in the 2020s by American Axle & Manufacturing.

[13] Naval New Construction Programmes. CAB 102/527

[14] Moss & Hume, *op cit*, p433

# CHAPTER SIX

# Vickers-Armstrongs Ltd

## THE FIRST DECADE

After the merger, effective on 1 January 1928, the Vickers-Armstrongs' Board decided to concentrate gun manufacture at Elswick, leaving only a nucleus facility at the former Vickers Sheffield works – Barrow had never made guns. Gun mounting production ceased at the former Whitworth Openshaw works with its four pits for 6in and 8in mountings after completion of *London*'s twin 8in mountings in 1928. The works were retained in the rump of the Armstrong Whitworth organisation, but its armour production was transferred to English Steel Corporation at the former Vickers Sheffield works, which company was a merger in 1929 of Vickers, Cammell Laird and Armstrong Whitworth steel interests. At Elswick, ESC operated the forge shops, used in gun making. The Erith works were closed in 1931 with transferable gun mounting construction moved to Barrow (also from Elswick) and director firing gear moved to Crayford, although Barrow made the director structures and assembled them. It still retained an interest in the Japanese and Spanish ordnance works.

Under the ten-year rule, no new battleship construction was envisaged, so no heavy gun mountings would be ordered for over a decade, although design studies were continued for the Admiralty for mountings of 12in and over. With the end of 8in cruiser construction in the 1930 Programme, the only heavy gun mounting orders that could be expected for Barrow and Elswick would be for cruiser 6in. The first order for twin 6in Mark XXI for the *Leander* class was placed in 1930 (see page 128). Throughout the 1930s, a fairly steady stream of 6in mountings was ordered by the Admiralty, with the triple Mark XXII ordered from 1933 then the Mark XXIII, with 46 being completed between 1934 and 1939.

Transferable mountings continued to be produced for destroyers, sloops and submarines plus anti-aircraft guns for all types of ship. Some export orders were received from Chile, Portugal, Brazil, Siam (Thailand), Sweden and Holland and even Japan, of calibres from 40mm to 4.7in, usually on credit terms. The most substantial of these was for four 15in coast defence single mountings from Barrow for Spain in 1925, followed by fourteen more, at about £70,000 each, a highly profitable order worth nearly £2M including the guns at about £18,000 each, all too large to be manufactured in Spain. These were installed at Ferrol, La Coruña, Cartagena and Mahon, although trials of the later mountings were delayed by the Spanish Civil War and had not all been completed before WW2 broke out. Several mountings still exist, including two at La Coruña and four at Cartagena.

One twin 8in was supplied from Barrow for the Spanish cruiser *Canarias*, plus the design allowing manufacture of a further seven in Spain – see page 123. Such orders enabled a core of designers and skilled workers to be employed throughout the depression years. But in the early 1930s, employment at Elswick had dropped to about 3500, compared with about 13,000 (excluding the shipyard) in 1914, and 8000 at Barrow, which also included the shipyard. Charles Craven, a former RN Commander who had joined Vickers in 1912, was appointed managing director of Vickers-Armstrong in 1931, strongly believing in maintaining the company's facilities and even investing in new, including the 250-ton crane at the Naval Yard, still in use in the 2020s.

When the major reconstruction of battleships started with *Warspite*, in 1934 Elswick were awarded the contract to upgrade her four twin 15in mountings to a new Mark I/N standard. This entailed increasing the elevation from 20° to 30° but retained the old arrangement of magazines above the shell rooms. Mounting components were shipped to Elswick for rebuilding, which was completed at the end of 1936. Elswick also received orders to reline a number of existing 15in guns at about £5500 each.

Not before time, re-armament started in 1936, requiring a big increase in output, also of armour production which included gunhouse protection. After exploring many alternative designs of battleship mounting guns from 12in to 16in calibre, the Admiralty settled on 14in. Ten were to be fitted in each ship, the most possible within the 35,000-ton Treaty limit if adequate speed and protection were to be maintained. V-A started the design of 14in guns and mountings in 1935, each ship to receive two quadruple Mark III and one twin Mark II. The order for the mountings for the first two ships were placed on 15 May 1936, for *King George V* at Elswick and *Prince of Wales* at Barrow, with delivery planned for 1939. But the quad turntable at 39ft diameter weighed 200 tons with lifting gear, so cranes had to be upgraded both at the works and at the shipyards. A fourth large pit was needed at Elswick in an extension to Nº 24 Shop in 1937. The new pit would be 48ft deep and 30ft inside diameter, with the upper part 46ft

The new North Shop at Barrow with three pits was built just before WW2 for erecting the 14in mountings. The turntable for *Howe's* A mounting was the heaviest component at about 187 tons, seen probably in 1941. Also visible are the turret buffer stops to prevent training beyond ±143°. Below is the turntable for Y mounting. *(Dock Museum 5060)*

diameter to contain the turntable. The 60ft wide main bay with the four pits then had two 100-ton and one 90-ton cranes, two of which were needed to handle a 14in turntable. Barrow needed a whole new erecting shop, so North Shop was added close to the Boiler Shop with three pits N^os^ 17–19, but to be paid for by V-A unless expected orders did not mature in four years when the Admiralty would contribute. It was also necessary to erect a 250-ton crane close by at Devonshire Dock to lift the heavy components into ships for transport to the shipyards; it was built by Arrol in 1939. But the Admiralty did contribute 60% of the £424,000 for other plant extensions at Barrow. Barrow was also equipped to make director towers for destroyers up to battleships but with internal components supplied from their Crayford works. At Elswick new tank building and shell shops were built on the site of the former shipyard.

Under re-armament from 1936 the Government was prepared to help finance capital expenditure to build warships and their equipment. There were three schemes where Vickers-Armstrongs could benefit:[1]

[1] Cambridge University Library, Manuscripts. Vickers Archive 1435

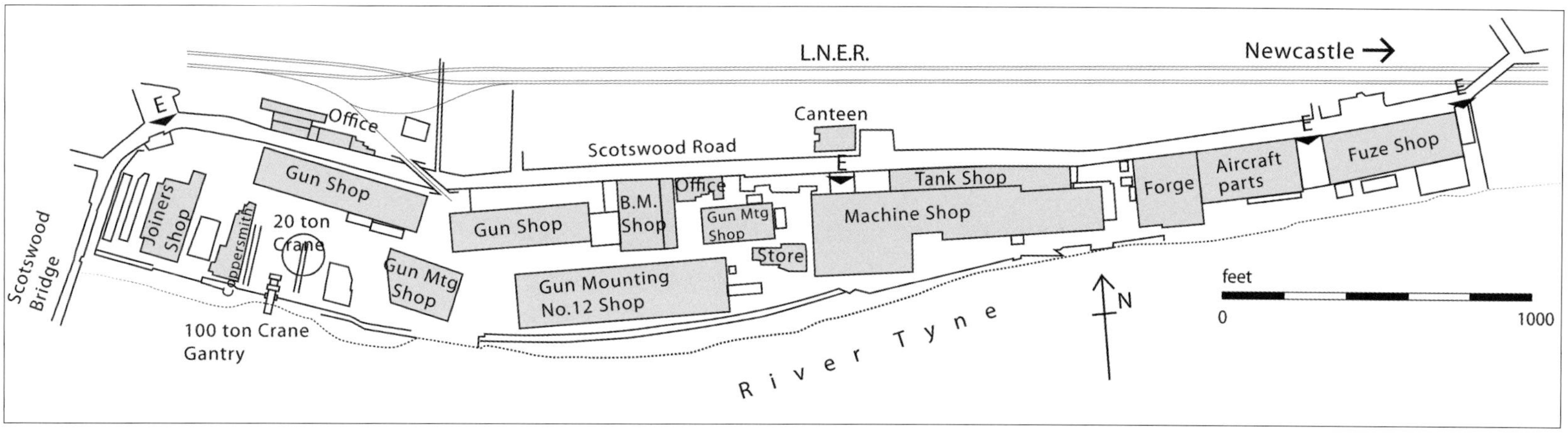


Armstrong Whitworth had set up its Scotswood works, 1.5 miles west of Elswick, in 1899 to make munitions. It was not included in the Vickers-Armstrong merger, continuing to make engineering products including locomotives. It was taken over in 1936 for ordnance work, managed by Vickers-Armstrongs. Its layout in WW2 included shops for medium calibre and anti-aircraft guns. The site became a new tank factory for Vickers in 1982. *(Ian Johnston)*

The Government paying all costs: Scotswood re-equipping for gun mountings £754,956, English Steel Corporation for armour plate plant £1,056,000, Openshaw for anti-aircraft guns £167,500, Crayford for 2pdr guns and fire control £53,599.

The Government paying 60% of the total costs: Vickers-Armstrongs £883,941, ESC £200,625.

Vickers-Armstrongs paying all costs but the Government to pay if insufficient orders were received in the next four years: New gun mounting shop at Barrow £240,500, extension to Elswick gun mounting shop and a new pit £50,800.

ADM 1/17863 records total Government capital expenditure during 1936–1943 as Barrow £518k, Elswick £2444k, Scotswood £2730k with an overall total of £11.681M. But since the companies were not using their own capital, the allowed profit rate on armaments was only 5.5% in 1944–1945.

But the Government did start to bring in general engineering firms to make the simpler transferable mountings, with Harland & Wolff at Scotstoun starting to make 4.7in Mark XVIII singles. In 1936, both Admiralty and War Office eyes were cast on Armstrong Whitworth's works at Scotswood just a mile and a half west of Elswick and not included in the Vickers and Armstrong merger, which had been used to make munitions in WW1. In the 1930s it was producing locomotives and marine diesel engines. Its workshops could be extended and re-equipped with new machine tools to make ordnance items. The Admiralty gave them a trial contract in July 1936 to make forty of the new 4in twin Mark XIX HA mountings.[2] But they found the firm inefficient, so they decided to purchase the whole works for £1,308,000 and hand the management to Vickers-Armstrongs, seen initially as an overflow works from Elswick. Some £600,000 was initially spent by the Admiralty on extending the shops and re-equipping the former crane shop for erecting mountings. A further order for 150 4in Mark XIX and 78 4.5in Marks II and III was received. Scotswood was planned to take manufacture of medium calibre mountings from the parent works including 4.5in BD and 5.25in and anti-aircraft weapons for the War Office.

DNO reported that armament contracts for ships laid down between 1936 and 1939 included:[3]

| | *Elswick* | *Barrow* |
|---|---|---|
| 14in quadruple | 5 | 5 |
| 14in twin | 3 | 2 |
| 6in triple | 32 | 24 |
| 5.25in twin | 18 | 28 |
| 4.7in | 21 | 76 |
| 4.5in twin | | 74 |
| 4in twin | | 30 |

Barrow also received orders for 2pdr mountings, both 8-barrel Mark VI and 4-barrel Mark VII and quadruple 0.5in; see Appendix 1.

In November 1938 the Admiralty assessed the gun mounting industry's capacity and concluded that annual production would be limited to:[4]

Battleships: Two and a half ships 14 to 16in (say 7–8 mountings)
Aircraft carriers: Two 4.5in to 5.25in
6in to 8in cruisers: Four and a half
Light cruisers: Four 4.5in to 5.25in

This was just as much a bottleneck as shipyard capacity. They noted delays in production owing to:

Many new designs stretching drawing office staff
Shortage of and slow delivery of new machine tools
Shortage of experienced managers and skilled men
Much more complex mountings.

[2] Production of Gun Mountings. CAB 102/526.

[3] *Ibid*

[4] ADM 265/2

Nº 12 gun mounting erecting shop at Scotswood on 31 March 1941 showing in the foreground two of *Cleopatra's* twin 5.25in (one with lifting frame ready for shipping) and beyond several twin 4.5in Mark II for aircraft carriers. *(Vickers)*

**Elswick gun production 1936–1944**

| Gun Mark | 1936 | 1937 | 1938 | 1939 | 1940 | 1941 | 1942 | 1943 | 1944 | Total | Typical price £ | Approx Value £ |
|---|---|---|---|---|---|---|---|---|---|---|---|---|
| 14in VII | 0 | 3 | 9 | 11 | 10 | 2 | 5 | 4 | 0 | 44 | 22,000 | 968,000 |
| 6in XXIII | 25 | 29 | 15 | 31 | 12 | 7 | 0 | 3 | 17 | 139 | 2,800 | 389,200 |
| 5.25in I | 0 | 0 | 12 | 39 | 27 | 43 | 0 | 0 | 0 | 121 | 2,700 | 326,700 |
| 4.7in IX, XI, XII | 38 | 35 | 47 | 57 | 102 | 141 | 48 | 41 | 14 | 523 | 2,200 | 1,150,600 |
| 4.5in I, II, III | 0 | 8 | 41 | 61 | 22 | 16 | 0 | 47 | 164 | 359 | 2,340 | 840,060 |
| 4in XVI | 51 | 37 | 106 | 135 | 120 | 88 | 300 | 206 | 92 | 1,135 | 1,600 | 1,816,000 |
| Total | 114 | 112 | 230 | 334 | 293 | 297 | 353 | 301 | 287 | 2,321 | | 5,490,560 |

Also 37 8in Mark VIII 1940–1943 at about £6,500 each and several 16in Mark II in WW2 at about £25,000 each.

*Sources:* Cambridge University Vickers archive; Elswick Works Quarterly Reports; Elswick Accounts and WW2 Output. See also Appendix 2.

Barrow was particularly hard pressed: the Admiralty found only 370 fitters and turners working on *Prince of Wales*' 14in while Elswick had 640 on *King George V*.[5] But Barrow was also working on mountings for the Army, on export orders and on commercial work, these latter two deemed important by the British Government. The Admiralty encouraged V-A to do more sub-contracting of components, although that increased costs.

Other firms brought in included Marshall & Sons of Gainsborough, who normally made agricultural machinery, taking on twin 4in Mark XIX, 117 such being on order in 1939, later making single 4in Mark XXIII for escorts. Harland & Wolff at Belfast and Garrett at Leiston in Suffolk were also used for 4in. In 1939, Openshaw was brought back into gun mounting manufacture under Vickers-Armstrongs management but only for smaller mountings including the 4-barrel 2pdr Mark VII and later 40mm twins.

## WARTIME PRODUCTION

Wartime demand for gun mountings of all types put great pressure on V-A, with production limited by skilled staff – unskilled labour, dilutees and women could only be used for a limited range of tasks. Both at Elswick and Barrow, the works were also used for a wide range of other products. At Barrow the same facilities and staff were used to make marine engines and submarine diesels, while Elswick also produced guns and breech mechanisms, tanks, hydraulic machinery and Army equipment. James Callander, general manager of the Barrow works in his third quarter 1940 report to the V-A Board stated: 'The controlling factor in production in both the engineering and shipbuilding departments continues to be the supply of men both skilled and unskilled. Our manpower is insufficient for our contract commitments and delays are accumulating.' There was a particular shortage of fitters, affecting engineering. At that time Barrow was employing 18,600 people, about two-thirds in engineering which included gun mountings. The pool of labour in this relatively remote area of the Furness peninsula was much less than around Tyneside serving Elswick, then employing 14,000. The trade unions were reluctant to accept many 'dilutees', *ie* semi skilled men who could be trained to do repetitive work. However, the number of women employed continued to increase, both in the works and the offices.

[5] CAB 102/526

Elswick was a major supplier of guns to the Admiralty, as well as the War Office, making over 2000 of 4in and above as shown in the tables, which excludes 3.7in and below, War Office, foreign, relinings, spare guns and loose barrels. Prices include breech mechanism typical of *c*1937.

Elswick Nº 7 Shop with at least four triple 6in in course of erection. *(Vickers)*

**Right:** Scotswood built mainly medium calibre mountings such as these 4.5in Mark IV, which could be erected on the shop floor without pits. Dated 11 April 1944 these are probably the first two for *Barfleur* building at Swan Hunter. *(Vickers)*

During the war production priorities were continually changing, which affected supply of materials and allocation of labour. Early in the war anti-aircraft mountings had priority. Inevitably production of heavy gun mountings slipped, often affecting the delivery of ships. Work on 6in mountings for *Uganda* and *Ceylon* was suspended in 1940, while only nine out of the first eleven of the *Dido* class completed with their planned five 5.25in mountings. Cancelling some orders and reducing the number of spare mountings did ease things a little. While six triple 16in mountings had been ordered in 1938 for the first two *Lion* class battleships (three at Elswick, three at Barrow) little progress was made, partly due to the suspension of construction of the ships. However, a start was made in 1942 on assembling one of *Lion*'s 16in triple mounting at Elswick – see page 116.

The Elswick accounts from 1931 to 1947 are usefully broken down between the various products by sales value, with annual naval armament sales shown in the graph; 'Hydraulic mountings' covered 6in and over.

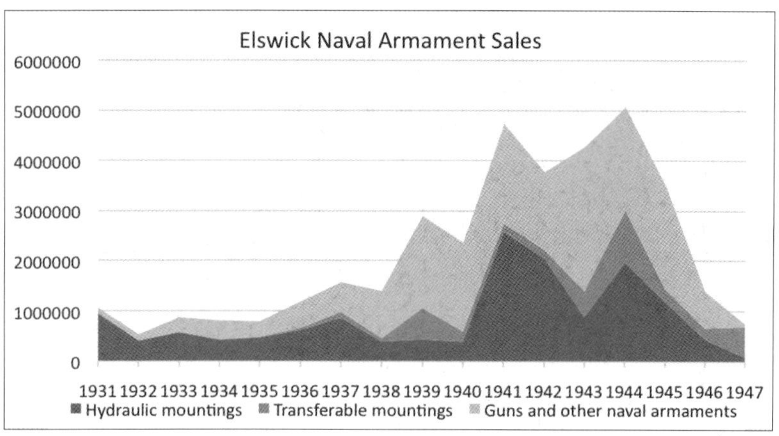


For most of the period they amounted to 30–40% of total Elswick sales. Over the period 1939–1945 they averaged around four times the annual peace time figure:

| *Annual average 1939–1945* | £ | % |
|---|---|---|
| Hydraulic mountings 6in and over | 1,362,230 | 35.80 |
| Transferable mountings | 423,239 | 11.12 |
| Guns and other naval armaments | 2,019,916 | 53.08 |
| **Total** | **3,805,385** | **100** |

Scotswood concentrated on transferable mountings. Elswick profit margins 1937–1939 were about 20% of costs for hydraulic mountings, but only 9% for transferable mountings.

A number of Vickers-Armstrongs annual accounts and works quarterly reports have been analysed to give a general picture of the range of products and their relative value for both Elswick and Barrow.[6] Figures

[6] Cambridge University Library, Manuscripts. Vickers Archive 208–211

An aerial view of Vickers-Armstrongs' works at Elswick on the north bank of the Tyne post WW2. The 150-ton crane is the black erection just below the two funnelled ship in the river and Nº 24 Shop for erecting the heaviest mountings is the three bays immediately below the crane. *(Author's collection)*

exist for expenditure to date at the end of 1943 on contracts in hand, although for major contracts like heavy gun mountings and ships, costs would be incurred over several years. Around half of Barrow's output by value was in ships and their machinery. Although Elswick produced neither, it did make submarine parts, while V-A's Naval Yard on the Tyne built hulls. The latter's equivalent figure was about £10M with Scotswood a similar amount. For both works the total expenditures are similar at about £47M. Generally, Barrow reported higher profits as a proportion of sales than Elswick despite higher wages, suggesting higher productivity. In the table, the Ministry of Supply procured both Army and RAF ordnance, while the Admiralty retained its own procurement organization.

Gun mounting work at Barrow constituted 37.9% of the total with Elswick at 26.3%, but adding guns to the latter takes the percentage to 39.3, close to Barrow.

In 1941 8in mountings had been planned for a new class of cruiser, but resources were simply not available to take those forward. At the end of the war construction of triple 6in Mark XXIV was suspended (see page 138) while bulk orders for the new 4.5in Mark VI were reduced.

Neither Elswick nor Scotswood seem to have suffered serious damage

**Expenditure at 1943 £k**

| | Elswick | Barrow |
|---|---|---|
| Gun mountings for Admiralty | 8,242 | 12,603 |
| Gun mountings for Ministry of Supply | 4,194 | 5,242 |
| Guns & breech mechanisms for Admiralty | 4,255 | |
| Guns & breech mechanisms for Ministry of Supply | 1,892 | |
| VSG pumps | 3,058 | |
| Tanks and tractors | 15,391 | 5 |
| Shells, cartridge cases, fuzes & primers | 5,931 | |
| Aircraft parts | 2,196 | 4,106 |
| Torpedo tubes | 239 | 259 |
| General engineering and gun proofing | 992 | 1,454 |
| Submarines | 882 | 9,302 |
| Surface warship hulls & machinery | | 9,122 |
| Merchant ship hulls & machinery | | 2,854 |
| Fire control & directors | | 1,159 |
| Mines and depth charges | | 269 |
| Foundry and repair work | | 650 |
| Total | 47,272 | 47,025 |
| Average number of employees 1943 | 18,971 | 19,569 |
| Average weekly wage £ | 5.08 | 5.60 |

**Right:** The main gun mounting production for Elswick post-war was 6in and 3in twins for the *Tiger* cruisers and 4.5in Mark 6 twins. Here one of *Hampshire*'s two 4.5in mountings is waiting outside Nº 24 Shop for loading into a ship on 2 October 1961 for delivery to Clydebank. *(Tyne & Wear Archives)*

FRONT

from air raids during the war, despite their obvious location on the north bank of the Tyne. Barrow reported two raids: the one on 15 April 1941 included two unexploded bombs but had little effect on production; but the one on 4 May with HE bombs demolished three workshops, suspending some production.[7]

Employment at Elswick rose from about 11,500 in 1937 to about 19,000 in 1943, and that at Barrow from 14,000 to 20,000 with Scotswood about 8600 in 1943. Unfortunately, no breakdown of those figures was given by department, only some figures for apprentices and women. The figures for Barrow also include the shipyard. Shipbuilding was more labour intensive than engineering which latter had a higher proportion of bought-in materials and components. At a rough estimate the total split there was probably about one-third each to gun mountings, to marine engines and other engineering, and to ship and submarine hulls.

Post-war, Scotswood completed its 6-barrelled 40mm mountings orders about 1950, reverting to civilian manufacturing such as industrial tractors, but those activities ceased in 1979. The site was then rebuilt to become a major tank manufacturing facility in 1982 for Vickers Defence Systems. But with a dearth of orders for the British Army and export, it was finally closed by BAE Systems in 2014.[8]

The Admiralty continued to place modest gun mounting orders with V-A post-war. Apart from the 4.5in Mark VI (now 6), the most important of these were the twin 6in Mark 26 and twin 3in Mark 6 for three cruisers unfinished in 1945. Although designs for a new 5in HA/LA mounting were prepared, none were built. 5.25in HA mountings continued to be produced for the Army, four of which were still at Gibraltar in 2017.

In December 1954 Vickers-Armstrongs Ltd split into Vickers-Armstrongs (Shipbuilders) Ltd at Barrow and Naval Yard and into Vickers-Armstrongs (Engineers) Ltd at Barrow and Elswick, with guns and gun mountings coming under that company, as well as marine engines, tractors and foundries. V-A (E) also took over Scotswood, Crayford and Whitehead torpedo works at Weymouth.

In the late 1950s Elswick was tasked with developing the horizontal magazine and loading arrangements for the Seaslug missiles in the first four 'County' class destroyers, laying out the 220ft long rail mechanisms in N° 24 Shop. They completed their orders for the 4.5in in 1969. Tank building continued into the 1970s – it had started in 1917 – before switching to Scotswood. The rambling Elswick site was then run down, closing in 1982 and after demolition became a business park.

The main gun mounting shops at Barrow continued to make them including the 4.5in Mark 8 and land artillery into the twenty-first century under the BAE Systems Armaments Division, although the shops to their west have been cleared to make way for a new fabrication hall for the *Dreadnought* class submarines.

## BRITISH PRODUCTION OF GUN MOUNTINGS IN WW2

**Value of gun mounting equipment £k**[9]

| Year | UK Gun Mountings | UK Revised | UK Fire Control | US at UK prices | Canada at UK prices | Total value |
|---|---|---|---|---|---|---|
| 1939/40 | 6,700 | 7,200 | 2,104 | | | 9,304 |
| 1940/41 | 10,500 | 12,000 | 2,800 | | | 14,800 |
| 1941/42 | 12,600 | 14,600 | 3,000 | 315 | 60 | 17,975 |
| 1942/43 | 14,500 | 16,500 | 3,110 | 3,850 | 2,598 | 26,058 |
| 1943/44 | 17,300 | 19,300 | 3,232 | 3,684 | 2,400 | 28,616 |
| 1944/45 | 16,000 | 18,000 | 3,086 | 2,694 | 2,311 | 26,091 |
| 1945/46 | 24,000 | 14,000 | 2,718 | 1,264 | 2,415 | 20,397 |
| Total | 101,600 | 101,600 | 20,050 | 11,807 | 9,784 | 143,241 |

The 1945/46 gun mounting figure includes a substantial amount for claims settled long after completion, estimated value £10M, distributed by the author over six previous years to give the revised column. The US supplied £35.420M of gun mounting and fire control material, but at prices about three times higher than UK, so only one-third of US figures are shown. Similarly, Canada's £19.567M are halved as prices were twice UK. These were all of 5in calibre or less, mostly close range weapons. Thus UK production amounted to about 85% of the £143M total. UK warship hull and machinery production over the same period amounted to about £555M or with Admiralty supply items like armour, ordnance and electrical equipments about £915M.

Another analysis recorded from 1939 to 1945 73 triple 6in, 114 5.25in (which includes mountings for the Army) and 137 twin 4.5in, peaking in 1940–1943.[10]

[7] CAB 102/526

[8] *The Tank Factory*. Tyne & Wear Archives & Museums, 2018

[9] CAB 102/532 and Buxton, Ian. *Warship Building and Repair during the Second World War*. Centre for Business History in Scotland, 1998

[10] CAB 102/526

CHAPTER SEVEN

# 12-inch Mountings

## THE FIRST BATTLESHIPS WITH MODERN 12-INCH ARMAMENT

The ordering of nine new battleships of the *Majestic* class resulting from the Naval Defence Act was a watershed as it ushered in ships with more powerful 12in guns in twin hydraulic gun mountings. By using wire winding, stronger barrels could be made capable of using the new more powerful cordite propellant, resulting in improved ballistics. Harveyed armour with a hard face and soft resilient back offered the same level of protection as compound armour 50% thicker, thus reducing thickness and weight.[1] The new triple expansion engines and steel boilers allowed higher pressures and increased efficiency, reducing fuel consumption and increasing speed to 17 knots.

The Royal Gun Factory at Woolwich designed the 35-calibre gun designated Mark VIII by the RN, the first of which was ready in December 1893, incorporating an improved breech mechanism. Two ships had just been ordered, with the other five to follow at intervals intended to spread expenditure over the next four years; the other two had a later mounting. The configuration with one turret forward and one aft, with 6in secondary armament, became the norm for the next decade, albeit with improvements to the gun mountings, protection and machinery. The 12in calibre was used for British battleships until 1909 when the 13.5in was developed, having gone through four Marks of gun and ten Marks of twin mounting.

As well as inviting Armstrong at Elswick to prepare a design of twin 12in gun mounting, somewhat surprisingly Palmers Shipbuilding & Iron on Tyneside was also invited, despite having no record of making heavy gun mountings, plus Woolwich whose experience was mostly in smaller gun 'carriages' for the Army. Vavasseur who had made some previous hydraulic mountings had been taken over by Armstrong and Vickers had yet to enter the market, so Armstrong was the only serious contender. Its design was accepted, with fourteen mountings designated Mark B.II being ordered for the first seven ships at a price of around £30,000 each excluding guns. It had improved features over previous mountings with an enclosed armoured gunhouse giving the gun crews better protection as well as reducing the chance of an enemy shell setting off a chain of ammunition explosions reaching down to the magazine. The gunhouse – the term turret seems to have been introduced at this time – was supported on a turntable which rotated on tapered rollers, with its centre of gravity at the centre of rotation, so permitting a smaller training engine and in emergency hand operation, even when the ship was rolling. The gun itself was also balanced with its CG at the trunnions, allowing a less powerful elevating mechanism as well as hand operation in emergency. The hydraulic pumps (one per turret) powered the ammunition hoists, training and elevating mechanisms, rammers and breech operation.

Two loading methods were fitted. Eight ready-use shells per gun could be stowed at the rear of the gunhouse, loaded at a fixed elevation of 1° using a rammer in line with the breech. This enabled the turret to remain trained on the target during reloading. Once those shells had been used, the turret had to train fore and aft for it to line up with mechanisms in the loading chamber below in the pear-shaped barbette, which was protected by 14in curved armour; initially it was called a redoubt. The narrow rear of the chamber housed a second fixed rammer, requiring the gun having to be brought to its maximum elevation of 13½°. The gun could then be loaded from a shell and cordite hoist delivering from the magazine below to the side of each gun, or from the 20 ready-use shells stowed in the same compartment. The gunhouse ready-use shells could be replenished through a hatch in the gunhouse floor from the loading chamber once the turret had been trained fore and aft. The first system resulted in a loading cycle of 70 seconds, the second 100 seconds, before replenishment was required, faster than the 130 seconds for the *Royal Sovereigns*' 13.5in. In gunlayers tests around 1908 reported in Brassey the time for the first method had been brought down to 50 seconds. Total stowage was 80 rounds per gun: 10 AP shot, 48 common, 22 Palliser. The mounting crew was made up of 23 in the gunhouse and chamber and 24 in shell room and magazine. With so much equipment on the fixed structure, the revolving weight was only 264 tons (including 92 for the guns), half the weight of the B.XI mounting of 1910.

The first two ships were ordered at the end of 1893: *Majestic* from Portsmouth Dockyard and *Magnificent* from Devonport, completed quickly in only two years. In January 1894 ten shipbuilders were invited to tender, but only two were awarded contracts: *Mars* to Laird at Birkenhead and *Jupiter* to Thomson at Clydebank, *Prince George, Victorious* and *Hannibal* going to dockyards, although commercial

[1] Johnston, Ian & Buxton, Ian. *The Battleship Builders*. Seaforth Publishing, 2013, p212

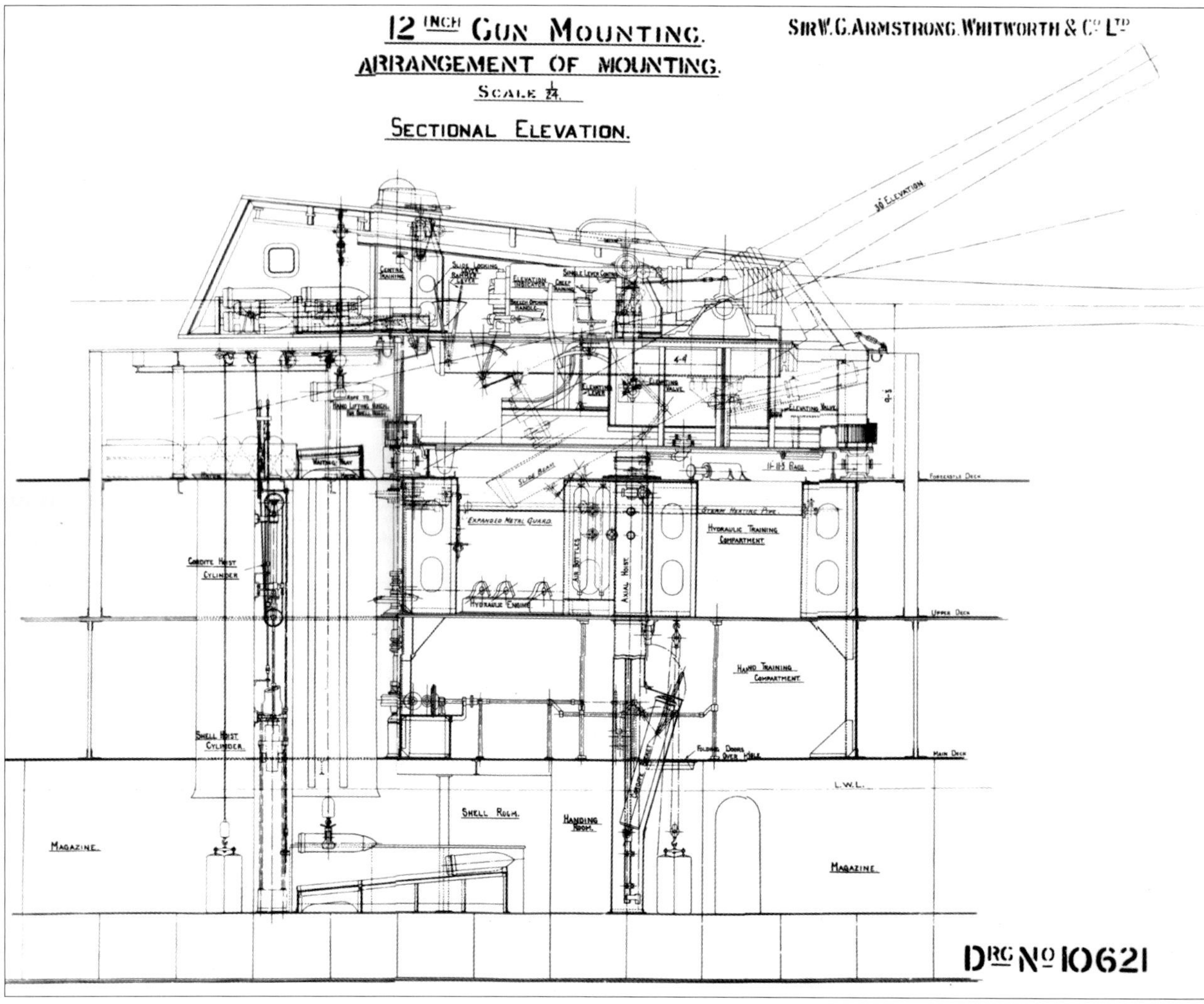


Early in 1915, four *Majestic* class had their two 12in B.II mountings removed to arm eight monitors for shore bombardment. To get a greater range, Armstrong were asked to modify the design to give 30° elevation. Their new plan showed the removal of the loading position in the narrow end of the barbette to make space for intermediate stowage supplied from magazine and shell room, leaving only the loading position in the gunhouse itself. Only shells already stowed in the gunhouse could be fired on target, before having to train the turret fore and aft to reload through hatches in the gunhouse floor. Cordite charges could also be supplied via a centreline hoist. *(NMM)*

### 12-inch BL Mark VIII Gun

| | |
|---|---|
| Calibres, bore | 35 |
| Length overall, in | 445.5 |
| Weight incl BM, tons | 46 |
| New gun muzzle velocity, ft/sec | 2375 |
| Shell, lb | 850 |
| Charge cordite Mark I, lb | 174 |
| Charge cordite MD45, lb | 200 |
| Life, efc cordite MD | 400 |

firms supplied machinery for all five dockyard ships. While no delivery dates for the gun mountings have been found, a document in the *Majestic* class Ships Cover gives the gunshield delivery period of the first two ships as the last quarter of 1895, *ie* shortly before completion, and the other five as the second half of 1896 – although by mid-1895 the latter had yet to be ordered. The front armour was 10½in, sides 5½in, rear 4in, roof and gunhouse floor 2in, all nickel steel except sides Harveyed. Those are nominal thicknesses, as armour was ordered by weight in terms of lb/sq ft. For steel, 1in = 40.8lb/sq ft, but in quoting thicknesses, 40 was usually assumed, *ie* 4in = 160lb, although the latter's true thickness was 3.92in. As the rolling tolerance was usually 0% over, ⅛in under, actual thicknesses were even less.

Initially 75 Mark VIII guns were made, 49 being manufactured by Woolwich, 10 by Armstrong, 10 by Whitworth and 6 by Vickers, taking between 16 and 26 months to make.[2] Service showed that the guns were liable to 'choke' where copper from the shell driving bands was deposited in the bore causing a constriction at the shoulder between inner A tube (liner) and A tube, which could retard the shell sufficiently to start the fuze and cause a premature detonation damaging the gun. Later Elswick and Vickers each made five new guns with thicker A tubes which prevented a choke building up.

Most deployed initially in the Channel Fleet, but once the dreadnoughts began to arrive, they were relegated to secondary roles, usually in local home squadrons. Only two of the class were deployed to support bombardment operations at the Dardanelles in 1915, *Prince George* from 3 March and *Majestic* from 26 February, but she was torpedoed on 27 May.

Broadly similar mountings were supplied by Armstrong in 1897 for the Elswick-built Japanese *Yashima* and Thames Iron Works' *Fuji,* although their guns were 40 calibres and could elevate to 15°.[3]

[2] Breer, *op cit*, p286. Friedman, Norman. *Naval Weapons of World War One*. Seaforth Publishing, 2011, p52

[3] Noble, Andrew. 'The Rise and Progress of Naval Artillery'. *Trans Inst of Naval Architects* 1899.

## THE 12-INCH GUN MONITORS

Early in the First World War it became evident that the *Majestic* class pre-dreadnoughts would be of limited use. In December 1914, Churchill, First Lord of the Admiralty, proposed further shallow draft coast bombardment vessels to get close inshore, armed with heavy long-range guns, following on from the four 14in gun monitors just ordered using US-made gun mountings. Admiral Percy Scott, who had done so much to improve RN gunnery pre-war, was called out of retirement and quickly established with Elswick that it would be possible to increase the elevation of the 12in B.II mounting to 30°. This would increase range using 2 crh (calibres radius of head) shell and the now current cordite MD from 13,700 to 21,300 yards. However, to accommodate the 36in recoil at maximum elevation, it would be necessary to modify the turntable and remove the rammer and loading position at the rear of the barbette. Depression angles would also be sacrificed, previously 5°. The only loading position retained was in the gunhouse, where 20 rounds could be stowed, enough for 20 minutes firing at one round per minute from the turret, long enough to get spotting corrections. Removing the rammer from the barbette increased stowage there to 34 rounds. A hoist from the shell room lifted the shells horizontally via a waiting tray in the barbette to the gunhouse when the turret was trained fore and aft. Another lifted the cordite charges vertically from the after magazine. The main cordite hoist was on the centre line so delivered from the forward magazine direct to the gunhouse, where the quarter charges were loaded manually into the breech. The unduly heavy curved barbette armour was replaced by 8in flat plates.

Four of the *Majestic*s were selected to surrender their 12in mountings, allowing the construction of eight monitors of 5900 tons but only 7 knots speed. *Magnificent* arrived at Belfast in February 1915, where Coventry Ordnance Works carried out the conversion work in situ. The mountings were shipped by floating crane in the first two Harland & Wolff ships in July, *Lord Clive* and *General Craufurd.* Elswick converted *Victorious* in the Tyne, shipping one to Hamilton's *Prince Rupert* using John Brown's 150-ton crane and one to *General Wolfe* at Palmers in October. *Mars'* mountings were converted by Vickers at Belfast before shipping in *Earl of Peterborough* in September and *Sir Thomas Picton* in October. *Hannibal* was sent to Dalmuir where Vickers used Beardmore's 150-ton crane to install one in Scott's *Sir John Moore* in July and COW the other in Harland & Wolff (Govan)'s *Prince Eugene* in August.

Four *Majestic* pre-dreadnoughts had their twin 12in B.II mountings removed to arm eight monitors in 1915. *Sir John Moore*, seen here, received one of *Hannibal*'s, converted to 30° by Vickers. *(Author's collection)*

Rushed into service off the German occupied Belgian coast, the new mountings with their increased pressures gave trouble initially from leaking joints, requiring careful maintenance. The introduction of 4 crh shells increased range by a further 15%, while in 1918 a 878lb 8 crh shell became available capable of reaching 26,000 yards using a super-charge – double the original range. After the 15in monitors *Erebus* and *Terror* arrived in 1916, they were used less but still fired well over 2000 rounds during the war. Full details of their operations are given in *Big Gun Monitors*.

## *CANOPUS* CLASS WITH TWIN 12-INCH B.III GUN MOUNTINGS

The *Canopus* class battleships comprised six ships, all with 12in Mark VIII guns but with three different twin hydraulic mountings: three with the Mark B.III from Openshaw, two with B.IV from Elswick and one with B.V from newcomer Vickers at Barrow. The last two of the *Majestic* class, the dockyard-built *Caesar* and *Illustrious,* also carried the B.III mounting designed by Sir Joseph Whitworth & Co of Manchester. This featured what was later called a working chamber beneath and rotating with the turntable for the final transfer to the gunhouse, thus permitting all-round firing. A fixed ammunition trunk on its centreline delivered shells and cordite charges from magazine and shell room to this circular chamber. The shells were unloaded there by overhead crane with 24 per gun being stowed around the perimeter. When required one was transferred via a waiting tray to a gun loading cage which also held two half charges loaded by hand, which was then raised into the gunhouse alongside each breech. Each gun had been brought to its fixed loading angle of 13½° so the 850lb shell was rolled sideways onto the loading tray to be rammed into the breech by a telescopic hydraulic rammer. The charges were then moved to the loading tray for ramming and the breech closed.[4]

With the rammer on the rotating structure rather than on the fixed structure as in the B.II, the armoured barbette could be circular instead of pear-shaped and thus be lighter. Three shells per gun were stowed in the gunhouse, so an alternative manual loading position could be used when the gun was brought to 1° elevation. As in the B.II, the hydraulic training engine was on the fixed structure driving a vertical shaft with a pinion engaging on the training rack fitted outside the turntable; hand operation was also possible. Later mountings had the engine on the revolving structure with the training rack fitted inside the supporting ring bulkhead. *(See colour plates on page 152)*

Having a central ammunition trunk rather than the direct path in the

[4] ADM 186/872. *Addenda to Manual of Hydraulics. 'Caesar' Class. 1898*

B.II mounting reduced the risk of an explosion in the gunhouse reaching the magazine. The 7ft diameter trunk contained a 'shot' hoist for each gun with the shell carried vertically and three hoists for two half charges carried at 45° plus a vertical ladder. As the trunk did not rotate, transfer of shells stowed horizontally in the shell room bins was by grab running on an overhead rail, delivered to a bucket at the base of the hoist to be tilted vertically. Magazine capacity was 80 rounds per gun, 2 crh.

The mounting had the same maximum elevation of 13½° and depression of -5° as the B.II and gun and ammunition so ballistic performance was the same. Its firing cycle was 65 to 80 secs, as fast as the B.II only as long as the latter's ready-use shells permitted. There were two recoil cylinders under the gun and two run-out cylinders above operated at 1050lb/sq in. A series of lever-operated interlocks ensured that each movement could only be undertaken in the correct sequence. The mounting crew was slightly larger than the B.II with more operations having to be performed.

Armstrong Whitworth's Nº 7 Shop at Elswick was used for erecting heavy mountings. This 1902 view shows two of *Cornwallis'* B.VI mountings. The shop had just been rebuilt after a fire. *(Library of Congress LC-DIG-ggbain-00178)*

The first three ships of the 13,000-ton 18-knot *Canopus* class were ordered in August 1896, fitted with Openshaw mountings: *Canopus, Goliath* and *Ocean,* one from each of the main dockyards, by which time the first two B.III ships *Caesar* and *Illustrious* were ready to be launched. Their central hoists were slightly different, with two cages, each with a shell and two half charges lifted horizontally as one unit, rather than separately. The ships featured the new Krupp cemented armour so thickness could be reduced by about 25%, with the gunhouse front now 8in.

The ships had been designed to operate on the China station but were later in the Mediterranean or Home Fleets. *Ocean* took part in bombardments at the Dardanelles before being mined on 18 March 1915 while *Goliath* was torpedoed there on 13 May. *Canopus* had been at the Falklands in December 1914, firing at Von Spee's cruisers at maximum range. *Caesar* saw no significant action in WW1, nor did *Illustrious.* She was used in harbour service from 1916, so her two 12in turrets were removed. These were converted by Elswick for coast defence near the mouth of the Tyne.

At the end of 1916 the Government became concerned at the threats to Tyneside's shipyards and ordnance factories from long range

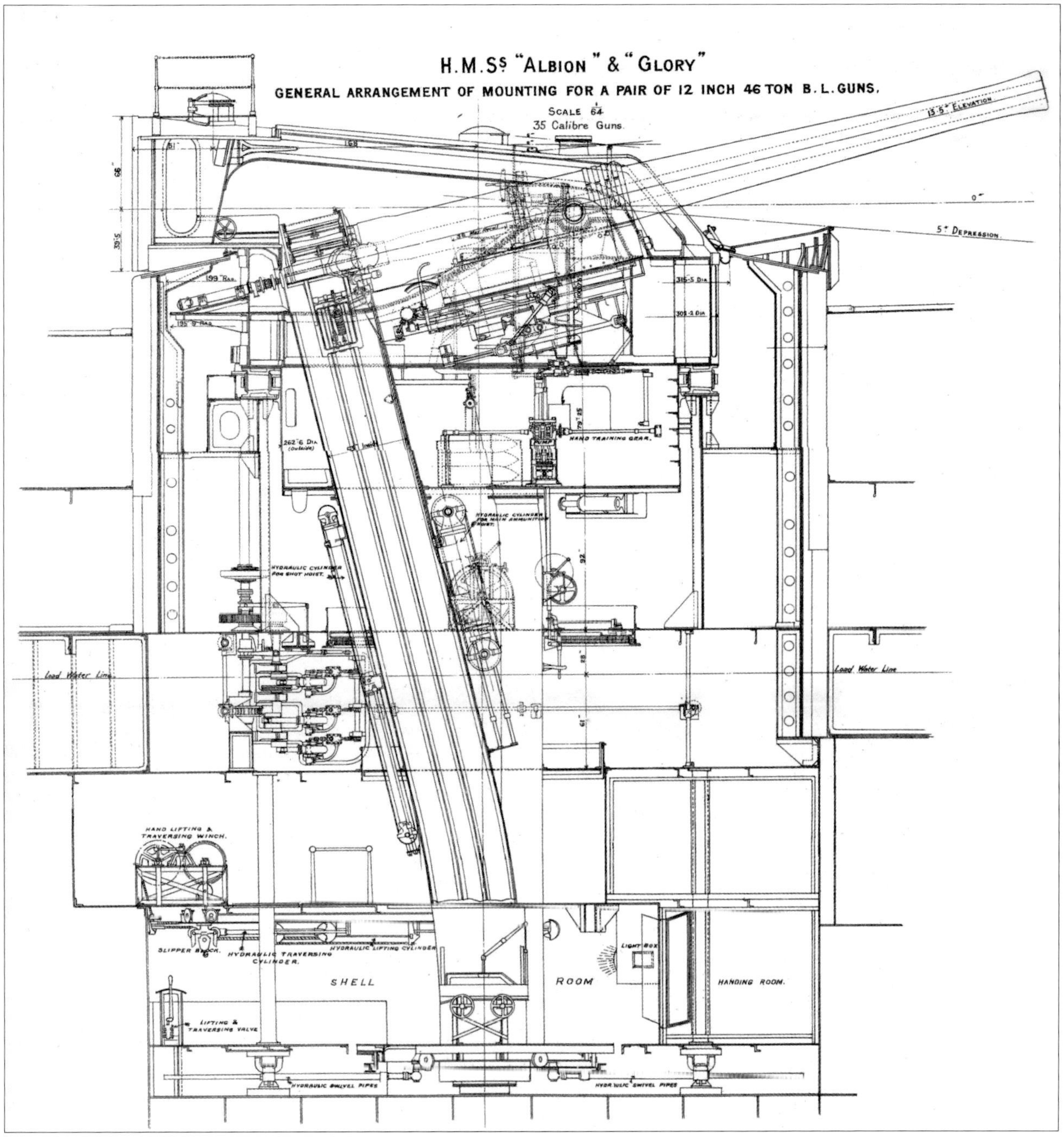


*Albion* and *Glory* were fitted with the Elswick designed twin 12in B.IV mounting. In order to speed up loading, the hoist extended right from gunhouse to magazine, revolving with the mounting, but this increased the risk of fire or flash in the gunhouse reaching the magazine. *(*Trans Inst Naval Architects*, 1899)*

fire from hostile ships. The existing coast defence guns were up to 9.2in but with low elevation limiting their range. The twin 12in mountings with Mark VIII guns from disarmed *Majestic* class ships could be upgraded to 30° elevation as had been done in eight monitors giving a range of about 24,000 yards. The intention was to mount one turret at Marsden south of the Tyne entrance and one at Hartley to the north. *Illustrious* arrived in the Tyne in December 1916 for Elswick to remove her two B.III mountings for conversion. Meantime the emplacements were being constructed, but the mountings were only delivered to Hartley on 15 April 1919 and Marsden on 21 May, after the war was over.[5] Trial firings of 12 rounds each were carried out in September 1921, Marsden being known as Kitchener Battery and Hartley as Roberts. The ammunition outfit would have consisted of 30% APC and 70% CPC. The mountings were scrapped by Blyth shipbreakers Hughes Bolckow in 1926, the two yielding 2227 tons of saleable scrap worth £9700 – £4.35 per ton, a tiny fraction of manufacturing cost. Most of the underground concrete installations were probably left intact.[6]

## TWIN 12-INCH B.IV GUN MOUNTINGS

Two more *Canopus* class were also ordered in August 1896, but with the B.IV mounting designed by Armstrong's Elswick Ordnance Co, *Albion* from Thames Iron Works and *Glory* from Laird. They had the same Mark VIII guns and performance as the others, with 13½° maximum

[5] MUN 4/2943 & 2850. 12-inch Mountings for Tyne Coast Defence Batteries. 1918

[6] Marine Technology Special Collection, Newcastle University 12040/25

elevation and -5° depression. The Director of Naval Ordnance wanted an ammunition supply direct from magazine level to the gunhouse, without a transfer at the working chamber potentially slowing the rate of fire. This required the full length of the hoist mechanism to rotate with the turret, with the gun loading cage carrying both shell and the two half charges. A turntable was required at the bottom of the trunk, initially fixed to the hull on which to deposit the shell lifted from the shell room bins, then released to rotate to the loading door at the bottom of the hoist. The cage then delivered to the rear of the gun at its fixed elevation of 13½° to be rammed into the breech. Having to lower the cage down the full 42ft distance to reload slowed the operation, while the direct route from gunhouse to magazine posed a serious explosion risk. A secondary loading method used largely hand methods with eight shells per gun stowed in the working chamber and three in the gunhouse. See drawing on p57.

At that time commercial shipbuilders delivered the hull and machinery of a warship to the Admiralty for armament and other equipment like boats to be fitted at one of the dockyards. Laird delivered *Glory* in December 1899 for her guns to be shipped at Portsmouth; she was commissioned in November 1900. That extended period suggested problems installing the new mountings. *Albion*'s guns were shipped at Chatham in August 1900, commissioning in June 1901 three years after launching. She had been delayed by her machinery contractors (Maudslay, Sons & Field) going bankrupt in 1899. She was at the Dardanelles bombarding on 19 February 1915, later sustaining hits from Turkish artillery, before taking up guardship roles at home. When she arrived to break up at Morecambe in January 1920 her guns had been removed but mountings retained. *Glory* was also at the Dardanelles in 1915 before becoming flagship in North Russia in 1916; scrapped in 1922.

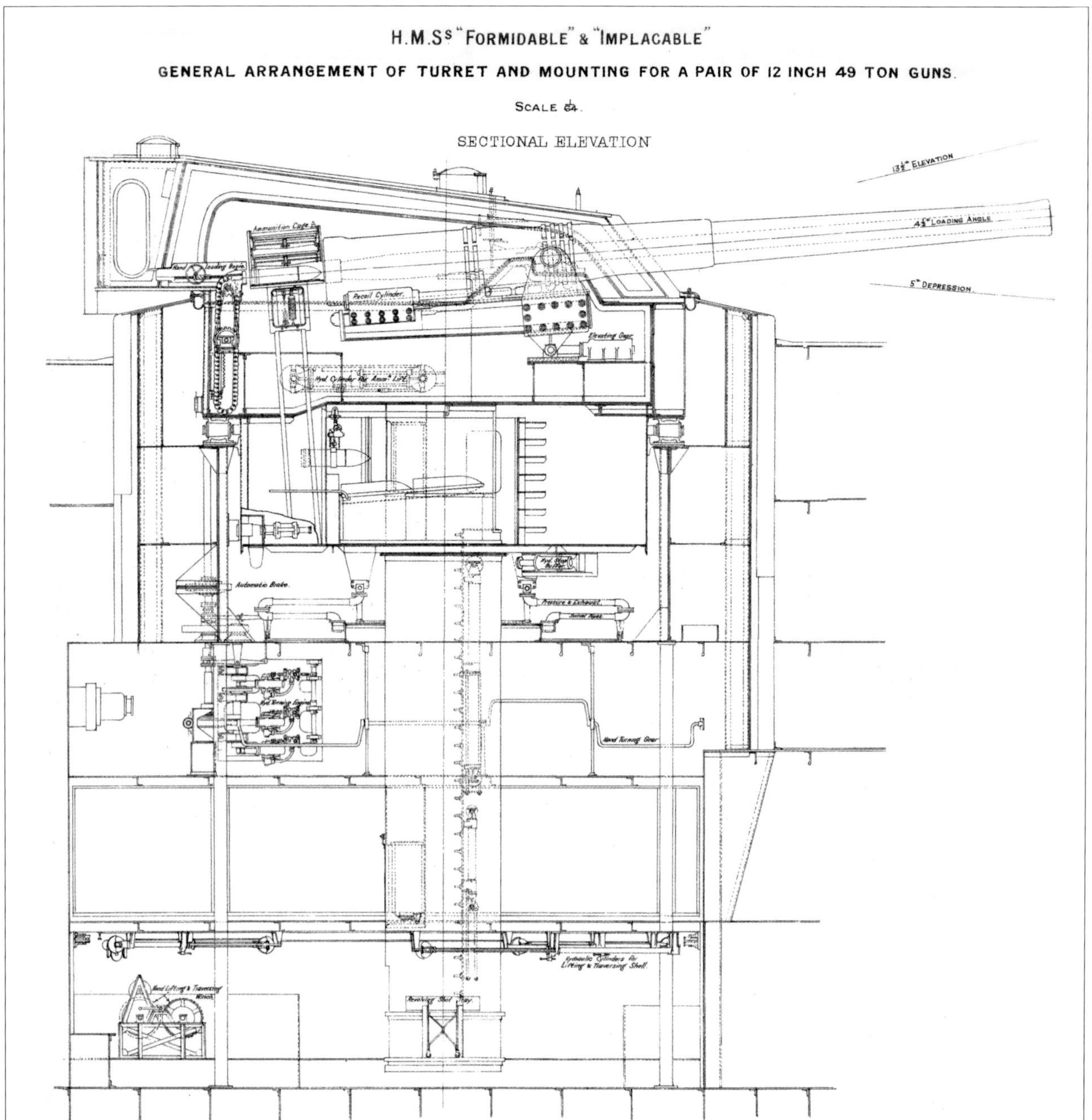


*Formidable* and *Implacable* and the *Duncans* were fitted with the Elswick designed twin 12in B.VI mounting, which reverted to the working chamber arrangement. (Trans Inst Naval Architects, *1899)*

## TWIN 12-INCH B.V GUN MOUNTINGS

Having acquired the shipyard at Barrow, Vickers with its record of making guns and armour for the Admiralty no doubt lobbied hard for the order for the sixth *Canopus* class battleship, to be ordered under the 1897 Programme. The new company, now Vickers, Sons & Maxim, was awarded the hull and machinery contract for the 12,950-ton *Vengeance* on 6 July 1897, plus some of the armour. Having no doubt advised the Admiralty of its plan to build a new gun mounting manufacturing plant at Barrow, it proceeded to develop its own design of a 12in mounting using the existing Mark VIII gun. This featured a major innovation, all-angle loading, from 13½° to -5°. By fitting the rammer on a loading arm extending from the rear of the gun slide, the gun loading cage running vertically on curved rails could follow in line with the breech as the gun elevated. In place of a linear hydraulic shaft rammer, a chain rammer was used which could be extended from and retracted into a housing on the gun slide, making for a shorter mechanism. The cage was loaded with shell and charges in the working chamber, which incorporated another novel feature: the central ammunition trunk now revolving with the mounting, so simplifying the operation of shell transfer in the working chamber. A turntable was fitted at the bottom of the trunk as in the Mark IV, so that shells could be transferred from the shell bins on the fixed structure via a bogie to the hoist. These arrangements introduced a break in the ammunition supply making it safer and offered all-angle loading featured in subsequent 12in mountings.

*Exmouth* was fitted out with her two Vickers twin 12in B.VII mountings in 1902 at her builders, Laird of Birkenhead, rather than at a dockyard, hitherto the norm. Her forward 90-ton turntable and working chamber was lifted by the Mersey Docks Board 100-ton crane in West Float. *(Wirral Archives)*

Vickers put in a competitively priced tender for two twin 12in mountings. Probably also desirous of bringing in a competitor to the new monopoly of Armstrong Whitworth, the Admiralty awarded them the contract on 19 March 1898, designating the mounting B.V. The price of £52,128 excluded the gunshield armour and guns, both procured separately. The delivery date of June 1899 would prove a challenge for the newcomer, with the shops and plant still under construction at Barrow.

| *Item* | *Tons* | *£* |
|---|---|---|
| Turntable, loading chamber, roller path, trunk | 191.5 | 10,978 |
| Turning gear and engine | 28.0 | 7,180 |
| Slides, carriages, cradles and presses | 72.2 | 15,300 |
| Steam pumping engines | 24.0 | 5,700 |
| Hydraulic pressure and exhaust piping on revolving structure | 10.0 | 1,330 |
| Magazine and shell room machinery | 14.0 | 3,900 |
| Gear between magazines and shell room for ammunition transfer and gun loading | 18.2 | 2,800 |
| Hydraulic pressure and exhaust piping outside revolving structure | 12.0 | 4,940 |
| Total | 369.9 | 52,128 |
| Gunshield armour | 121.9 | |
| Excludes guns, 46 tons each.[7] | | |

The work took longer than expected, with launch not until 25 July 1899, and mountings not installed until 1900. Although with extras, the final price was increased to £63,455, costs were much higher on this pioneer project, making a loss of £35,912. Hull and machinery final price was £444,205 also resulting in a loss of £19,475.[8]

## CONTINUING EVOLUTION OF THE 12-INCH GUN MOUNTING

British naval policy in the late nineteenth century was based on the Royal Navy being as large as the next two biggest navies combined, among them France, Germany and Russia. A steady programme of battleship building continued over that first decade of the 1900s. During the first years the programme continued to build what later became known as pre-dreadnoughts, each with two twin 12in mountings. After *Dreadnought* appeared in 1906, that pattern changed to five twin mountings in battleships and four in what became known as battlecruisers. Forty 12in gunned ships were built, armed with three Marks of 12in guns with four Marks of 12in hydraulic mountings, plus two unsuccessful Marks of electric mountings.

The guns evolved to meet the demands of longer range at which it was thought future battlefleets would fight. The 35-cal Mark VIII gave way to the 40-cal Mark IX of 1898, the 45cal Mark X of 1904 and then the 50-cal Mark XI of 1907. Of the 110 Mark IX initially made, 44 came from Woolwich, 36 from Vickers and 30 from Elswick. These longer and heavier guns demanded heavier mountings and faster ammunition supply. To supplement production from the existing works at Elswick, Sheffield and Woolwich, Beardmore at Glasgow and Coventry Ordnance Works were brought in, making their first guns in 1909. Common features to all guns were wire winding and the new Welin breech mechanism whose stepped screwed threads could be closed in one-twelfth of a turn, usually hydraulically operated. Initially they all fired 2 crh 850lb shells, but later a more aerodynamic 4 crh shell was introduced, giving a range increase of about 14%.

**12-inch gun data**

| 12in BL Gun | Mark IX | Mark X | Mark XI |
|---|---|---|---|
| Designer | Vickers | Vickers | Vickers |
| Calibres, bore | 40 | 45 | 50 |
| Length overall, in | 496.5 | 556.5 | 617.7 |
| Weight incl BM, tons | 50.8 | 57.6 | 66.7 |
| Muzzle velocity nominal ft/sec | 2560 | 2725 | 2825 |
| Shell, lb | 850 | 850 | 850 |
| Charge cordite MD45, lb | 246* | 258 | 307 |
| Life on MD, efc | 280 | 280 | 220 |
| Number made | 129 | 135 | 85 |
| Range at 13½° elevation 2 crh, yd | 15500 | 16500 | 21000 ** |

* *King Edward VII* class had heavier 254lb charge 2610ft/sec requiring a strengthened mounting.
** 4 crh at 15°

Marks IX*, X*, XI*, XII varied in extents of wire winding and internal tube geometry. Two reserve gun bodies were generally carried for every five afloat, and one breech mechanism for every four. Seven Mark X guns were shipped to France as land artillery and nineteen Mark IX for nine railway mountings.

The mountings all had some common features: a central ammunition trunk from magazine to working chamber and rotating with it, manual transfer of cordite charges between hoists, and the shell rooms below the magazine. Hodges discusses the differences; the table opposite summarises their features.[9]

The *Formidable* class were the first to be ordered, all dockyard-built, the name ship and *Implacable* each getting two Elswick B.VI mountings (ordered in March 1898) and *Irresistible* getting the Vickers B.VII. Vickers had tendered for all three ships but received an order for only one, on 24 March 1898 for delivery in 21 months. The contract price was £53,250 with a weight of 372 tons for the two mountings (£143 per ton), but these figures excluded gunhouse armour and guns, both ordered by the Admiralty under separate contracts to be delivered to Barrow as a free issue for erection. The 8in to 10in thick armour contract was worth about a further £37,600.

Elswick armed ten more pre-dreadnoughts, erecting the mountings in their rebuilt workshops, then shipping the components to the shipbuilders for final erection, but few contract details have survived. For Vickers and its ten other pre-dreadnoughts, a selection of contract information survives in the Vickers archive at Cambridge University and in the BAE shipyard archive at Barrow. (*See colour plates on pages 148-149)*

[7] Cambridge University Library, Manuscripts. Vickers Archive 1116
[8] Vickers, Sons & Maxim Annual accounts. BAE Systems, Barrow shipyard archives

[9] Hodges, Peter. *The Big Gun*. Conway Maritime Press, 1981, pp41–54

*Britannia* was one of the *King Edward VII* class with two twin 12in B.VIIS mountings built by Vickers. She also had two single 9.2in Mark VS on each broadside plus ten 6in. *(World Ship Society)*

**12-inch gun mountings**

| Mark | B.VI | B.VII | B.VIIS | B.VIII | B.VIII* | B.XI |
|---|---|---|---|---|---|---|
| Designer | Elswick | Vickers | Vickers | Vickers | Vickers | Elswick |
| Gun Mark | IX | IX | IX & IX* | X | X | XI & XII |
| Number of ships fitted | 9 | 5 | 8 | 5 | 6 | 6 |
| Loading angle, degrees | 4½ | 13½ to -5 | 13½ to -5 | 13½ to -5 | 13½ to -5 | 15 to -5 |
| Elevating speed, deg/sec | | | | 3 | | 3 |
| Training speed, deg/sec | | | | 2 | | 2 |
| Where training engines fitted | Fixed structure | Rotating | Rotating | Rotating | Rotating | Rotating |
| Firing cycle, sec | 50 | | | 30 | 30 | 30 |
| Revolving weight incl guns, tons | 345 | | 433 | 480† | 470 | 530 |
| Gunhouse armour, front, KC in | 8 | 8 | 9 | 12† | 11 | 11 |
| Sides, KC in | 8 | 8 | 8 | 12† | 11 | 11 |
| Rear, KC in | 10 | 10 | 12 | 13† | 12 | 11 |
| Roof, KNC in | 2–3 | 2–3 | 3 | 3 | 3 | 3–4 |
| Features | Working chamber. Chain rammer on turntable. Shell & charges separate main hoist | Developed from B.V. Chain rammer on loading arm | Strengthened for heavier charge | Chain rammer on loading arm | Auxiliary hoists power worked | Single cage main hoist. Swash plate training engine |

† *Agamemnon.* Weight about 10 tons less in *Dreadnought* as gunhouse armour 1in thinner, 40 tons less in battlecruisers as front, sides and roof 7in.

For B.IX and B.X electric, see text.

**Right:** *Lord Nelson* had the same twin 12in B.VIII mountings as *Dreadnought* but only two (by Elswick) instead of five. On each broadside she also mounted two twin 9.2in Mark VII and one single 9.2in Mark VIII by Elswick, being Tyne built. *(World Ship Society)*

## Two 12-inch Mark B.VIII Gun Mountings for *Dreadnought*: Contract Estimate

| Item | Tons | £ | £/T |
|---|---|---|---|
| Turntable, loading chamber, roller path, trunk | 187 | 18,038 | 96.5 |
| Turning gear and engines | 35 | 12,538 | 358.2 |
| Gun slides, carriages, cradles and elevating gear | 96 | 29,733 | 309.7 |
| 2 steam pumping engines* | 31 | 3,950 | 127.4 |
| Hydraulic pressure and exhaust piping on revolving structure | 11.5 | 4,583 | 398.5 |
| Magazine and shell room machinery and rails | 9 | 3,549 | 394.3 |
| Hoists and gear for ammunition transfer between magazines and shell room and gun loading* | 44.5 | 14,528 | 326.5 |
| Hydraulic pressure and exhaust piping outside revolving structure* | 18.5 | | |
| Brackets, rails and guards | 5 | 2,242 | 448.4 |
| Gunshield structure, roof and sighting hoods | 125 | 17,355 | 138.8 |
| Gunshield armour. Front, side, rear. Estimated | 205 | 33,067 | 161.3 |
| Sub-total | 767.5 | 139,583 | 181.8 |
| Drawings | | 562 | |
| Spare mounting | | 8,812 | |
| Delivery of two mountings | | 1,480 | |
| Delivery of spare mounting | | 140 | |
| Erection of two mountings | | 9,300 | |
| Total contract | | 159,877 | 208.3 |
| Mounting revolving weight | 709 | | |
| Four guns | 230.4 | *c*46,000 | 191 |
| Total revolving weight | 939.4 | *c*205,877 | 217 |
| Per mounting | 470 | *c*102,938 | |

*excluded from revolving structure

The story of how *Dreadnought* was built in only fifteen months (not twelve) is well known. Less well known is how her five 12in mountings, normally taking two years to build, were delivered in July 1906 and ready for trials in October when they had only been ordered in September 1905. This was only possible by splitting the order, with Barrow building X and Y and Elswick A, P and Q. But as *Agamemnon*'s two had already been ordered from Barrow in December 1904 and *Lord Nelson*'s two from Elswick, all of the same Mark B.VIII, it was possible to take much of their material for *Dreadnought* and rush a fifth mounting. Vickers passed jigs and gauges to enable Elswick to build their mountings. As a result the two pre-dreadnoughts were completed a year later than planned, using material which would have gone into *Dreadnought*; *Indomitable*'s Elswick mountings were delayed over a year. The manufacturers of the mountings of the other 12in battleships and battlecruisers are shown in the table on p65.

Vickers were criticised by *Excellent*'s team at *Inflexible*'s gun trials on 18 June 1908 as follows:[10]

> It must be remarked that the erection and care and maintenance since erection, of the 12-inch mountings, as in the case of the

[10] Johnston, Ian. *Clydebank Battlecruisers*. Seaforth Publishing, 2011, p16

## Some representative Vickers 12-inch contracts

| | Contract | Mountings | Ordered | Tender price £ | Final price £ | Cost £ | Profit £ | Profit/ cost % | Price £ per mtg | Notes |
|---|---|---|---|---|---|---|---|---|---|---|
| *Vengeance* | 1G | 2 B.V | 19/3/98 | 52,128 | 63,455 | 99,367 | -35,912 | -36.1 | 31,728 | Excludes gunshield armour, delivery & erection |
| *Irresistible* | 2G | 2 B.VII | 24/3/98 | 53,250 | 64,739 | 84,659 | -19,920 | -23.5 | 32,370 | Excludes gunshield armour, delivery & erection |
| *Exmouth* | 10G | 2 B.VII | 28/7/99 | 78,756 | 86,063 | 82,716 | 3,347 | 4.0 | 43,032 | Erected at Birkenhead |
| *Prince of Wales* | 58G | 2 B.VII | 9/1/01 | 83,319 | 97,270 | 72,994 | 24,276 | 33.3 | 48,635 | Includes erection £12,000 and spare mounting |
| *King Edward VII* | 67G | 2 B.VIIS | 22/7/02 | 108,585 | 115,715 | 80,418 | 35,297 | 43.9 | 57,858 | Price excludes armour. Priced per ton adding *c*£21,000 per mounting |
| *New Zealand* | 70G | 2 B.VIIS | 8/11/02 | 103,280 | 111,276 | 57,540 | 53,736 | 93.4 | 55,638 | Price excludes armour. Priced per ton adding *c*£21,000 per mounting |
| *Hibernia* | 76G | 2 B.VIIS | 13/1/04 | 100,804 | 109,413 | 64,970 | 44,443 | 68.4 | 54,707 | Price excludes armour. Priced per ton adding *c*£21,000 per mounting |
| *Agamemnon* | 87G | 2 B.VIII | 8/12/04 | 157,710 | 163,947 | 101,347 | 62,600 | 61.8 | 81,974 | Price probably includes armour |
| *Dreadnought* | 94G | 2 B.VIII | 2/9/05 | 126,817 | 137,713 | 80,511 | 57,202 | 71.0 | 68,857 | Some costs probably included under 87G. Armour adds *c*£18,000 per mtg |
| *Invincible* | 95G | 2 B.IX Elec | Mid-1905 | | 148,371 | 132,798 | 15,573 | 11.7 | 74,186 | High cost for new electrical mounting |
| *Inflexible* | 99-102G | 4 B.VIII | 23/11/05 | 202,580 | 226,452 | 127,562 | 98,890 | 77.5 | 56,613 | Price probably excludes armour, *c*£22,000 per mounting |
| *Temeraire* | 108G | 5 B.VIII* | Mid-1906 | | 277,014 | 179,497 | 97,517 | 54.3 | 55,403 | Price probably excludes armour, *c*£22,000 per mounting |
| *Vanguard* | 110G | 5 B.XI | End 1907 | | 300,449 | 205,283 | 95,166 | 46.4 | 60,090 | Price probably excludes armour, *c*£23,000 per mounting |
| *Colossus* | 112G | 5 B.XI | Early 1909 | | 320,772 | 212,095 | 108,677 | 51.2 | 64,154 | Price probably excludes armour, *c*£23,000 per mounting |

The battlecruiser *Inflexible* had four twin 12in B.VIII mountings by Vickers. Two single 4in on P.I* mountings are sited on top of A turret. *(UCS 118-364-35)*

**Pages 65-68:** Four detailed views of installing *Colossus'* twin 12in B.XI mountings at Scott's Greenock yard, built by Vickers as Contract 112G; another is on page 11. *(All NRS GD-3-323-12)*

**Above:** On 18 August 1910, A barbette was ready to receive the cylindrical main hoist trunk then the working chamber and turntable. The training rack has been fitted inside the supporting ring bulkhead on top of which is the roller path partly covered over. Turret buffer stops and two arms of a walking pipe lie in the well.

## Manufacturers of 12-inch gun mountings

| Mounting and Mark | Gun Mark | Manufacturer | Ships fitted (number of mountings) |
|---|---|---|---|
| Twin 12in B.VI | IX | Elswick | *Formidable* (2), *Implacable* (2), *London* (2), *Bulwark* (2), *Queen* (2), *Duncan* (2), *Russell* (2), *Montagu* (2), *Cornwallis* (2) |
| Twin 12in B.VII | IX | Barrow | *Irresistible* (2), *Venerable* (2), *Exmouth* (2), *Albemarle* (2), *Prince of Wales* (2) |
| Twin 12in B.VIIS | IX | Elswick | *Commonwealth* (2), *Hindustan* (2), *Africa* (2) |
| Twin 12in B.VIIS | IX | Barrow | *King Edward VII* (2), *Dominion* (2), *New Zealand/Zealandia* (2), *Britannia* (2), *Hibernia* (2) |
| Twin 12in B.VIII | X | Elswick | *Dreadnought* (3) A, P, Q, *Lord Nelson* (2), *Indomitable* (4) |
| Twin 12in B.VIII | X | Barrow | *Dreadnought* (2) X,Y, *Agamemnon* (2), *Inflexible* (4) |
| Twin 12in B.IX (electric) | X | Barrow | *Invincible* (2) A, Y |
| Twin 12in B.X (electric) | X | Elswick | *Invincible* (2) P, Q |
| Twin 12in B.VIII* | X | Elswick | *Superb* (5), *Indefatigable* (4), *New Zealand* (4) |
| Twin 12in B.VIII* | X | Barrow | *Bellerophon* (5), *Temeraire* (5), *Australia* (4) |
| Twin 12in B.XI | XI | Elswick | *St Vincent* (5), *Collingwood* (5), *Hercules* (5) |
| Twin 12in B.XI | XI | Barrow | *Vanguard* (5), *Neptune* (5), *Colossus* (5) |

> *Agamemnon*, compares unfavourably with mountings erected by Messrs Armstrong, as for instance in *Indomitable.* Many simple adjustments, among others, some of which were pointed out to Messrs Vickers' representatives at the preliminary hydraulic trials on the 12 May were still incorrect at Gun Drills and Trials, such as run-out cut off, fouling of sights against the armour, etc, and it is considered that more attention in this respect should be paid by Messrs Vickers. …

Some representative Vickers 12in contracts are summarised in the table.

The price per mounting doubled over the decade, partly reflecting the increased weight and complexity of the B.XI over the B.VII but also the scope of each contract. For the early orders, the Admiralty continued its normal practice of ordering gunhouse armour and guns separately from the mounting. At that time the dockyards installed a warship's armament after delivery of hull and machinery from the shipbuilder. This was simple enough when for most ships the armament consisted of transferable mountings which could be drawn from stock together with guns. But the new non-transferable mountings with ammunition trunks penetrating into the hull and heavy rotating parts broken down into components for transport were a problem for erection in the dockyards. About 1904 the Admiralty realised it was better for the work to be done at the shipyard, although in practice it was mostly done by the gun mounting contractor's staff. This also meant

By 6 September A mounting has had the turntable installed. It awaits the gunhouse floor, armour and guns, with the two gun slides, cradle, round rammer motors and control lever stands already fitted. Machined surfaces on rams and cradle are protected by wood.

Mounting A receives the first of her 66-ton 12in Mark XI guns on 5 October. The second gun lies on the dockside together with the turntable and working chamber of P mounting.

that the gunhouse armour had to be delivered to the shipyard, so the gun mounting contracts now began to include the armour as well, albeit supplied by the specialist armour makers. But armour was priced per ton so its cost depended on the actual weight of each plate, so a firm price for armour could not be included in the gun mounting contract, only the rate agreed between the Admiralty and armour makers, which also covered belt and barbette armour. Around 1907 those rates were about £160 per ton for front, side and rear KC plates, and £120 for the KNC roof, at a time when shipbuilding mild steel plates only cost about £7. The gun mounting price usually now included delivery from the works to the shipyard and its erection there, a significant sum. Furthermore, drawings and spare parts were not always included in the contract. Thus the figures in the table on p64 are not always directly comparable.

The figures are taken from the Vickers accounts spread over the several years of the contract, which do not always specify the scope, *eg* whether armour was included. The final price was usually greater than the tender price as there were modifications and adjustments made as work progressed. The first Vickers 12in mounting was quoted at a very low price resulting in a substantial loss. But it appears that for later contracts Vickers were able to negotiate higher prices, *eg* that for *King Edward VII* was double that for *Vengeance.* At the same time Vickers was able to reduce manufacturing costs as it gained experience – the learning curve. As a result the fourth contract (for *Exmouth*) showed a modest profit, and thereafter profit rates of 50–70% of cost were not unusual.

Where Vickers was also supplying gunhouse armour from Sheffield, they were able to make even higher profits, as all five armour makers (Vickers, Armstrong, John Brown, Cammell Laird and Beardmore) were members of a price fixing cartel. Profit rates were typically over

Both P (left) and Q (right) mountings have been installed and *Colossus* is ready to leave in July 1911.

100% of cost; a fact of which the Admiralty was unaware.[11] Similarly, making large calibre guns also yielded profits of over 100%, even when selling at the same price as the Royal Gun Factory at Woolwich, notoriously overmanned and inefficient. Vickers could be well satisfied with its investment in armaments manufacture.

The approximate make up of a twin 12in B.XI mounting was:

| | |
|---|---|
| Mounting | £64,000 |
| Armour | £23,000 |
| Two guns | £23,000 |
| Total | £110,000 |

[11] *The Battleship Builders*, Chapter 9 and Appendix 2.

Excluding the guns, that worked out at about £220 per ton. The Admiralty often used such gross figures in estimating the costs of new designs. On a typical battleship price of £1.7M, the main armament including guns constituted about one-third. Ammunition stowage for all Marks was the standard 80 rounds per gun.

## ELECTRIC MOUNTINGS

Although the Admiralty was broadly satisfied with its hydraulically operated heavy mountings, it noted that the US Navy was using electrically operated mountings, so decided on a trial in 1905. One of the three battlecruisers to be ordered was selected: *Invincible* to be built at Elswick. Her two centreline mountings were to be supplied by Vickers (A and Y as B.IX) and two wing mountings by Elswick (P and Q, as

B.X). The ship would need greater electrical generating capacity, four 200kW generators being required. The main operations would be electric: hoist winches, training, ramming, elevating (though the 10hp motor driving a screw system used proved inadequate).[12] For most of these actions the linear motion of hydraulic mechanisms was better suited than the rotary motion of electric motors. Gun run-out would have to be different from the usual hydraulic: Vickers chose springs, Elswick pneumatic. The mountings were no heavier than hydraulic ones when account was taken of the latter's tanks and piping.

In the event *Invincible* was delivered later than *Indomitable* (Fairfield) and *Inflexible* (John Brown). Indeed, in November 1907, seven months after launch, the shipyard was already complaining of late delivery of the mountings despite being manufactured only a few hundred yards away. Partly due to a shipyard strike in March 1908, the ship was only ready for gun trials in November by which time her two sisters had been delivered. Final gun trials were held in March 1909, the ship being handed over on the 15th.[13]

In service the electrical mountings proved unsatisfactory, with poor creep control, and were less reliable and less safe. Training was slower than the designed 4° per sec. The decision was taken in 1913 to convert the four mountings to hydraulic operation. Work was completed in August 1914 just in time for war, so *Invincible* was able to take part in the Battle of the Heligoland Bight on 28 August 1914 sinking *Köln*. Vickers charged the Admiralty £52,336 for converting their two, about half the price of a new mounting, excluding armour which could be re-used.

The Admiralty investigated the possibility of triple 12in mountings, with Elswick producing a design in 1907. But it was difficult to provide a fast ammunition supply to the centre gun, while firing a wing gun produced throw-off, nudging the mounting off target. The concept was deemed too complex so not pursued, although Elswick and Vickers did supply triple 12in mountings to Italy in 1913.

Over the decade, twin 12in mountings broadly similar to contemporary British ones were installed in the Japanese battleships *Shikishima, Asahi, Hatsuse, Mikasa, Kashima* and *Settsu*, all by Elswick, with *Katori* and *Kawachi* by Vickers. Elswick supplied twin mountings for the Italian *Regina Margherita* and *Benedetto Brin,* and parts for single mountings in four more building in Italy. In the 1910s triple and twin 12in and parts were made for five more Italian battleships and twin 12in for three Spanish battleships.[14] For the ex-Brazilian 12in see page 78.

At the Battle of the Falklands in December 1914 *Invincible* fired 513 rounds of 12in (128 APC, 259 CP, 126 HE) with 257 remaining, while *Inflexible* fired 661 (157 APC, 343 CP, 161 HE) with 210 remaining.[15] Most of the 12in dreadnoughts and battlecruisers were in the Grand Fleet, most now fitted for director firing. Campbell gives the ammunition expenditure at Jutland as about 1666 12in rounds with a discussion of general shooting performance.[16] The last of the pre-dreadnoughts, *Lord Nelson* and *Agamemnon*, spent most of WW1 in the Mediterranean, partly to bottle up *Goeben* in the Dardanelles.

A somewhat unsatisfactory 12in mounting with a single Mark IX gun was fitted in three submarines *M.1–M.3* towards the end of WW1. A twin 12in mounting, probably a Mark VIIS, was erected above ground at the Royal Marines Eastney establishment for training which survived into the 1950s.

[12] *The Big Gun*, p52
[13] *The Battleship Builders*, p125
[14] *Ibid*, p200
[15] Naval Staff Monograph, Vol IX, 1923
[16] Campbell, John. *Jutland: An Analysis of the Fighting*, pp346-7

## CHAPTER EIGHT

# Armoured Cruiser Mountings

### 6-INCH TWIN MOUNTINGS MARKS I AND II

The ten armoured cruisers of the *Monmouth* class were the first to be fitted with twin 6-inch mountings, electrically powered. One such mounting was fitted forward and one aft, plus ten more single 6in guns on pedestal mountings on the broadside. In November 1899, the Admiralty entrusted Vickers, new to the gun mounting business, to design and supply the two twin mountings for each of the first four ships, *Monmouth, Kent, Essex* and *Bedford,* at a price of £22,000 each including gunshield armour but excluding the guns. The mounting was designated Mark I, with chain driven hoists reaching down to the magazine. *(See colour plates on page 153)*

A new 6in 45-calibre gun was also designed by Vickers, designated Mark VII, with the lefthand gun designated Mark VIII. It went on to be made in large numbers, 928 for the Admiralty alone, fitted mostly in single mountings in battleships and cruisers and during WW1 in monitors, gunboats, armed merchant cruisers and defensively armed merchant ships.[1] The Army also acquired several hundred, many for coast defences, remaining in service into the 1950s.

[1] Campbell, John. 'British Naval Guns 1880-1945 Nº 11'. *Warship* 28, 1983

The 9800-ton armoured cruiser *Monmouth* was built in 1903 by London & Glasgow Engineering & Shipbuilding, the former Napier yard. She had one twin 6in Mark I mounting forward and one aft made by Vickers, plus on each broadside five single 6in P.III BD mountings. *(World Ship Society)*

| **6-inch BL Gun** | **Mark VII** |
|---|---|
| Calibres, bore | 45 |
| Length overall, in | 279.2 |
| Weight incl BM, tons | 7.4 |
| Shell, lb | 100 |
| Charge MD16, lb | 23 |
| Muzzle velocity, ft/sec | 2570 |
| Range at 15° 2 crh, yd | 11,300 |
| Life, efc | 1200 |

Vickers shipped the first two mountings for *Monmouth* in 1903, not surprisingly making a loss of £10,400. In 1905, spare mountings were supplied for each ship at £12,900 each which suggests only the main gunhouse fittings, not the entire mounting. Two more outfits were ordered in December 1900 for £24,750 each for *Cornwall* and *Suffolk.* But Barrow now had developed the experience to manufacture them efficiently, with the last set generating a profit of £12,000, nearly 50% on cost.

In the meantime the mountings for the last four ships were ordered from Elswick, designated Mark II, for *Donegal, Berwick, Lancaster* and *Cumberland.* In all twenty mountings were made plus ten spares stored at the main dockyards.

In both Marks, the main motions were electrically powered, except for elevation. Since both guns were held in a single cradle (although recoiling separately), elevating some twenty tons by hand to 15° was a considerable effort. The training motor was of 6hp at 100 volts, which could rotate the turret at angles of heel up to 10°. The two hoist motors were of 3hp each and could supply shells to alongside each gun or to a circular carrier space beneath the gunhouse platform, holding no fewer than 153 rounds. Normal magazine capacity was 120 rounds per gun (although 200 could be stowed), with rate of fire about 3.5 rounds per minute. Gunhouse armour was 4in front and sides, 2in rear and roof, with 3in surrounding the ammunition trunk. Revolving weight was 75 tons.[2]

Friedman reports that the mounting was considered cramped and

[2] *Handbook for the Twin 6-inch Gun Mounting, 1904.* IWM LBY 16965-1 (1100 copies printed)

The 10,850-ton armoured cruiser *Argyll* was built in 1905 by Scotts of Greenock (at a large profit, restoring the company's finances). She carried four single 7.5in Mark I mountings made by Elswick, plus six single 6in P.III BD mountings. *(World Ship Society)*

unwieldy with its single cradle, poor gunsights and prone to electrical faults.[3] *Jane's Fighting Ships* in 1914 reported that consideration was being given to replacing the 6in with 7.5in mountings but the outbreak of war prevented any such change. It was to be another twenty years before the Admiralty considered twin 6in mountings again for new ships.

Non-transferable twin 6in mountings were fitted by Vickers in the three ex-Brazilian river monitors, *Severn, Mersey* and *Humber* (one in each) and by Elswick in the two ex-Norwegian coast defence ships *Glatton* and *Gorgon* (four singles each) all as described in *Big Gun Monitors*; all saw service in WW1 but were short lived.

On later single mountings 30° elevation was provided with a heavier charge and 4 crh shell giving longer range.

## THE 7.5-INCH MOUNTINGS

For the 1901 armoured cruiser programme, the Admiralty sought a gun between 6in and 9.2in calibre. A shell of 200lb was reckoned the heaviest two men could handle, appropriate for a 7.5in gun. The Mark I was a Vickers design of 45 calibres. To get a reasonable rate of fire, mechanical ammunition supply was required. *(See colour plates on page 151)*

The six *Devonshire* class of 10,850 tons (*Devonshire, Antrim, Argyll, Carnarvon, Hampshire, Roxburgh*) mounted four hydraulically operated single 7.5in, plus six single 6in P.III BD mountings. Twenty-four Mark I mountings of Elswick design were made by them plus six spares,

| 7.5-inch BL Gun | Mark I | Mark II & V |
|---|---|---|
| Calibres, bore | 45 | 50 |
| Length overall, in | 349.2 | 386.7 |
| Weight incl BM, tons | 13.7 | 15.0 |
| Shell, lb | 200 | 200 |
| Cordite MD26, lb | 61 | 61 |
| Muzzle velocity, ft/sec | 2760 | 2840 |
| Range at 15° 2 crh, yd | 14,000 | 14,300 |
| Life on MD, efc | 600 | 600 |

[3] Friedman, Norman. *British Cruisers of the Victorian Era*. Seaforth Publishing, 2012, p342

| 7.5-inch single mounting | Mark I | Mark II |
|---|---|---|
| Revolving weight, tons | 87 | 98 |
| Maximum elevation, degrees | 15 | 15 |
| Maximum depression, degrees | -7 | -7½ |
| Rounds per minute | 3.5 | 4 |

although elevating portions only. The hoists reached right down to the magazines capable of supplying four rounds per minute; stowage was 120 rounds per gun. Hand operation could be used in emergency.[4]

The 1903 Programme saw four larger armoured cruisers at 13,550 tons ordered (*Warrior, Achilles, Cochrane, Natal*) armed with six 9.2in and four 7.5in. The latter were longer at 50 calibres, the Mark II and V guns being essentially similar. Some of the guns were used for coast defences overseas. These Mark II mountings by Elswick were mounted two on each broadside at upper deck level. The 1904 Programme saw three more vessels, the 14,600-ton *Minotaur* class, with four 9.2in and ten 7.5in Mark II mountings. Vickers at Barrow made the mountings for *Natal, Shannon* and *Defence* at about £13,000 each (excluding guns about £3900 each), while Elswick made those for the other three *Warrior*s and *Minotaur*.

The gunhouses of both Marks were heavily armoured: Mark I 7½in KC front, 5½in sides, 4½ rear with 2in nickel steel roof; Mark II 8in KC, 6in, 4½in, 2in KNC roof.

The 7.5in guns in *Swiftsure, Triumph* and *Hawkins* class cruisers were all in transferable mountings.

## 9.2-INCH NON-TRANSFERABLE MOUNTINGS

Although there had been 9.2-inch guns in RN cruisers from the 1880s, it was not until high velocity wire-wound guns using cordite became available that a powerful armament could be provided for armoured cruisers. With a 380lb shell, it would be possible to penetrate 6in Krupp cemented armour, now being increasingly used for protection. The 46.7-calibre 9.2-inch Mark X gun of 1898 went on to be widely used both by the Navy and the Army. Indeed, coast defence guns mounting it can still be seen at IWM Duxford, Gibraltar, Rottnest off

| 9.2-inch BL Gun | Mark X | Mark XI |
|---|---|---|
| Calibres, bore | 46.7 | 50.1 |
| Length overall, in | 442.4 | 474.4 |
| Weight incl BM, tons | 28.3 | 28.40 |
| Shell, lb | 380 | 380 |
| Cordite MD37, lb | 120 | 130 |
| Muzzle velocity, ft/sec | 2720 | 2880 |
| Range at 15° 2 crh, yd | 15,500 | 16,200 |
| Life on MD, efc | 450 | 300 |

Fremantle, Cape Town and Bermuda. *(See colour plate on page 156.)*

At 28 tons, such a heavy gun really needed full mechanical operation. Electrical operation was still regarded as 'ungovernable' whereas the well-tried hydraulics could be used for training, elevating, loading and ramming. Two *Cressy* class armoured cruisers were the first to be fitted with two Elswick single Mark V mountings, plus twelve 6in, completed from 1901. These 12,000-ton ships were fitted with 6in side armour and could steam at 19 knots. Four others of the class had two of the Vickers Mark VI design, ordered in April 1898 at £21,600 each excluding the 6in gunshields, with delivery in 30 months.[5] That was only a month after Vickers' first gun mounting contract – two 12in twin for *Vengeance* – suggesting that the Admiralty was keen to bring in another manufacturer as soon as possible. Although *Hogue* was the first to be ordered (Contract 3G), she was two years late after collapse of the dock entrance at Barrow trapped her; she made a loss on her gun mounting contract of £23,800. John Brown's *Sutlej* (4G) overtook her, although as was the custom at the time, her mountings were installed by Chatham Dockyard rather than by her builders (Vickers charge £3500). By Vickers' fifth contract, *King Alfred*, the company was making a bare profit on the 9.2in.

A protected cordite hoist on the mounting centre line delivered the charges direct from magazine to gunhouse. But shell supply was more complex. While full stowage was 105 rounds per gun, less than half could be fired in action. At the rear of the gunhouse eight ready-use shells were stowed. Thirty-two more were stowed within the 4ft deep barbette structure (6in armour) which also contained the turntable and

[4] *Handbook for the 7.5-inch B.L. Mounting (Mark I). 1905.* Imperial War Museum LBY 16970

[5] Cambridge University Library, Manuscripts. Vickers Archive 1146

| 9.2-inch Mounting | Mark V | Mark VS | Mark VI | Mark VII | Mark VIII |
|---|---|---|---|---|---|
| | 2 singles | Single | 2 singles | Twin | 2 singles |
| Designer | Elswick | Elswick | Vickers | Vickers | Vickers |
| Gun Mark | X | X | X | XI | XI |
| Revolving weight, tons | 118 | 128 | 115 | 250 | |
| Maximum elevation, degrees | 15 | 15 | 15 | 15 | 15 |
| Maximum depression, degrees | -7½ | -5 | -7½ | -5 | -5 |
| Gunhouse armour, front/sides/rear/roof in | 6/3/4/2 | 7½/5½/4½/2 | 6/3/4/2 | 9/6/5/ | |
| Made by Elswick | *Cressy, Aboukir, Drake, Good Hope, Leviathan* | *King Edward VII* 4, *Africa* 4, *Commonwealth* 4, *Hindustan* 4, *Warrior* 6, *Achilles* 6, *Black Prince* 6, *Cochrane* 6 | | *Lord Nelson* 4, *Minotaur* 2 | *Lord Nelson* |
| Made by Vickers | | *Dominion* 4, *Britannia* 4, *New Zealand* 4, *Hibernia* 4, *Duke of Edinburgh* 6, *Natal* 6 | *Hogue, Sutlej, Bacchante, Euryalus, King Alfred* | *Agamemnon* 4, *Defence* 2, *Shannon* 2 | *Agamemnon* |
| Total made | 10 + 5 spares | 68 + 14 spare | 10 + 5 spare | 14 + 6 spares | 4 + 1 spare |

The 13,550 ton-armoured cruiser *Warrior* was built in 1906 by Pembroke Dockyard, engined by Wallsend Slipway & Engineering. She carried six single 9.2in Mark V strengthened mountings made by Elswick (one forward, one aft, the others broadside) plus on each broadside two single 7.5in Mark II mountings, whose gunhouses were little smaller than the 9.2in. *(Marine Technology Special Collection)*

training engine. There was a hatch in the gunhouse floor served by an overhead crane for these shells to be brought into the gunhouse when required. Replenishment thereafter was more complicated and much slower. Two unprotected shell hoists per mounting delivered from the shell room above the magazine to hatches in the upper deck. The turret then had to be trained so that a hatch in the overhanging rear of the gunhouse plumbed each hatch allowing shells to be lifted first into the gunhouse, then lowered through another hatch in the gunhouse floor into the barbette stowage area. Rate of fire was typically 3.5 rounds per gun per minute, until the ready-use ammunition was exhausted.

The eight pre-dreadnoughts of the 16,300-ton *King Edward VII* class had a mixed armament of the usual two twin 12in mountings but with four single 9.2in Mark V strengthened mountings with 150 rounds per gun, plus ten single 6in. However, it was found as with the later *Lord Nelson*s that it was difficult to distinguish the fall of shot between the 9.2in and 12in.

The two armoured cruisers *Duke of Edinburgh* and *Black Prince* had six single 9.2in each plus ten 6in, while the four *Warrior*s of the 1903 Programme also had six 9.2in plus four 7.5in. All the 9.2s were single Mark VS costing around £20,000 each including gunshield armour. All carried 100 rounds per gun.

A new *Good Hope* shows off her forward single 9.2in Mark V mounting. The deflection rail prevents the gun from firing into the hull. *(Fairfield collection)*

The 1904 Programme ships were the last to mount 9.2in, with *Lord Nelson* and *Agamemnon* each having four twin Mark VII (roller path diameter 16.5ft) and two single Mark VIII (diameter 13ft), all Vickers designed, plus the usual two twin 12in. The three *Minotaur* class armoured cruisers had two twin Mark VII each plus ten 7.5in. By this time Vickers were able to make a profit of about 20% of cost on each mounting priced at about £46,500.[6] These all had the 50-calibre Mark XI gun of which 45 were made. The guns could be loaded between 6½° elevation and 5° depression.[7]

Two single 9.2in mountings were fitted in the ex-Norwegian coast defence ships *Glatton* and *Gorgon,* with guns linered down from the original 9.4in, as described in *Big Gun Monitors*.

[6] Vickers, Sons & Maxim Annual accounts. BAE Systems, Barrow shipyard archives

[7] Friedman, Norman. *Naval Weapons of World War One*, Seaforth Publishing, 2011, p72

CHAPTER NINE

# Heavy Mountings Ordered for Foreign Navies

### 10-INCH TWIN MOUNTINGS

Tension between Chile and Argentina in 1901 resulted in Chile ordering two 12,000-ton battleships of 19 knots armed with four 10in guns. The basic design came from Sir Edward Reed, former DNC, and was limited in size by the drydock at Talcahuano, the Chilean naval base. They were ordered on 26 February 1902, one from Armstrong Whitworth to be named *Constitucion* and one from Vickers, *Libertad,* to be delivered in eighteen months. Each company would supply the main armament and armour, with two twin 10in mountings in turrets fore and aft and fourteen single 7.5in centre pivot mountings in casemates.

Keen to defuse a South American naval arms race, Britain persuaded the Chileans in 1903 to put their ships up for sale, Russia and Japan showing interest. But instead, Britain decided herself to buy the two ships on 21 November 1903. *Libertad*'s turrets had been installed in July and September and she started trials in the Clyde on 28 November still under Chilean auspices. The Chilean authorities and Vickers had already decided to publicise the ship with an extensive technical write-up in the journal *Engineering*, probably as a sales pitch. The article only appeared in the issue of 15 January 1904, running to 25 pages with fold-out plates on her 10in mountings and machinery, with an extensive description of all her armament: four 10in, fourteen 7.5in BL, fourteen 14pdr QF and four 6pdr QF.

Both ships were commissioned into the Royal Navy on 21 June 1904, *Libertad* as *Triumph, Constitucion* as *Swiftsure.* The Admiralty had paid £956,000 for each, which presumably also included their special ammunition, with the Chileans being reimbursed. The RN

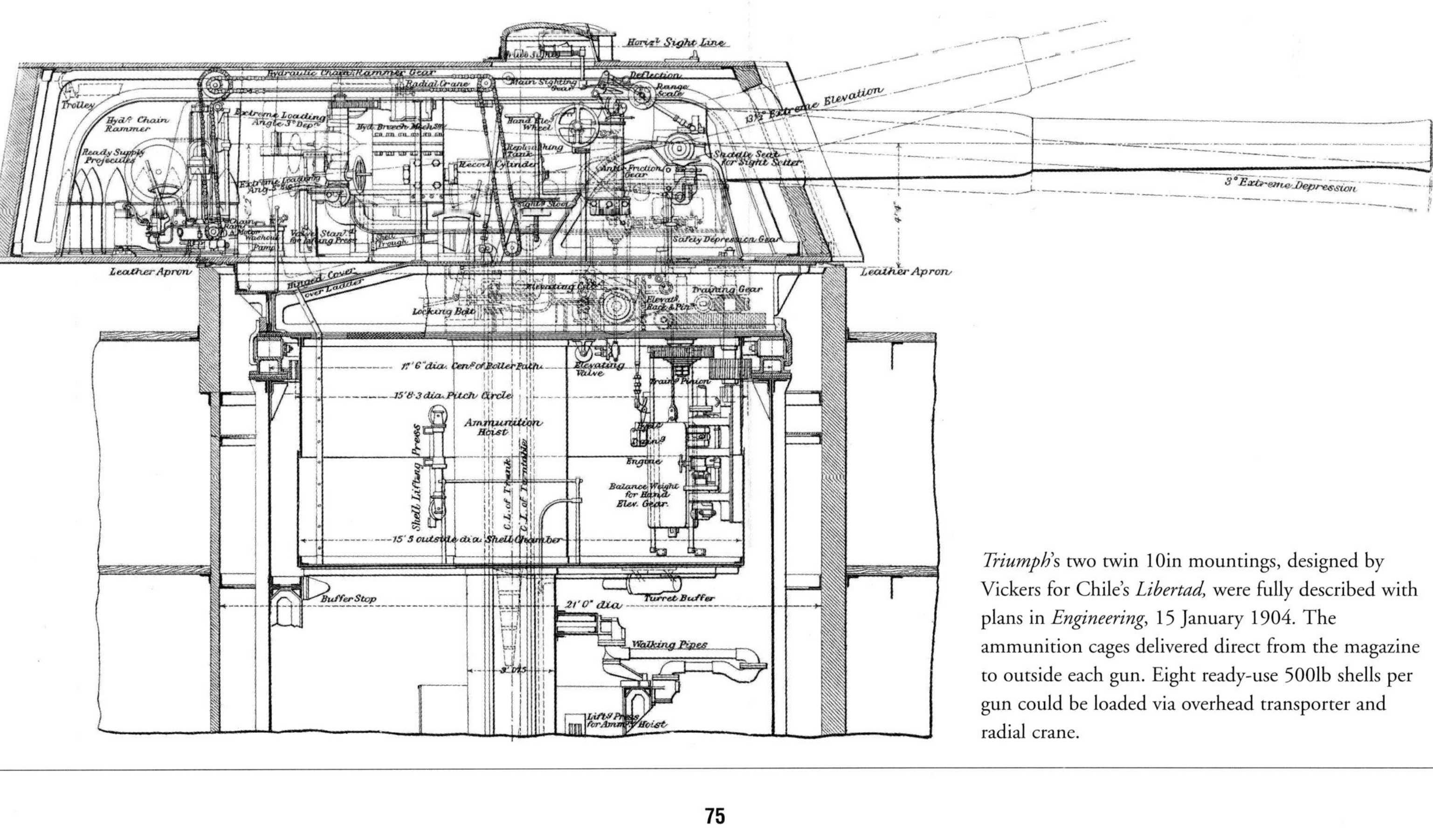


*Triumph*'s two twin 10in mountings, designed by Vickers for Chile's *Libertad,* were fully described with plans in *Engineering*, 15 January 1904. The ammunition cages delivered direct from the magazine to outside each gun. Eight ready-use 500lb shells per gun could be loaded via overhead transporter and radial crane.

The 12,000-ton battleship *Swiftsure* had been built by Armstrong for Chile as *Constitucion* but was bought by the RN in late 1903. She mounted two twin 10in made by Elswick, one forward, one aft. On the broadside were fourteen single 7.5in centre pivot mountings in casemates. Her near sister *Triumph* (ex-*Libertad*) had the Vickers 10in mounting. *(Author's collection)*

categorised Armstrong's 10in gun as BL Mark VI and mounting B.IV, with Vickers BL Mark VII and B.V.

The Vickers 45-calibre 10in gun was wire-wound over half its length, designed to use 210lb nitro-cellulose propellant. This gave a higher muzzle velocity of 2840ft/sec compared with cordite but at the expense of shorter life. But in RN service 146¾lb cordite was used giving 2656ft/sec. It fired a new armour piercing capped shell. Only five guns were made for each ship, one spare.

The mounting was simpler than for larger calibres and was designed to fire up to three rounds per minute from each gun, faster than a contemporary 12in gun. No working chamber was used, the hoists from the magazines delivering the cage direct to the outer side of each gun. Here the shell was transferred by a radial crane to a loading tray which was swung in line with the breech. A chain driven rammer could move vertically to load the 500lb shell into the gun at angles between 5° elevation and 3° depression. The four quarter charges were light enough to be handled manually outside the cage. To accommodate the recoil at the maximum elevation of 13½°, a panel had to be opened in the gunhouse floor. Sixteen ready-use shells were stowed in the rear of the gunhouse served by an overhead rail. The main magazines held 90 rounds per gun.

The Vickers gun shield plates were flat (curved in Armstrong); front 9in KC, sides and rear 8in KC, roof 2in – a total of 117 tons. The rest of the mounting weighed 98.5 tons, the two guns 61.5 tons, making a revolving weight of 277 tons. Two hydraulic pumps were fitted, with one enough to power both mountings at 1000lb/sq in, to work hoists, training, elevating and ramming. The training engine could train the turret at 4° per second and operate up to 10° heel; hand drive could be used in emergency. The Elswick mounting was broadly similar, but their gun was heavier at 37 tons and the breeches were different. Although Vickers records do not show the price or profit on the two

| **10in BL Mark VII Gun** | |
|---|---|
| Calibres, bore | 45 |
| Length overall, in | 462.75 |
| Weight incl BM, tons | 30.7 |
| Nominal muzzle velocity, ft/sec | 2656 |
| Shell, lb | 500 |
| Charge in four quarters MD45, lb | 146.75 |
| Life, efc | 350 |

| **10in Twin B.VI Mounting** | |
|---|---|
| Rotating weight, tons | 277 |
| Max elevation, degrees | 13.5 |
| Max depression, degrees | -3 |
| Rounds per gun per minute | 2.5 |
| Range at 13½° 2 crh, yards | 14,800 |

A view inside *Libertad*'s (soon to become *Triumph*) twin 10in mounting with the left-hand breech open, right closed. (*Engineering*)

10in mountings, they do give their cost as £52,458 excluding guns, a somewhat higher cost per ton than for a 12in.

Although a well-balanced design for a medium sized navy, the RN regarded them as second class battleships so relegated them to second line duties. Early in 1915 both ships were sent to support operations at the Dardanelles, although their low elevation guns with low trajectories were only effective when targets were clearly visible (and many were not) and spotting possible. *Triumph* fired 14 10in over two hours at Cape Helles fort from 7700 yards on 19 February with no success.[1] She supported the Anzac landing on 25 April but was torpedoed off Gaba Tepe on 25 May. *Swiftsure* also undertook shore bombardments between 2 March and 7 May although stocks of her 10in ammunition would have been limited. She remained in the Aegean until early 1916, paid off in 1917 and was scrapped in 1920.

### 12-INCH TWIN MOUNTINGS FOR BRAZIL

Brazil's economy was thriving in 1905 based on a near monopoly of rubber exports for tyres for the new motor vehicles. Naval rivalry had developed between its near neighbours Argentina and Chile, so a large naval programme was proposed. This included three 13,000-ton battleships which were ordered from Armstrong Whitworth in July 1906. But after the powerful *Dreadnought* was completed, they had second thoughts, wanting larger ships of 19,000 tons with twelve 12in 45-calibre guns and 21 knots. The order was revised to two ships in February 1907, with *Minas Geraes* to be built at Elswick with steam reciprocating machinery from Vickers and *São Paulo* from Vickers. Both would get six twin 12in mountings (one more than *Dreadnought*) supplied by Elswick, with superimposed turrets fore and aft and two wing turrets. The lengthy discussions about the many ship designs for Brazil are fully described by David Topliss in *Warship International* Nº 3, 1988. While there are detailed costs for *Minas Geraes'* hull, none have been found for armament, although the latter including secondary armament amounted to about one-third the total cost of £1,821,000. *Minas Geraes* carried out gun trials on 5 October 1909 and was handed over on 5 January 1910, leaving the Tyne on 5 February.[2] *São Paulo* departed from the Clyde on 16 September 1910. Both ships went on to have lives of over 40 years.

[1] Corbett, Julian. *Naval Operations*, Vol II, 1921, p145

[2] Cambridge University Library, Manuscripts. Vickers Archive 1106

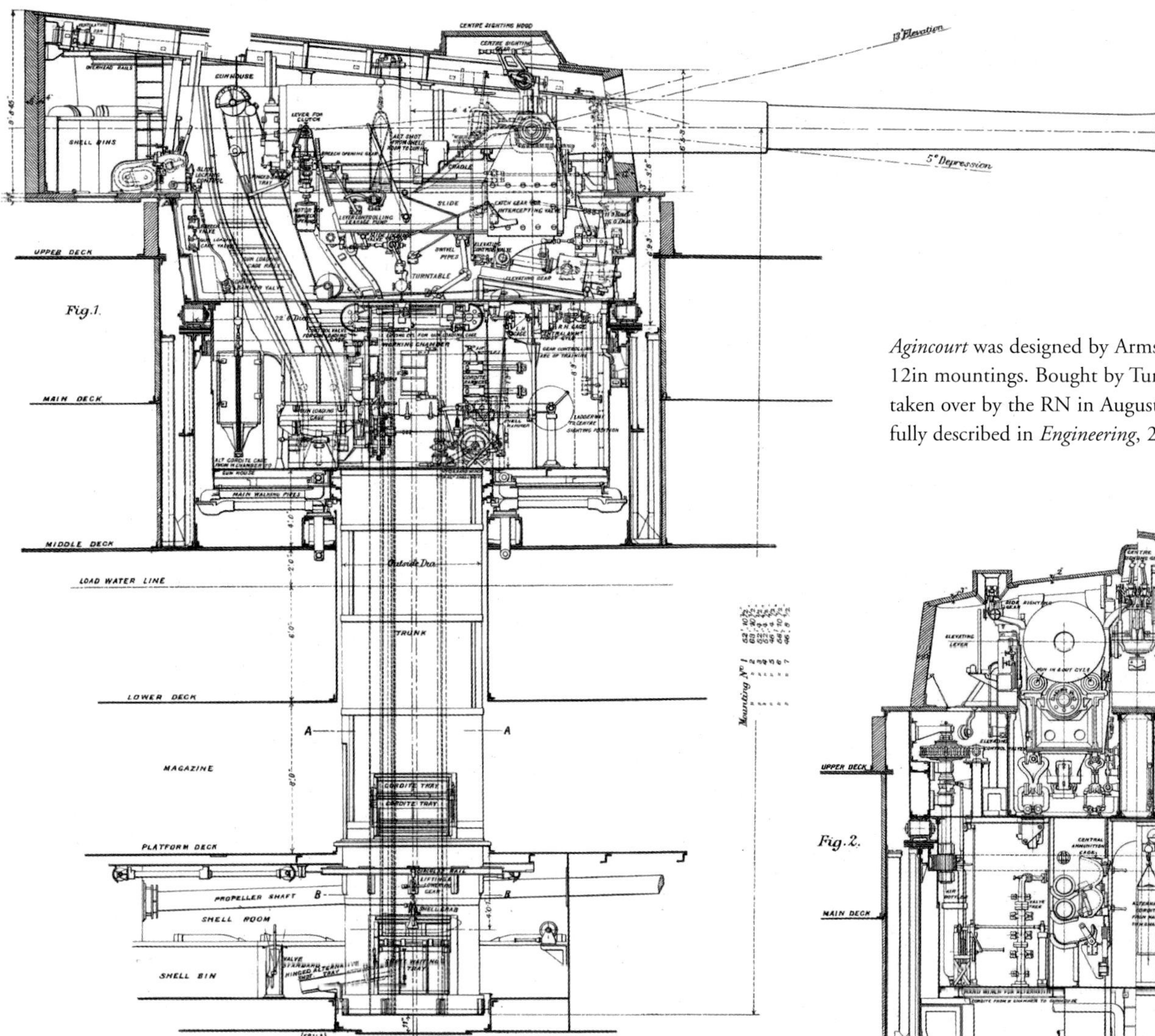


Fig.1.

*Agincourt* was designed by Armstrong for Brazil with seven twin 12in mountings. Bought by Turkey before completion she was taken over by the RN in August 1914. Her mountings were fully described in *Engineering*, 28 November 1919.

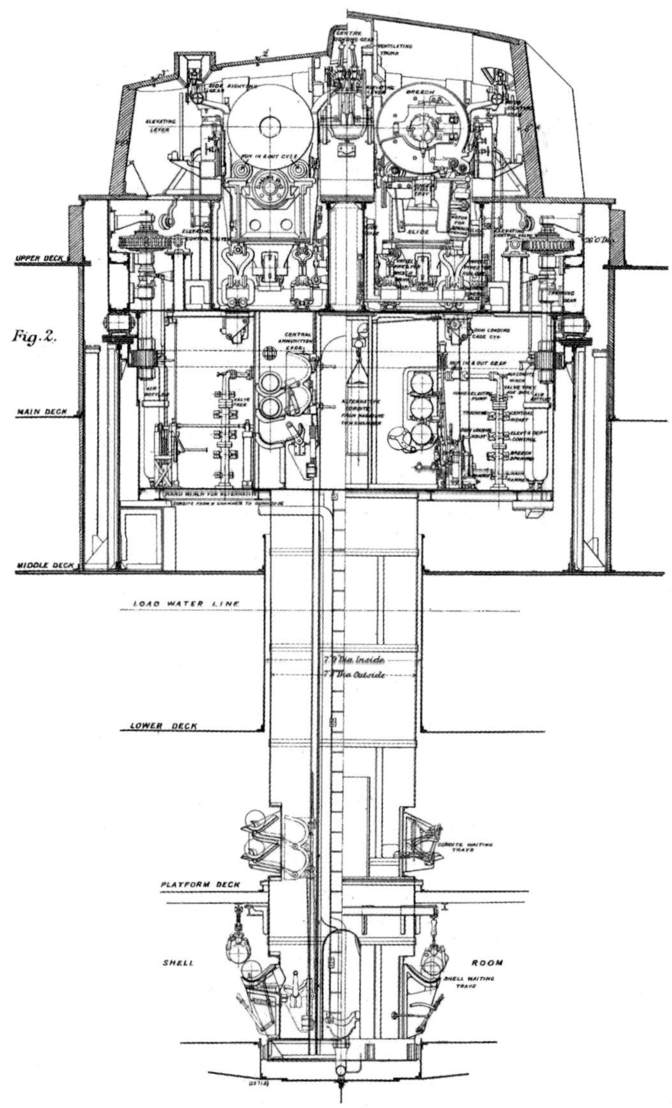


Fig.2.

Fig.3.
HALF PLAN ON TOP.

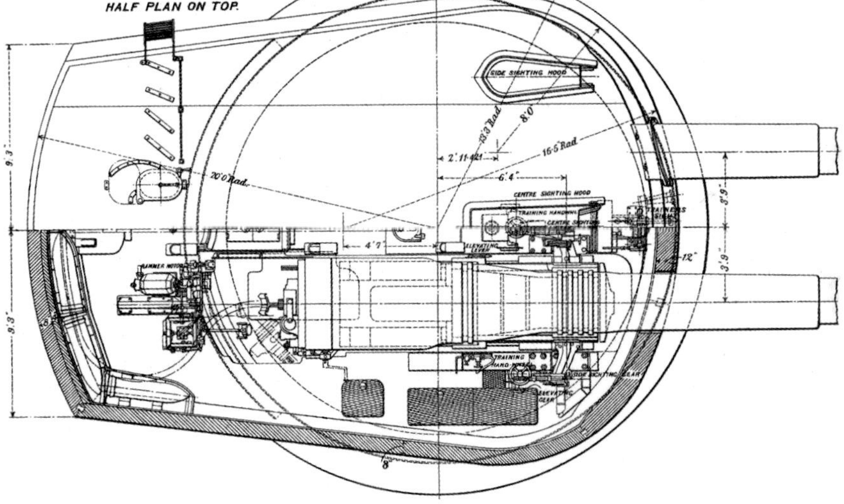

Armstrong had reserved Yard number 792 for the third battleship, so Brazil continued discussions as soon as the first two were completed. Even more powerful designs of up to 30,000 tons with guns from 12in to 16in calibre were investigated. Armstrong's chief naval architect Tennyson d'Eyncourt, who went on to become DNC in 1912, was closely involved in the negotiations. He finally clinched the order in Rio de Janeiro on 3 June 1911 for a ship of 27,500 tons, 22 knots and mounting seven twin 12in mountings, the greatest number ever fitted in a battleship. Two were mounted forward, two amidships and three aft, with Y superimposed over X and Z. Both mounting and gun were of Armstrong's design, the 45-calibre gun able to use the same ammunition as the two earlier ships. Elswick now had no fewer than 32 heavy mountings on order.

*Rio de Janeiro* was launched from Elswick on 22 January 1913, its longest ship to date at 672 feet overall, fully described in *Engineering* on 24 January 1913. But by now Brazil was facing a financial and political crisis, with rubber exports undercut by new plantations in Malaya. So Rothschilds, who had extensive financial interests in Brazil, sought buyers worldwide for the part-built ship. *Reshadieh* (also rendered as *Reşadiye*) had recently been launched in September 1913, so Turkey agreed in December to buy a second battleship. By this time the main machinery had been shipped as well as most of the gun mountings. Renamed *Sultan Osman I* she went on trials on 7 July 1914, although lacking both guns in X mounting.[3] Meantime in April 1914 Turkey had also ordered *Fatih* from Vickers and Brazil *Riachuelo* from Armstrong both with 15in guns, although in the event both were cancelled. On 3 August 1914 the Royal Navy took over the virtually completed ship in the Tyne and renamed her *Agincourt,* the Admiralty reimbursing Turkey its costs – the final cost to Britain was £2.725 million.

The RN designated the Armstrong Pattern W 12in gun as Mark XIII. After the original fourteen, seven spares were made. The mounting was an Armstrong 'special', which was fully described in *Engineering*, 28 November 1919 edition. The mountings were originally designated from forward to aft, 1, 2, 3, 4, 5, 6, 7 but they may have been changed by the RN to A, B, P, Q, X, Y, Z. A special feature was its single lever in the gunhouse controlling four motions: opening the breech, locking the gun at its loading angle, raising the cage and ramming. Its maximum elevation was only 13°, depression -5°, but in 1918 this was altered to 16°/-2° presumably by modifying the trunnions. A fixed loading angle of 5° enabled a 20-second firing cycle to be achieved on trials, although with only three hydraulic pumps serving seven mountings, unlikely to be maintained in action. Gunhouse

[3] Dodson, Aidan. *The Windfall Battleships*. Seaforth Publishing, 2023, p16

**12in BL Mark XIII Gun**

| | |
|---|---|
| Calibres, bore | 45 |
| Length overall, in | 556.7 |
| Weight incl BM, tons | 60.7 |
| Nominal muzzle velocity, ft/sec | 2725 |
| Shell, lb | 850 |
| Charge in four quarters MD45, lb | 260 |

**12in Special Twin Mounting**

| | |
|---|---|
| Rotating weight, tons | 452 |
| Max elevation, degrees | 13 |
| Max depression, degrees | -5 |
| Rounds per gun per minute | 2.4 |
| Range at 16° 4 crh, yards | 20,670 |

*Agincourt* was designed by Armstrong for Brazil with no fewer than seven twin 12in mountings by Elswick, two more than contemporary Royal Navy battleships. She was resold to Turkey but taken over by the RN in August 1914. This view shows the two mountings forward, two amidships and three aft. *(World Ship Society)*

When *Agincourt* was broken up at Rosyth in 1923, the interiors of her mountings were revealed. P and Q are nearest, with aft X, superfiring Y with part of her guns cut off and Z. *(Rosyth Dockyard)*

armour was 12in KC front, 8in KC sides and rear, with 3in at roof sides, 2in in the middle. The rear had an additional 4in mild steel balance weight plate fitted.

The original outfit comprised armour piercing capped, common pointed capped and shrapnel (with 7766 bullets) projectiles. The handbook prepared in July 1914 advised that reduced charges should be used unless the full power of the gun was required, *eg* for piercing thick armour, with sixteen half charges taken as equivalent wear to one full charge.[4] The handbook showed the mounting weights as:

| | |
|---|---|
| Turntable and working chamber | 133 tons |
| Gun carriages, slides and pipes | 44 |
| Ammunition trunk and hoists | 17 |
| Gun shield and fittings | 137 |
| Two guns | 121 |
| Total | 452 |

At Jutland *Agincourt* fired 144 rounds as part of the First Battle Squadron, all CPC.

## 13.5-INCH TWIN MOUNTINGS IN *ERIN*

During 1910–11 Turkey discussed a variety of battleship designs with Armstrong with guns ranging up to 16in. Eventually they settled on a design with five twin 13.5in mountings, 23,000 tons and 21 knots. Although an Armstrong design, the order for the hull and machinery and X and Y mountings was placed with Vickers in June 1911. The other three mountings and some armour were ordered from Armstrong, with John Brown making some armour, shafting and forgings and Cammell Laird making some armour. The companies already had a work- and profit-sharing agreement for any Turkish orders, having to pay commission to the other partners.[5]

The 13.5in gun was a Vickers design of 45 calibres; Vickers building five and Elswick five. The mountings were basically similar to the British Mark II, weighing about 600 tons and able to take the larger 1400lb shell. The gun profile and cradle were the same as the RN Mark V, so no spares were ordered. Indeed, two Mark V guns were actually fitted in

[4] Tyne & Wear Archives DS.VA/6/18/11/1

[5] *The Battleship Builders*, p229

*Erin* was launched by Vickers as *Reshadieh* for Turkey but taken over by the RN in August 1914. She mounted five twin 13.5in of Elswick design, although Vickers built X and Y, Armstrong A, B & Q. *(World Ship Society)*

one mounting, with charges adjusted to give the same muzzle velocity as the Turkish guns which were designated Mark VI by the RN.[6]

*Reshadieh* was launched at Barrow on 3 September 1913 and the mountings installed in 1914. The ship was about to be handed over to Turkey when war was declared. As permitted under the shipbuilding contract, the Admiralty could take over ships building for foreign navies, so she was taken over on 2 August and renamed *Erin.* After drydocking at Liverpool (25–30 August), she joined the Grand Fleet after gunnery trials off Loch Ewe.[7] Vickers accounts record their share of the contract as £1,325,754, slightly more than half the total value, but resulting in a modest loss of £23,655. Vickers' two mountings cost £155,396. Later spares including two cradles were ordered from Vickers for £75,649.[8] Little was done on her sister *Fatih* before cancellation in November.

The Admiralty reckoned that her guns were not up to their standards. She fired no 13.5in rounds at Jutland, due to her position in the fleet and not having director firing. The mountings were scrapped with the ship at Queenborough in 1923.

[6] *Naval Weapons of World War One*, p52
[7] Dodson, *op cit*, p21
[8] All such figures extracted from Vickers Barrow accounts

**13.5in BL Mark VI Gun**

| | |
|---|---|
| Calibres, bore | 45 |
| Length overall, in | 625.9 |
| Weight incl BM, tons | 77 |
| Muzzle velocity nominal ft/sec | 2445 |
| Shell, lb | 1400 |
| Charge MD45, lb | 265 |
| Range at max 20° elev 4 crh, yd | 23,100 |

## 14-INCH TWIN MARK I MOUNTING

Since *Libertad* and *Constitucion* had been taken over by Britain, Chile had been considering what capital ships to build in the light of what their South American neighbours were building. Eventually they settled on two large fast battleships with 14in guns but with modest protection. They placed the order for the first with Armstrong Whitworth in November 1911, to be named *Valparaiso,* then changed to *Almirante Latorre*, of 28,000 tons and 23 knots with five twin 14in turrets. On 28 July 1911 John Brown had also submitted a tender with three alternative armament suppliers, Armstrong, Vickers and Coventry Ordnance Works, of £2.294 million, which was slightly less than Armstrong's price of £2.335 million.[9] Vickers also probably tendered. The contract for a sister ship *Almirante Cochrane* was signed by President Barros Luco and Armstrong chairman Sir Andrew Noble on 29 July 1912 for £2,425,190. This included hull, machinery (from John Brown), armour (from Openshaw), armament as well as ammunition and spare guns, to be completed by 1 March 1915.[10]

Elswick had not built this calibre before, 13.5in being the then largest which they had just supplied to the RN, although Vickers already had an order for 14in armament for the Japanese battlecruiser *Kongo*. The 14in gun was wire-wound with 45 calibres bore length. Ten plus four spares were made for the Chilean ship. Each of the five turrets featured all-angle loading to their maximum elevation of 20° and was fitted with a 15ft rangefinder. The gunhouse was less heavily armoured than for RN battleships with 10in front and sides, 9in rear and 4in roof.

*Almirante Latorre* was launched on 27 November 1913, the largest ship to date from the cramped Elswick shipyard, having been delayed by late delivery of armour from Openshaw. Armstrong had just laid the keel of the first ship, *Malaya*, at its new more spacious Naval Yard down-

[9] *The Battleship Builders*, p281
[10] *Almirante Cochrane* contract in the Admiralty Library, Portsmouth

The Chilean battleship *Almirante Latorre* was taken over in 1914 to become HMS *Canada*. Her five twin 14in mountings were unique in the RN. She also had sixteen single 6in. *(Author's collection)*

stream of the Tyne bridges, but *Almirante Cochrane*'s had already been laid at Elswick. It is thought that installation of *Latorre*'s main armament had just started when war broke out. The Admiralty considered taking over both ships, after reviewing their suitability. It noted the lower level of protection, the larger magazine capacity of 100 rounds per gun and the lack of redundancy in the hydraulic pumps – only three to serve five mountings compared with four for four in the RN. The superstructure would need to be modified to accept new fire control positions and equipment. The Admiralty purchased the ship on 9 September 1914, renaming her *Canada,* and reimbursing the Chileans for the monies they had paid, with an option for them to repurchase after the war.[11] The Elswick Pattern A 14in guns were designated Mark I by the RN and her twin mountings also Mark I. The second ship was also purchased but construction was suspended. Work on her resumed in 1917 after she had been redesigned as an aircraft carrier, to be named *Eagle.* None of her battleship mountings seem to have been made, although three guns were completed for railway mountings.[12]

**14in BL Mark I Gun**

| | |
|---|---|
| Calibres, bore | 45 |
| Length overall, in | 684.4 |
| Weight incl BM, tons | 84 |
| Muzzle velocity nominal ft/sec | 2500 |
| Shell, lb | 1586 |
| Charge MD45, lb | 344 |
| Life, efc | 250 |

**14in Mounting Twin Mark I**

| | |
|---|---|
| Revolving weight, tons* | 660 |
| Max elevation, degrees | 20 |
| Max depression, degrees | -3 |
| Training speed, deg/sec | 3 |
| Elevating speed, deg/sec | 3 |
| Max range, yards, 4crh | 24,400 |

* includes 220 tons armour and 168 tons guns

*Canada* commissioned on 10 August 1915, and took part in the battle of Jutland as part of the 4th Battle Squadron, firing 42 rounds.

After the war, *Canada* was sold back to Chile with three destroyers also taken over, for £1.4 million the lot, less than half what they had cost the Admiralty, being handed over on 1 August 1920. As part of their major naval expansion which included six new destroyers and three submarines (all with Vickers Armstrongs armament), the Chileans decided to modernise *Almirante Latorre.* She was taken in hand at Devonport Dockyard in June 1929, where her machinery installation was replaced in 1930 with oil fired geared steam turbines by Vickers-Armstrongs from Barrow. Also in 1930, Elswick relined two of her 14in guns for £16,850 and supplied 412 14in forged steel practice shot for £10,840, and she received new fire control equipment.[13] All work was completed in March 1931; she was not scrapped until 1959.

The RN also had a 14in Mark II mounting, one of which armed each of the four *Abercrombie* class monitors in 1915. They had been built by Bethlehem Steel for the battlecruiser *Salamis* building in Germany for Greece. Unable to deliver them owing to the British blockade, they were sold to the Admiralty. They are fully described in *Big Gun Monitors.*

[11] Dodson, *op cit,* p29.
[12] *Warship* 17. 1981, p59
[13] Elswick accounts 1931. DS.VA/3/5/4

CHAPTER TEN

# 13.5-inch Mountings

### THE 13.5-INCH TWIN MOUNTING MARK II

After the 12in Mark XI high velocity gun had not fulfilled all its promise in 1909, the Admiralty sought a battleship main armament with greater destructive effect. Triple mountings or more than five 12in twin mountings or higher muzzle velocity were not considered desirable. Therefore, a heavier shell with a larger burster was desired which could defeat ever improving armour protection. The next step in gun calibre was 13.5in, last used in the 30-calibre Marks I to IV of the late 1880s. Vickers were asked to design a new wire-wound 13.5in, initially described as 12in A to disguise its true calibre. At 76 tons, both it and its 1250lb shell were about 50% heavier than the 12in Mark XI. Trials in 1909 of this 45-calibre Mark V gun proved encouraging, but required a new mounting. This was based largely on the 12in B.XI which was about to enter service. The new mounting retained the same general dimensions, which helped keep weight down of both mounting and its surrounding barbette armour, including the same 24ft 6in roller path diameter. The relatively narrow trunk meant that although the new shell could be loaded horizontally at the bottom of the hoist, it would have to be canted to 38° to travel up in the cage to the working chamber where it reverted to horizontal in the waiting tray, and then canted again in the gunloading cage. The gunhouse was too compact to allow the auxiliary loading hoists from the working chamber to deliver between the guns, so they had to be moved to the outer side of each gun. Eight ready-use shells per gun were stowed at the rear of the gunhouse, handled by a radial crane, with six more per gun in the working chamber. The shell rooms were planned to hold 80 per gun, a mix of APC, CPC and HE.[1] *(See colour plate on page 155)*

To increase range, the elevation was increased to 20° with all-angle loading, requiring a new single cylinder mechanism, able to attain 3° per second. Two of the new swashplate training engines were fitted on the turntable, only one driving at a time, with the training rack fitted inside the ring bulkhead supporting the rotating structure. Flash doors were fitted at the top and bottom of the main hoist and enclosing the gunloading cage, features which contributed to the survival of *Lion* at Jutland, despite taking a direct hit on Q turret. Repairs required the mounting to be removed from the ship for two months for repairs at Elswick. Other improvements included water squirts on the loading arm, one to damp any smouldering embers in the gun chamber and one to cool the mushroom head of the obturator.

The gunhouse armour consisted of eight KC 440lb (11in nominal) plates, front, side and rear, with three KNC plates on the roof, the front two of 160lb (4in) which could be unbolted to change guns and the rear one 120lb (3in). The less well protected battlecruisers had 9in front and sides, 8in rear and roof 3¼ and 2½in. Revolving weight with guns was 590 tons.

A provisional order for five Mark II mountings (all on the centre line) for battleship *Orion* to be built at Portsmouth was placed with Vickers on 20 August 1909 and also four at Elswick for battlecruiser *Lion* at Devonport, enabling them to make a start on drawing office work and ordering material in advance of a formal contract. *Orion*'s final price came out at £394,000 excluding guns.[2] On 10 December Vickers received the formal order for *Orion* and four (plus one spare) for battlecruiser *Princess Royal* to be built at Barrow and three for *Thunderer* at Thames Iron Works. On the same day, Elswick got the formal order for *Lion* and for *Thunderer*'s other two and five for *Monarch* to be built at Elswick. Coventry Ordnance Works also received its order on 10 December for five mountings for *Conqueror* but built to their own design as discussed below.

The Mark II* mounting was fitted in four ships of the 1910 Programme: *King George V, Audacious* (both Vickers), *Centurion* and *Queen Mary* (both Elswick). It had improved chain rammers and breech mechanisms and a new type of shell room bogie was introduced that eased moving shells from stowage into the hoists. The gun was strong enough to take a 4 crh 1400lb shell, whose greater momentum gave a range similar to the 1250lb shell despite a reduced muzzle velocity, so could be used in these mountings. Four pumping engines were fitted compared with the three in earlier ships, providing greater gun run-out capacity when fired simultaneously.[3]

For the 1911 Programme ships *Iron Duke, Marlborough* (both Elswick) and *Emperor of India* plus battlecruiser *Tiger* (both Vickers), the mounting was modified with a wider hoist trunk which allowed the shell

[1] *Addenda to Hydraulic Manual 13.5 inch Mountings Mark II, 1913*. ADM 186/190

[2] *The Battleship Builders*, p187

[3] *Addenda to Hydraulic Manual 13.5 inch Mountings Mark II**, 1914*. ADM 186/192

The first of the new 13.5in Mark II mountings were ordered in August 1909 for *Orion* (Vickers) and *Lion* (Armstrong). One of the latter's is shown in its pit in Elswick's new Nº 24 Shop. *(Tyne & Wear Archives)*

to be raised while horizontal, becoming Mark II**. This trunk could contain the new 1400lb shell, 110 of which were typically carried per gun in the battleships, although nominal capacity was 80. As usual, the shell was loaded into its part of its cage via a bogie which could rotate via a circular rack between the bins in the shell room and the revolving trunk. At the magazine level above four quarter charges could be loaded manually via flashtight hoppers into their part of the cage. Then these two parts were combined as they were lifted up the twin shafts at the sides of the trunk. The trunk also contained the lifting presses in the centre, plus auxiliary shell (raised vertically) and cordite hoists, a ladder from shell room to working chamber and flasks for high pressure air. The elevating rate was increased to 5° per second. The gunhouse armour was the same as the Mark II. *(See colour plates on pages 154, 155)*

*Tiger's* contract was worth about £238,000 plus £49,000 for her

**13.5in Twin Mark II Mounting**

| | |
|---|---|
| Revolving weight with guns, tons | 590 |
| Maximum elevation, degrees | 20 |
| Maximum depression, degrees | -5 |
| Elevating speed, deg/sec | 3 |
| Training speed, deg/sec | 2 |
| Rounds per gun per minute | 1.5-2 |

armour (30 tons less per turret), or about £72,000 per mounting; battleship *Orion* came to about £78,000. Excluding guns that amounted to about £178 per ton.[4]

There were now six plants manufacturing heavy guns: Vickers Ltd (its title from 1911) who made 53 of the 13.5in guns at Sheffield, Armstrong Whitworth at Elswick (49) and also at Openshaw (22), Coventry Ordnance Works (14), William Beardmore at Glasgow (10) and the Royal Gun Factory at Woolwich (58), a total of 206. Vickers

[4] *Orion* class Ships Cover. NMM

Bay N° 12 at Vickers' Barrow works showing 12in and 13.5in guns and mountings in 1911. The nearest 13.5in turntable and working chamber is for *Orion*'s 13.5in B mounting (Contract 115G) with roof plates from Sheffield in the foreground. Alongside are two 13.5in and one 12in gun. The cylindrical ammunition trunks beyond are for *Thunderer* (120G) and *Princess Royal* (119G). In Bay 13 to the left are components for *Australia*'s 12in (121G) and *Princess Royal. (Dock Museum 6512)*

| **13.5in BL Mark V Gun** | **Light** | **Heavy** |
|---|---|---|
| Calibres, bore | 45 | 45 |
| Length overall, in | 625.9 | 625.9 |
| Weight incl BM, tons | 76.1 | 76.1 |
| Muzzle velocity nominal ft/sec | 2582 | 2491 |
| Shell, lb | 1250 | 1400 |
| Cordite MD45, lb | 293 | 297 |
| Life, efc | 550 | 450 |
| Range at 20° elevation 4 crh, yd | 23820 | 23740 |

charged £11,300 for the gun bodies plus £1350 for the breech mechanism, giving a profit rate of 140% on cost.[5]

Elswick proudly announced that they had shipped all five of *Monarch*'s mountings in only four days from 23 July 1911. Their 150-ton crane made 250 lifts ranging from 145 tons to 1.5 tons, totalling 2800 tons including all ten guns.[6] Where Vickers was building both the hull and the mountings for 13.5in gunned ships such as *Princess Royal, Emperor of India* and *Erin,* the cost of delivering them was minimal – their Barrow ordnance shops were very close to their fitting out berths equipped with 150-ton cranes. But for shipyards further away, those same facilities had to be used to load the turret components into a transport ship for re-erection at destination. They had used the same technique when shipping submarine sections to Japan in 1908, even building a special ship called *Transporter.* In 1912–13 they used a newly built 1554 grt cargo ship also called *Transporter* to ship 12in and 13.5in mountings. Her first charter took 13.5in mountings for *King George V* to Portsmouth, leaving Barrow on 14 June 1912. Her next charter took two 12in mountings for *Australia* at Clydebank on 14 September, followed by the other two on 31 October. She was back at Barrow on 3 December taking four days to load probably one or two 13.5in mountings for *Audacious* at Birkenhead. She returned to Barrow on 11 December taking eight days to load 12in mountings (probably two) for Spanish battleship *España* at Ferrol. She was back at Barrow on 12 January 1913 to load more for *Audacious,* returning for another load on 8 February. In all such cases, the shipyard would use its heavy fitting out cranes to discharge the components on to their quayside in only one or two days, before starting erection on board, supervised by Vickers outside staff. Vickers gun mounting contracts included one sum for 'delivery' (typically £1000 for one heavy mounting) and another for 'erection', typically £4000 each.

Campbell gives the ammunition expenditure at Jutland and number of APC and CPC rounds fired, together with discussion of general shooting performance:[7]

[5] *The Battleship Builders*, pp175–7

[6] *Engineering*, 4 August 1911, p174

[7] Campbell, John. *Jutland: An Analysis of the Fighting*. Conway Maritime Press, 1986, pp346–7

**Right:** Battlecruiser *Queen Mary* fitting out at Palmer's on 18 March 1913 with her four twin 13.5in Mark II* newly installed by Elswick. *(NMM)*

1250lb ships: *Orion* 51, *Monarch* 53, *Thunderer* 37, *Conqueror* 57, *Lion* 326, *Princess Royal* 230. Total 754. *Lion* had two guns disabled, one on hand loading and one from failure of cordite supply.

1400lb ships: *Ajax* 6, *Centurion* 19, *King George V* 9, *Benbow* 40, *Iron Duke* 90, *Marlborough* 162, *Tiger* 303, *Queen Mary* *c*150. Total *c*779.

The story of how some of the mountings were to be used for Turkish coast defences is recounted in Appendix 4.

## COVENTRY ORDNANCE WORKS 13.5-INCH MOUNTINGS

For the 1909 Programme, the Admiralty took the opportunity to invite Coventry Ordnance Works to tender for their first non-transferable mountings, partly in the hope of getting a reduction in price from that normally charged by Vickers and Armstrong as well as to expand production capacity. The general situation at the time at COW is discussed on page 40. COW submitted a design featuring curved gunhouse armour plates, claimed to be lighter than flat, improved training and elevating gear, and independent shell and cordite hoist mechanisms.[8]

The formal contract letters for 13.5in twin mountings were sent out to the firms who had tendered on 10 December 1909.[9] That for the five for *Conqueror* differed from the others in having a slightly later delivery date and no penalty for late delivery – £50 per day for the others. It excluded hydraulic pumping engines, presumably ordered from Elswick; the Admiralty was probably reluctant to trust these key components to an untested design. The letter to COW included:

> With reference to your letters dated the 21st September 1909 and 22nd September 1909, I am commanded by my Lords Commissioners of the Admiralty to inform you that they are pleased to accept your tender for the supply of the following items at the prices named for the 12 inch A [ie 13.5in] hydraulic gun machinery required for a battleship having five turrets, the order for which machinery was provisionally placed with you by Admiralty letter dated 20th August 1909.
>
> A further communication will be made to you in regard to the price to be paid for the erection of the component gun machinery on board and for its delivery to the ship. It is preferred that you on receipt of that communication quote a lump sum price for the delivery of the machinery.
>
> My Lords desire me to invite your attention to the contract dates stated in the second paragraph of Admiralty letter dated 20th August 1909 and to the necessity of these dates being adhered to viz:
>
> For complete erection on board of the component gun machinery 20th October 1911.
>
> For complete delivery of the spare mounting 20th July 1911.

[8] *Addenda to Hydraulic Manual 13.5 inch Mountings for H.M. Ships Conqueror, Ajax and Benbow.* ADM 186/193

[9] *Orion* class Ships Cover. NMM

QUEEN
KEEP OFF PROPELLER
W. PARRY & SON,
SOUTH SHIELDS.

Elswick installed all five of *Monarch*'s 13.5in Mark II mountings in only five days in July 1911. The turntable and working chamber of X is being lifted aboard by the 150-ton hydraulic crane. Her topmast is hinged down ready for passage downriver under the Tyne bridges. To the left is Nº 6 Shop, centre Nº 7 and far right Nº 24. *(Author's collection)*

I am to add that your tenders for the gun machinery of additional ships received with your letter dated 22nd September 1909 are declined with thanks.

The detailed prices in the COW letter did not include gunhouse structure, floor and roof armour nor delivery and erection, presumably all the subject of later letters. The price breakdown for *Monarch* (Elswick) and *Conqueror* differed appreciably:

| *Item* | *Description* | *Monarch £* | *Conqueror £* |
|---|---|---|---|
| 1 | Turntable, working chamber & roller path | 51,030 | 60,585 |
| 2 | Training gear and rack | 26,405 | 17,335 |
| 3 | Gun cradles & slides, elevating gear & rammers | 65,134 | 54,080 |
| 4 | 3 Hydraulic pumping engines | 10,922 | |
| 5 | Piping on revolving structure | 11,106 | 7,425 |
| 6 | Shell room handling gear | 8,014 | 5,268 |
| 7 | Transporting gear, hoists & cages | 45,681 | 25,510 |
| 8 | Piping on hull structure | 9,245 | |
| 9 | Brackets, rails, covers, voice pipes etc | 2,945 | 6,270 |
| 10 | Gunhouse structure, floor & roof | 47,030 | |
| 11 | Drawings and models | 950 | 830 |
| 12 | Spare 'mounting' (one cradle etc.) | 7,363 | 6,280 |
| **Sub-total** | | **285,825** | **183,583** |
| 10A | Front, side, rear gunhouse armour per ton | 137.7 | 137.7 |
| Delivery date | | 26 Sep 1911 | 20 Oct 1911 |

Comparing those items common to both ships shows that COW quoted a price 16% less than Elswick, notably in the complex hoist mechanisms which were barely half that of the more experienced company. Openshaw supplied the 585 tons of gunhouse armour for *Monarch* for £80,500 so *Conqueror*'s would have been similar in cost. Her five sets were ordered from John Brown at Sheffield in January 1910, at the highly profitable standard armour prices per ton.[10] Gunhouse structure, pumping engines and piping would have added about another £65,000. Delivery and erection would amount to about £22,000, while there was about £8000 for minor items such as spare plates and lamps. Although there are no final Scotstoun figures, the above suggest a total of about £359,000 for five mountings. The price per turret that COW received thus came to about £71,800, less than Elswick or Barrow. With a revolving weight of about 430 tons that came to about £167 per ton, excluding the two guns which were Admiralty supply items (152 tons worth £25,700).

Vickers' figures for their 13.5in mountings show substantial profits with prices around £84,000 per mounting including armour.[11] Given the then level of their profits (about 45% of costs excluding armour) suggests not only that they had low manufacturing costs but also that *Conqueror*'s costs were about 50% higher than those of their competitors.

COW clearly struggled to build their first heavy gun mountings, as *Conqueror* only went out for gun trials in June 1912, when they should have been ready in October 1911. There is a fascinating report on those

[10] Boursnell, David. *Forging the Fleet*, Chapter 11. Sheffield Museums Industrial Trust, 2016

[11] *The Battleship Builders*, p190

trials and its many problems by a senior member of Elswick's ordnance department. The writer's name is not given although there are initials RP at the end. He was clearly very experienced, making many observant, detailed and critical comments. The report is dated 19 June, so typed the day after he left the ship at Spithead, probably in Armstrong's London office. Appendix 3 includes the bulk of the report.

Bacon acknowledges that the trials were a failure. He wrote: 'This was a great blow. Only two things were wrong, but these two affected all five turrets. The one which was more apparent was due to a shaft having a hexagonal head instead of a square head … There was one other defect which was more difficult to run to earth, but needless to say, run to earth it was, and both troubles were rectified; so that on the next occasion the trials were quite satisfactory.'[12] This glosses over the many shortcomings in the whole hoist and cage mechanisms identified in the Elswick report.

The report underlines how complex were the components of a heavy gun mounting, with many things which could go wrong, and indicating the problems that the inexperienced company evidently encountered in designing and manufacturing a satisfactory non-transferable mounting. They had been specifically requested by the Admiralty not to make any arrangements with Vickers or Armstrong regarding technical information, although quite likely Bacon had copies of the older gun mounting handbooks from his time as DNO. The many failures during the trials suggest that the mountings did not spend long enough in the pits at Scotstoun to iron out all the faults – the easiest place to make changes. Perhaps the shipbuilders Beardmore were pressing for early delivery (and the resulting stage payments) to get the mountings lifted into place by the 150-ton crane at their Dalmuir fitting out basin. While there would have been some finishing work on hull and machinery, final delivery to the Admiralty on 17 November 1912 suggests that most of those five months was spent fixing the mounting faults. It is quite possible that one whole turret was lifted out and sent back to Scotstoun to be re-erected in a pit where the many modifications to correct the failures could be more readily made.

The COW experience was much the same as Vickers underwent when they tried to break into the heavy gun mounting market in 1898, quoting a loss leading price and making a substantial loss. Lacking experience and unable to profit from refining previous designs, COW had much to learn – previously they had only made the much simpler transferable mountings and field artillery.

The Mark III mounting now officially described as 13.5in was also fitted in *Ajax* and *Benbow* but firing the 1400lb shell – *Conqueror* fired the 1250lb shell. All had the standard magazine capacity of 80 rounds per gun. That these mountings were designated Mark III* and III** suggests that further modifications had to be made. They seem to have given trouble in service; there is anecdotal evidence that their gunnery officers called them 'cows'. An official manual criticised the gun loading cages: 'The whole system is heavy and clumsy and involves many almost unavoidable accidents occurring during loading.'[13] COW only built eight more heavy mountings, all in WW1, but to the successful Elswick 15in design.

[12] Bacon, Reginald. *From 1900 Onward*. Hutchinson, 1940, p193

[13] *Manual for Power-worked Mountings Book 1, 1921*, p38. ADM 186/255

One of *Conqueror's* twin 13.5in in a pit at the new Scotstoun works of Coventry Ordnance Works in 1912. It was their first heavy gun mounting order, proving troublesome on trials. The left gun is at maximum 20° elevation, the right at 5° depression. Sighting hoods and turntable rollers are visible, the rest of the mounting is below in the pit. *(Author's collection)*

# CHAPTER ELEVEN

# 15-inch Mountings

## THE 15-INCH TWIN MOUNTING MARK I IN BATTLESHIPS

For the battleships of the 1912 Programme, the Admiralty sought a fast well protected ship able to deliver a heavy broadside at long range. It noted that the United States (*New York*) and Japan (*Kongo*) were already building ships with 14in guns, so considered the 15in calibre. On a well-balanced design capable of 24 knots, only eight guns in four twin mountings could be carried compared with ten 13.5in, but they would have greater hitting power. The gun was initially called 14in Experimental with two prototypes ordered from Elswick in August 1912.

The gun was of 42 calibres length, wire–wound, weighing exactly 100 tons and was designated Mark I. At 20° elevation it could fire a 1920lb 4 crh shell 23,400 yards using 428lb of MD45 cordite. In all 186 guns were made between 1913 and 1919. *(See colour plate on page 160)*

That conformed to the Admiralty standard of two spare guns for every five in ships (one to four for breech mechanisms). That margin also allowed for worn guns to be relined, which was not possible in the German 15in. Indeed, some guns were relined twice over their life (including Nos 10 and 14), costing in the 1930s about £5500. A typical Vickers price new was £14,000 for the gun body (profit £7940) plus £1600 for the breech mechanism.[2]

The mounting was an Elswick design, broadly similar to the 13.5in, with differences mainly due to its larger size. The roller path diameter was increased to 27ft from 24ft 6in, with barbette inside diameter 30ft 6in. The larger gunhouse provided more space for loading from the cages and had a screened off officer's local control position at the rear with sights, protected by 3in armour, which also strengthened the gunhouse itself. There was better support at the bottom of the side armour plates, which as well as being keyed into the gunhouse floor had supporting chocks. The front plates were slightly thicker at 520lb (13in nominal) KC with sides and rear remaining at 440lb (11in) KC, typically with 3.7% nickel and 1.8% chromium. Roof plates were 200lb (5in) KNC. [3]

The gun was designed for all-angle loading from 20° elevation to -5° depression. In practice it was found that running the gun out at higher elevations strained the hydraulic system and there was a tendency for the loaded shell to slip back if the driving bands did not get a good grip on the rifling, so often the gun was brought down to 5° for hydraulic loading or 2° for hand loading. Four 500hp steam driven hydraulic pumps were fitted, which were cross-connected but also with two independent pipe runs to each mounting (pressure 4.26in diameter, exhaust 6in). Training was by two swashplate engines fitted below the gunhouse, although only one was clutched in at a time. The heavier gun required a more rigid slide of solid forged beams. The mountings had training angles of ±150° but superfiring B and X could not fire on bearings ±30° ahead and astern due to blast effects on A and Y.

As was usual, the magazines were above the shell rooms, to give better protection in the event of an underwater explosion from mine or torpedo. The main ammunition hoist had the shell cage at the lowest level, then at the level above, the cordite cage was loaded via the flashtight handing room with two quarter charges on two levels in its cage. When the hydraulic press started the hoist mechanism, the shell cage picked up the cordite cage on its way up and both were delivered to the working chamber together. A secondary loading arrangement was

[2] *The Battleship Builders*, p177

[3] *Addenda to Hydraulic Manual. 15-inch Mountings Mark I, 1916.* ADM 186/314 & 315

**15in Mark I Gun Manufacturers and Registered Numbers** [1]

| | | |
|---|---|---|
| Elswick Ordnance Co, Newcastle | 1–14, 59–66, 141–148, 183–186. | 34 completed. 187–189 cancelled. |
| Armstrong Whitworth, Openshaw | 15–20, 79–84. | 12 completed |
| William Beardmore, Glasgow | 21–25, 45, 47–50, 87–96, 121–126, 131–133, 149–152, 179–182. | 37 completed |
| Coventry Ordnance Works | 26–30, 111–115, 139–140, 157–160, 165–167. | 19 completed. 168–170 cancelled |
| Vickers, Sheffield | 31–42, 67–78, 101–110, 134–138, 153–156, 171–178. | 51 completed |
| Royal Gun Factory, Woolwich | 43–44, 46, 51–58, 85–86, 97–100, 116–120, 127–130, 161–164, 190–192. | 33 completed, 193–195 cancelled |

[1] World Ship Society. *Warships Supplement* No 67, 1981

*Barham*'s twin 15in Mark I X and Y turrets at John Brown's shipyard on 20 April 1915. Y looks complete apart from the roof plates while X awaits gun cradles and guns. *(NRS)*

*Barham*'s right gun (Reg Nº 67 or 68) being shipped in X turret on 7 May 1915 by Clydebank's 150-ton cantilever crane. The planned 6in P.IX mountings in the two after casemates were never shipped. *(NRS)*

provided, with nine shells per gun stowed in the working chamber and lifted vertically into the gunhouse. 6pdr sub-calibre guns were stowed in the gunhouse for training purposes.

The first four ships were ordered at the end of 1912: *Queen Elizabeth* (Portsmouth), *Warspite* (Devonport), *Barham* (John Brown) and *Valiant* (Fairfield) with mounting orders placed shortly after, the two dockyard ships from Elswick and the two Clyde yards from Barrow. COW was still working on *Ajax* and *Benbow* and could only erect one battleship set at a time. A fifth ship, *Malaya,* financed by the colony, followed a few months later, both hull and mountings from Armstrong Whitworth, the first ship to be laid down at their new Naval Yard on the Tyne.

With a typical battleship taking about 27 months from contract to delivery, the shipyard had to carefully plan for the installation of the mountings. With both shipyard and gun mounting staff working together and launch at about Month 18, a typical timetable in months would be:

| | |
|---|---|
| Ship ordered | 0 |
| Fit ring bulkheads | 16 |
| Machine lower roller paths | 17 |
| Fit hydraulic pumps | 18 |
| Fit training rack to ring bulkhead | 19 |
| Fit shell room gear | 20 |
| Install X and Y mountings | 20 |
| Install A and B mountings | 21 |
| Gun trials | 23 |
| Delivery | 27 |

The five 'R' class battleships formed the 1913 Programme, with three more in the 1914 Programme. But Fisher's return as First Sea Lord in October 1914 brought about many changes in the 15in manufacturing programme, as spelled out in the timeline below. The manufacturers were asked to 'accelerate' production, taking on more staff and paying for additional overtime, Elswick ordnance department numbers doubling in twelve months to mid-1915. One spare mounting per ship was usually ordered at the same time, although it was only the elevating portion of cradle and slide, not the entire mounting.

The tenders for gun mountings (excluding gunhouse armour and guns) from Armstrong and Vickers were often very close: *eg*, for two 'R' class battleships Elswick tendered £353,706 each and Barrow £354,518, while for two *Queen Elizabeth*s there was 0.6% difference and for *Tiger* only 0.3%.[4] It is inconceivable that for such complex pieces of engineering they should be consistently so close without collusion. The advantage of collusion to the two companies was to maintain high prices producing large profits but also to ensure that the Admiralty

[4] From class Ships Covers, NMM

**Typical price and weight breakdown for one battleship 15in mounting**

| | Item | One mounting weight tons | Approx price 4 mountings £ |
|---|---|---|---|
| 1 | Turntable and working chamber | 141 | 54,000 |
| 2 | Turning gear and training rack | 22 | 9,000 |
| 3 | Gun slides and cradle | 81 | 76,000 |
| 4 | Hydraulic pumps * | | 14,000 |
| 5 | Piping on revolving structure | 8 | 22,000 |
| 6 | Magazine and shell room fittings * | 9 | 18,000 |
| 7 | Transporting gear and hoists | 30 | 43,000 |
| 8 | Piping on fixed structure * | | |
| 9 | Brackets, covers etc | 4 | 2,000 |
| 10 | Gunshield roof, floor & fittings | 132 | 65,000 |
| 10A | Front, side, rear gunhouse armour | 142 | 74,000 |
| | Sub total | 560 | 397,000 |
| | One spare mounting | | 12,000 |
| | Danger signals | | 8,000 |
| | Director firing gear | | 9,000 |
| | Delivery | | 6,000 |
| | Erection | | 22,000 |
| | Total | | 454,000 |
| | Two 15in guns | 200 | 125,000 |
| | Total incl guns | 760 | 579,000 |

* Not in revolving weight
The approximate price per revolving ton excluding guns was £202.

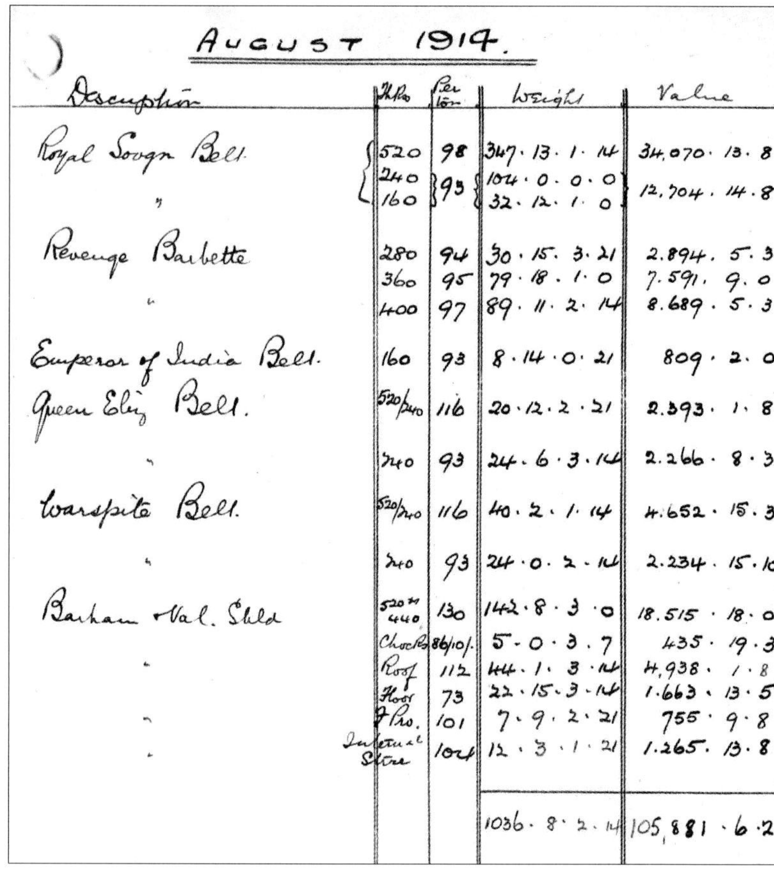

AUGUST 1914.

| Description | lbs | Per ton | Weight | Value |
|---|---|---|---|---|
| Royal Sovgn Belt | 520 | 98 | 347 · 13 · 1 · 14 | 34,070 · 13 · 8 |
| " | 240 / 160 | 93 | 104 · 0 · 0 · 0 / 32 · 12 · 1 · 0 | 12,704 · 14 · 8 |
| Revenge Barbette | 280 | 94 | 30 · 15 · 3 · 21 | 2.894 · 5 · 3 |
| | 360 | 95 | 79 · 18 · 1 · 0 | 7.591 · 9 · 0 |
| " | 400 | 97 | 89 · 11 · 2 · 14 | 8.689 · 5 · 3 |
| Emperor of India Belt. | 160 | 93 | 8 · 14 · 0 · 21 | 809 · 2 · 0 |
| Queen Eliz Belt. | 520/240 | 116 | 20 · 12 · 2 · 21 | 2.393 · 1 · 8 |
| " | 240 | 93 | 24 · 6 · 3 · 14 | 2.266 · 8 · 3 |
| Warspite Belt. | 520/240 | 116 | 40 · 2 · 1 · 14 | 4.652 · 15 · 3 |
| " | 240 | 93 | 24 · 0 · 2 · 14 | 2.234 · 15 · 10 |
| Barham & Val. Shld | 520 & 440 | 130 | 142 · 8 · 3 · 0 | 18.515 · 18 · 0 |
| | Chocks | 86/10/- | 5 · 0 · 3 · 7 | 435 · 19 · 3 |
| " | Roof | 112 | 44 · 1 · 3 · 14 | 4,938 · 1 · 8 |
| | Floor | 73 | 22 · 15 · 3 · 14 | 1.663 · 13 · 5 |
| " | F Pro. | 101 | 7 · 9 · 2 · 21 | 755 · 9 · 8 |
| " | Internal Stove | 104 | 12 · 3 · 1 · 21 | 1.265 · 13 · 8 |
| | | | 1036 · 8 · 2 · 14 | 105,881 · 6 · 2 |

Vickers kept careful records of all armour plates made at Sheffield. This page details output for August 1914, including 142 tons of *Barham*'s first gun shield front. The weights are in Tons, Hundredweights (20 per ton), Quarters (4 per cwt) and Pounds (28 per qtr). The prices are in Pounds, Shillings (20 to the £) and Pence (12 to the shillling). The month also includes part of the belt for *Royal Sovereign* and barbette for *Revenge*. *(Vickers Armour Notebook, Sheffield Industrial Museums Trust)*

divided orders equally, with no big price advantage by concentrating on only one company. The two companies also had price sharing agreements for export of warships to foreign navies.

Gunhouse armour formed part of the gun mounting contract but was priced per ton, so not included in the tender price. Each year the Admiralty would agree a schedule of prices with the armour makers, based upon thickness, quality, shape (flat, curved, tapered), and location (belt, barbette, bulkhead). While the Admiralty could estimate the cost based on calculated weights, the final price was based on the actual weight of each individual plate when finished. For the *Queen Elizabeth* and 'R' classes, the prices were £130 per ton for the Krupp cemented front, side and rear plates, £112 for the Krupp non cemented roof and £73 for the high tensile steel gunhouse floor.

Vickers kept detailed records of the plates they supplied to the Admiralty or to the shipbuilder. They also kept records of the output of the other four makers who formed the armour pool. They had an agreement that 20% of each order would be paid into a pool which would then be shared out, so that those who made less than the average output for that year would get additional payment.[5] The prices paid were far in excess of manufacturing costs at the time, as discussed in detail in *The Battleship Builders*. While the Admiralty was aware that there was an armour 'ring' (*ie* cartel) fixing prices, they had no idea that the profit margin was so high. Between 1907 and 1911, Vickers made 1838 tons of gunhouse armour with a sales value of £235,012 (£127.86 per ton) which cost only £93,683 (£50.97 per ton) to make, a profit margin of 150% on cost.[6] Corker reports that the sales value of John Brown's armour over that period was less at about £107 per ton (mostly side belt), with costs about £85, a profit rate of only about 25% – suggesting a less efficient producer?[7]

The gun mounting contractor chose the armour maker, normally an associated company (*eg* Elswick choosing Openshaw, thus keeping the high profits within the group). Vickers produced the gunhouse armour for *Valiant* between March and July 1914 to be railed to Barrow for erection on the mountings being built there before shipping to the Clyde. The table on p97 shows the average for *Valiant*'s four gunhouses.[8]

After *Renown* and *Repulse* had been changed from 8-gun battleships to 6-gun battlecruisers at the end of 1914, two twin mountings were notionally available to arm two monitors for shore bombardment with one 15in mounting each, to be completed in great haste. But with mountings taking a good two years from order to delivery, those two would not be available until 1916. So mountings would have to be diverted from a 1913 Programme ship to get mountings in time.

[5] Boursnell, David, *Forging the Fleet*, Chapter 9

[6] *The Battleship Builders*, p218

[7] Corker, C J. 'The Business and Technology of the Sheffield Armaments Industry, 1900–1930'. Sheffield Hallam University thesis, 2016, p119

[8] Armour Plate Pool Ledgers. HNVK0003, Sheffield Industrial Museums Trust

**Gunhouse armour: average of *Valiant*'s four mountings**

| Item | Thickness lb | Price £/ton | Average tons | Price £ |
|---|---|---|---|---|
| Shield front KC | 520 | 130 | 7.8 | 1,014 |
| Shield sides and rear KC | 440 | 130 | 134.4 | 17,472 |
| Roof KNC | 200 | 112 | 67.1 | 7,515 |
| Floor HT steel | 80 | 73 | 23.3 | 1,701 |
| Internals | 120 | 104 | 12.2 | 1,269 |
| Chocks, doors etc | | | 12.8 | 1,104 |
| Total | | | 257.6 | 30,075 |

Furthermore, ordering in early 1915 the three new battlecruisers with two mountings each further disrupted the manufacturing programme. The timeline below shows how deliveries were affected.

**End 1912** Four twin mountings ordered for each: *Queen Elizabeth, Warspite* (Elswick); *Barham, Valiant* (Vickers).

**March 1913** Four ordered for *Malaya* (Elswick).

**August 1913** Four ordered for each: *Resolution, Royal Oak* (Elswick); *Revenge, Ramillies* (Vickers); *Royal Sovereign* (COW).

**May 1914** Four ordered for each battleship: *Repulse* (Elswick), *Renown* (COW); *Resistance* (Vickers).

**January 1915** *Renown* and *Repulse* reduced to three mountings each, releasing two for monitors, although these were actually fitted in *Courageous* mid-1916.

**Early 1915** Two planned for *Ramillies* diverted to complete monitors *Marshal Ney* (August 1915) and *Marshal Soult.*

**March 1915** Two for *Ramillies* diverted to *Glorious* (3rd qtr 1916). One for *Renown* diverted to become *Courageous* A. One for *Repulse* diverted to become *Courageous* Y. *Resistance* at Devonport cancelled and four mountings diverted to *Ramillies.* Two ordered from Elswick for *Furious* should her two single 18in need replacing.

**May 1915** Four ordered for four new monitors (3 Vickers), although four would need to be diverted from *Royal Oak* to get early delivery.

**June 1915** Four monitors cancelled.

*Resolution* with all turrets now installed, after the original A and B were diverted to *Repulse.* She is alongside Palmer's Jarrow shipyard on 14 October 1916. *(Author's collection)*

**15in Gun Mounting Manufacturers***

| Ship | Shipbuilder | Mark | Manufacturer | Gunhouse Armour | Reg Nos** | Gun Reg Nos |
|---|---|---|---|---|---|---|
| *Queen Elizabeth* | Portsmouth | I | EOC | Openshaw | 1–8 | 1, 2, 21, 23, 24, 32, 35, 37 |
| *Warspite* | Devonport | I | EOC | Openshaw | 10–17 | 22, 25, 38, 39, 40, 41, 42, 50 |
| *Barham* | John Brown | I | Vickers | Vickers | 37–44 | A 87, 88, B 71, 74, X 67, 68, Y 69, 70 |
| *Valiant* | Fairfield | I | Vickers | Vickers | 28–35 | A 46, 47, B 29, 54, X 16,19, Y 44, 73 |
| *Malaya* | Armstrong | I | EOC | Openshaw | 19–26 | 4, 6, 7, 8, 9, 10, 59, 61 |
| *Royal Sovereign* | Portsmouth | I | COW | John Brown | 46–53 | A 30, 77, B 82, 92, X 51, 76, Y 17, 18 |
| *Royal Oak* | Devonport | I | EOC | John Brown | 55–58, 66–69 | 60, 79, 85, 86, 91, 95, 96, 103 |
| *Revenge* | Vickers | I | Vickers | Vickers | 73–80 | A 101, 104, B 93, 94, X 55, 56, Y 8, 49 |
| *Resolution* | Palmers | I | EOC | Openshaw & Beardmore | 64–5, 70–1 X & Y | A 114, 124, B 118, 120, X 57, 58, Y 102, 105 |
| | | I* | EOC | John Brown | 23–26 A & B | |
| *Ramillies* | Beardmore | I | Vickers | Vickers | 82–89 | A 63, 66, B 45, 62, X 83, 125, Y 34, 53 |
| *Renown* | Fairfield | I* | COW | John Brown & Cammell | 1–6 | 3, 12, 97, 112, 116, 117 |
| *Repulse* | John Brown | I | EOC | Openshaw | 59–62 A & B | 81, 98, 100, 113, 121, 122 |
| | | I* | EOC | John Brown | 8–9 Y | |
| *Marshal Ney* | Palmer | I | Vickers | Vickers | 91–92 | 52, 75 |
| *Marshal Soult* | Palmer | I | Vickers | Vickers | 93–94 | 27, 28 |
| *Courageous* | Armstrong | I* | A COW, Y EOC | Cammell & Openshaw | 10–13 | 13, 14, 108, 109 |
| *Glorious* | H&W, Belfast | I* | Vickers | Vickers | 19–22 | 11, 80, 90, 107 |
| *Erebus* | H&W, Glasgow | I* | EOC | Openshaw | 17–18 | 15, 36 |
| *Terror* | H&W, Belfast | I | Ex-*Ney* | | 91–92 | 52, 75 |
| *Hood* | John Brown | II | Vickers | Vickers | 1–8 | 31, 84, 111, 115, 128, 154, 155, 172 |

* Quarterly Appropriation List of Gun Mountings, November 1917. ADM 186/225

** Each elevating portion was given a Registered Number rather than the complete mounting.

Spare mountings. Mark I: 9, 18, 27, 36, 45, 54, 63, 72, 81. Mark I*: 7, 14–16. Mark II: 9

Total mountings completed: Mark I and variants 54, Mark II 4.

**Mid 1915** Two partially complete mountings for *Ramillies* sent to Elswick for modifications with shorter trunk for *Marshals* building on the Tyne.

**September 1915** Two planned for *Furious* diverted to two new 15in monitors.

**December 1915** Two diverted from *Resolution* A & B to *Repulse* A & B (2nd qtr 1916). Two diverted from *Repulse* to *Resolution* A & B (3rd qtr 1916).

**January 1916** Mounting removed from *Marshal Ney* (unreliable machinery) and converted by Elswick to 30° for *Terror. Marshal Soult*'s mounting converted by Elswick to 30°.

**February 1916** First mounting for *Furious* diverted to *Erebus*.

*Queen Elizabeth* was the first RN battleship to mount 15in guns in four twin Mark I mountings by Elswick. *(World Ship Society)*

Second mounting for *Furious* to reserve (fitted in monitor *Abercrombie* in 1943).

**Mid-1916** Four Mark II mountings ordered each for *Hood* and *Howe* (both Vickers), *Anson* (Elswick), *Rodney* (COW)

**Early 1917** Mountings installed in *Ramillies* (Vickers). Mountings for *Howe, Anson, Rodney* suspended, later cancelled.

A 15in gun was condemned when the bore wear at 1in from the commencement of rifling had reached 0.737in. By that time, muzzle velocity had fallen by about 100ft/sec, reducing maximum range by about 1500 yards. But guns would be replaced before that level of wear, as a battleship's guns should be able to fire off a full magazine outfit safely. With the longer HE shell, trajectory stability was lost earlier from bore wear reducing efc to about 260. But if the range or target was appropriate to using ¾ charges, erosion was reduced to only about a quarter of one full charge.

Typical ammunition outfit per gun:

WW1 battleships 100 APC
WW1 battlecruisers 120
WW1 monitors 90 HE and 10 CPC
WW2 monitors 30–60 APC, 80–50 HE, 8 practice
WW2 modernised battleships 106

*Queen Elizabeth* was sent out to the Dardanelles in January 1915 for shore bombardment firing at forts but this was a misuse of such a valuable ship and she was withdrawn in May. Campbell gives the ammunition expenditure (mostly APC) at Jutland, together with discussion of general shooting performance:[9] *Barham* 337, *Valiant* 288, *Warspite* 259, *Malaya* 215, *Royal Oak* 38, *Revenge* 102. Total 1239.

He reported comparatively few failures of guns and heavy mountings, with only eight guns put out of action due to hits or premature detonations of shells in the barrel, while *Malaya* had to use hand loading to the right gun of X turret due to a shell bogie out of alignment.

The 15in mounting remained on board an RN ship for half a century (1915–1965) with a well-earned reputation for accuracy, consistency and reliability. *(See colour plates on pages 157-159)*

[9] Campbell, John. *Jutland: An Analysis of the Fighting*. Conway Maritime Press, 1986, pp346–7

| **15in BL Mark I Gun** | |
|---|---|
| Calibres, bore | 42 |
| Length overall, in | 650.4 |
| Weight incl BM, tons | 100 |
| Muzzle velocity nominal, ft/sec | 2467 |
| Shell, lb | 1920 |
| Charge MD45, lb | 428 |
| Life, APC, efc | 335 |

| **15in Mounting** | **Mark I** | **Mark I/N** |
|---|---|---|
| Revolving weight, tons | 760 | 815 |
| Maximum elevation, degrees | 20 | 30 |
| Maximum depression, degrees | -5 | -4.5 |
| Elevating speed, deg/sec | 5 | 5 |
| Training speed, deg/sec | 2 | 2 |
| Recoil, in | 46 | 46 |
| Recoil at 30°, sec | | 0.36 |
| Gun run out time at 30°, sec | | 4.5 |
| Firing cycle, sec | 36 | |

### 15-INCH MOUNTING VARIANTS

With the 15in mountings needed for battlecruisers and monitors early in 1915, design changes were necessary. Vessels with less deep hulls than battleships needed shortened ammunition trunks. Faster vessels needed lighter mountings with thinner gunhouse armour. A revised design of shell room bogie which transferred shells from shell room to trunk resulted in the Mark I* design. Post Jutland, flashtightness was improved with new interlocks to prevent dangerous operations.

Such details were described in the 15in handbook, with the prosaic title *Addenda to Hydraulic Manual: 15-inch Mountings Mark I*. This confidential publication was produced by Elswick for the Admiralty to describe the principal components, illustrated by some ninety drawings, many fold-outs. As well as sections on the *Queen Elizabeth* and 'R' classes, there were two separate Appendices, one for *Royal Oak, Renown, Repulse, Glorious* and *Courageous*. Instead of shell bogies, these had an overhead rail and Kenyon doors to load shell into the hoist. Another section covered monitors *Erebus* and *Terror*, whose maximum elevation had been increased to 30° requiring stronger elevating gear. But even these two were not identical, with *Terror*'s mounting removed from the slow and unsuccessful *Marshal Ney*, with shell room arrangement as

**Gunhouse protection**

| Ships | Front KC lb | Side KC lb | Rear KC lb | Roof lb |
|---|---|---|---|---|
| 4 *Queen Elizabeths* (polygonal sides) | 520 | 440 | 440 | 200 KNC |
| *Malaya*, 5 *Royal Sovereigns* | 520 | 440 | 440 | 170 Van |
| *Renown, Repulse, Courageous, Glorious* (semi- circular sides) | 360 | 280 | 440 | 170 Van |
| *Marshal Ney, Marshal Soult, Terror, Roberts* | 520 | 440 | 440 | 200 KNC |
| *Erebus* | 520 | 440 | 440 | 170 Van |
| *Warspite, Queen Elizabeth, Valiant* Mark I/N | 520 | 440 | 440 | 200 NC |
| *Abercrombie* | 520 | 440 | 440 | 240 NC |
| *Vanguard* | 520 | 360 & 280 | 440 | 240 NC |

It was found that if vanadium alloy roof plates were hit obliquely, they were more likely to be dished than torn.

*Repulse*'s right gun being shipped in A turret, without its breech mechanism, at Clydebank on 12 April 1916. The Elswick-built mounting was planned for *Resolution*. The cruiser across the basin is *Canterbury*. *(UCS 118 443 22)*

*Royal Sovereign*'s A and *Erebus*' planned for *Furious* so had *Repulse*'s shell room arrangement.[10]

This 1916 handbook was overly detailed in its description of the mounting operation, while the drawings were of the working drawing type which could obscure important components. Between the wars, it was realised that this style of handbook was not sufficiently user-friendly for the ship's staff, so the Admiralty produced its own version with more emphasis on maintenance. The drawings were redrawn in a simpler style, usually in colour; the 1943 15in handbook is a good example. The handbook for each Mark of mounting was supplemented by a drill book which spelled out the duties of each member of the team. A typical crew consisted of the turret captain and the second captain, two gunlayers,

[10] ADM 186/314 & 315

one trainer, three sightsetters, six loading crew for each gun, five control and communication numbers (25 in gunhouse and working chamber) plus 19 in the magazine and 19 in the shell room, a total of 63 plus an ordnance artificer. There were also director and transmitting station staff serving all four mountings.

### 15-INCH MARK I/N MOUNTINGS *(See colour plates on pages 159-160)*

With no new battleship construction permitted, the Admiralty was faced with modernising the *Queen Elizabeth* and 'R' class battleships to better meet the ever increasing threats from aircraft and submarine attack. Bulges had been added to improve underwater protection and they got some extra thickness of deck protection, while further anti-aircraft weapons were added, 4in guns and multiple pom-poms. But by the early 1930s, more drastic reconstruction was called for. New improved higher pressure steam turbines had become available, reducing the number of boilers required to maintain 24 knots, at the same time reducing oil fuel consumption. The reduced machinery weight could be used for increased deck protection. The 15in mountings could be modified to 30° elevation, as had already been done in the WW1 monitors, whose minimum was +2°. That required the trunnions to be repositioned to allow recoil at higher elevation, also requiring a 12-ton balance weight to be fitted around the breech area to keep the centre of gravity in line with the trunnions. Pneumatic run-out replaced hydraulic where not previously done. Protective bonnets with mantlets were fitted over the now enlarged gun ports.[11] With a new 6 crh shell with improved ballistics and SC cordite charge, maximum range could now be increased 35% to 32,200 yards.

Vickers-Armstrongs at Elswick was given the contract in August 1934

[11] *Handbook for the 15-inch B.L. Mark I Gun on Twin Marks I and I/N Mountings, 1943*. ADM 234/168

The end for *Revenge*'s 15in guns as she is broken up at Ward's Inverkeithing yard in August 1949. Inboard is *Nelson* with B and X 16in mountings nearly demolished. *(Marine Technology Special Collection)*

to redesign the mounting to Mark I/N standard and modify *Warspite*'s four mountings. However, the original arrangement of shell rooms below magazines was not changed, although that had been reversed in the 16in mounting. Flashtightness was also improved and many minor modifications applied. *Warspite* was taken in hand at Portsmouth in October 1934, with her mountings dismantled and dispatched to Elswick for modifications. They were re-erected at the end of 1936 at a final cost of £182,594, around half that of new mountings. New guns were also supplied including four from *Courageous* after relining, with Reg N^os: A 162 & 163, B 13 & 108, X 14 & 109, Y 80 & 107. Revolving weight increased to 815 tons. Similar changes were made to *Renown*'s three twin mountings, completed at Portsmouth in 1939, and to *Valiant* (Devonport 1939) and to *Queen Elizabeth* (Portsmouth 1940).

The 15in battleships remaining unreconstructed in 1940 (*Malaya, Barham,* four 'R's and *Repulse*) kept their 20° mountings but were supplied with supercharges which increased their range to 28,700 yards at the expense of increased gun wear.

When Churchill returned as First Lord on the outbreak of WW2, he was keen to get the WW1 monitors back into service for shore bombardment. But *Marshal Soult* was in poor condition, so instead her twin 15in mounting was transferred to a new hull, the monitor *Roberts.* With her newly refurbished mounting and stronger anti-aircraft armament, she was completed in October 1941. She saw extensive operations in the Mediterranean in 1943 before being redeployed for Operation Neptune in June 1944.

When *Terror* was lost off North Africa in February 1941, a new monitor was ordered. *Furious'* second 15in mounting had been stored at Chatham in 1917. It was converted by Vickers-Armstrongs at Elswick to Mark I/N standard with 30° elevation but retained the orig-

**Below and opposite:** Two shots evocative of the big gun era: *Ramillies* having her guns replaced at Portsmouth in July 1944 having worn them out firing over 1000 rounds in shore bombardment at Normandy. Two new guns and a roof plate are stored at bottom left under the 250-ton crane while what looks like an old gun is shunted away on rail wagons, maybe A turret's right gun as the left has already been replaced. In the distance is monitor *Marshal Soult*'s mast, her 15in having been removed for *Roberts.* (The Times. *News Licensing)*

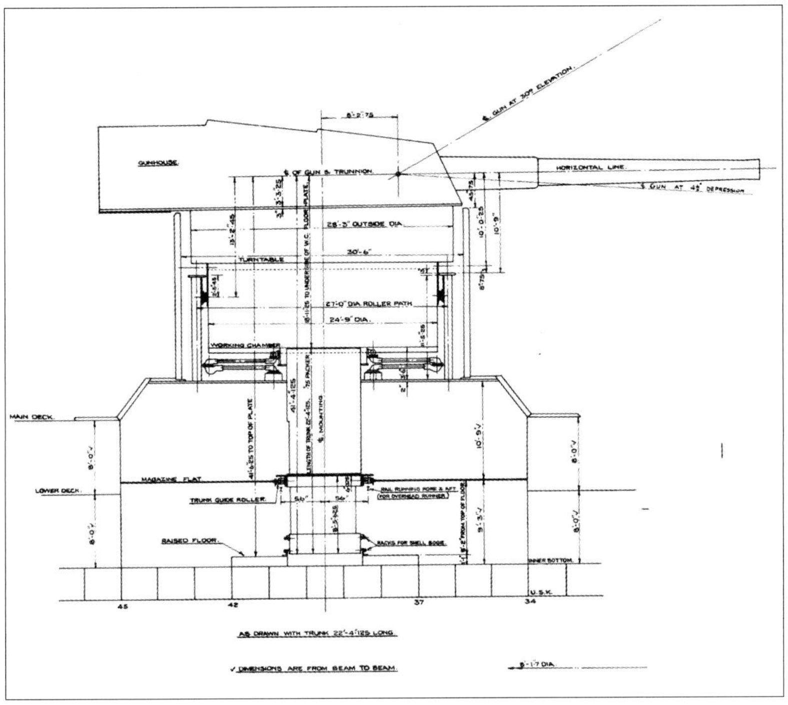


**Left:** This simplified drawing of the twin 15in Mark I/N fitted in monitor *Abercrombie* is an example of guidance information supplied to the shipbuilder by the ordnance manufacturer, showing key dimensions (although omitting the upper deck at about the level of the bottom of the working chamber). *(NMM)*

inal magazine above shell room arrangement. It was shipped in *Abercrombie* in January 1943; see the sequence of photographs included in *Big Gun Monitors* and on page 219.

*Warspite's* operations at Narvik in April 1940 and Matapan in March 1941 are well documented and her shore bombardments at Normandy and Walcheren in 1944 are covered in *Big Gun Monitors*. In July 1940 she obtained a hit on Italian battleship *Giulio Cesare* at 26,000 yards, probably the longest range by a big gun on a warship. Hits by glider bombs off Salerno in September 1943 so damaged her X turret that it

**Below:** *Ramillies'* new right gun in B turret being installed, the left gun cradle being readied to accept its new gun. Removed were Reg N$^{os}$ 16, 73, 87 and 150, replaced by 27 (ex-*Marshal Soult* relined), 41 (ex-*Warspite* relined), 58 (ex-*Resolution* relined) and 120 (ex-*Resolution* relined). The guns in X and Y were not replaced. Her mountings retained their original 20° elevation. *(*The Times. *News Licensing)*

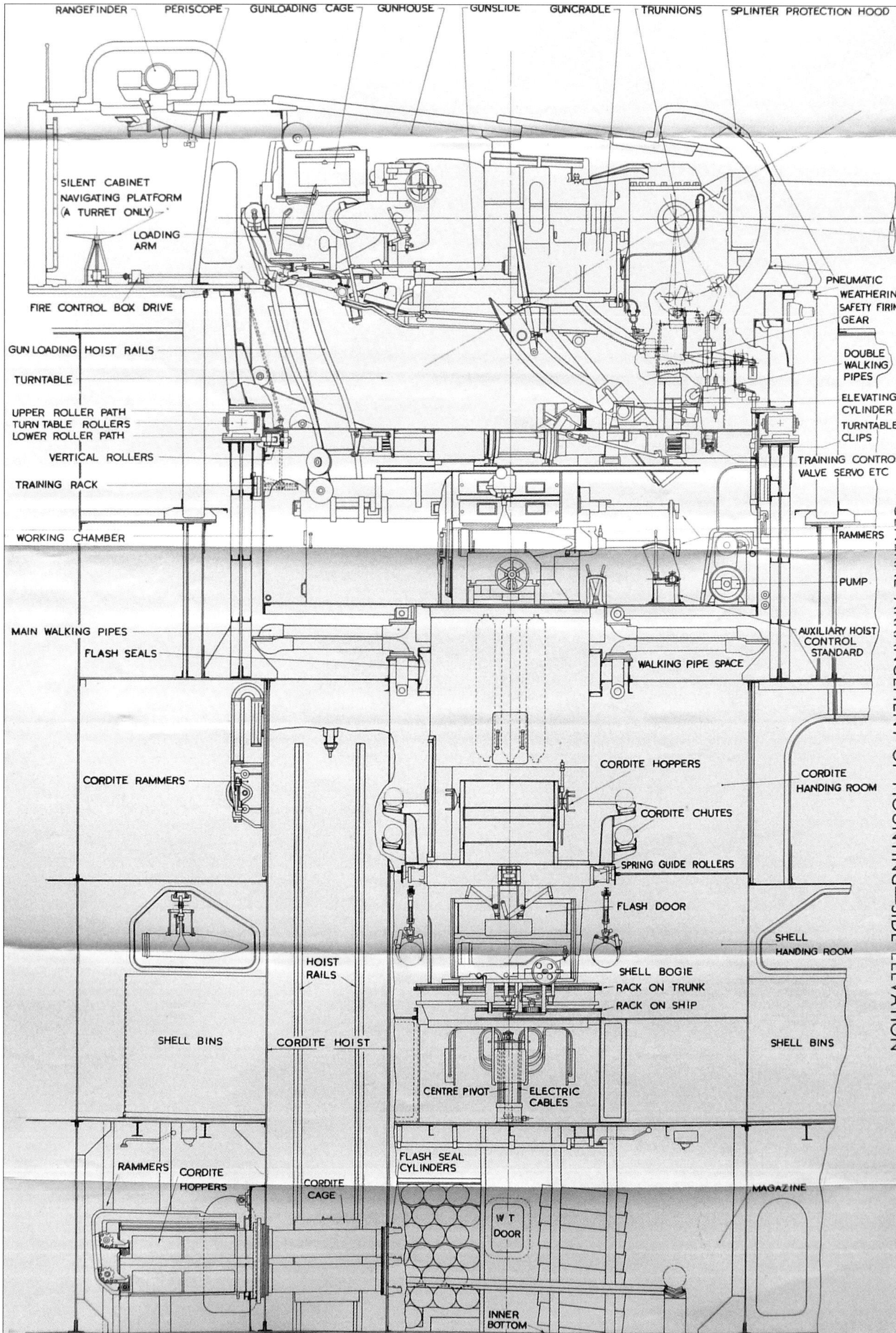


The last 15in mountings for *Vanguard* were converted to Mark I/N from *Courageous* and *Glorious* but now had the magazines below the shell rooms which required an additional hoist to raise the charges above the shells for loading in the main hoist cages. *(ADM 234/169)*

was never used again. Four 15in ships were active in shore bombardment at Normandy: *Warspite, Ramillies, Roberts, Erebus* all wearing out guns which needed replacing. *Roberts* fired 692 rounds wearing out two sets of guns. At Walcheren in November 1944 *Warspite, Roberts* and *Erebus* fired 636 rounds.[12]

[12] *Big Gun Monitors*, p208

**Right:** *Vanguard* was only built because 15in mountings removed from *Courageous* and *Glorious* had been stored after the ships were converted to aircraft carriers. As well as general refurbishment they needed conversion to Mark I/N standard with 30° elevation. While Vickers-Armstrong supplied new parts, Harland & Wolff's Scotstoun works carried out much of the work, here machining the upper roller path on X's turntable. *(Harland & Wolff)*

**Below:** *Vanguard*'s four twin 15in at Scotstoun in mid-1945 nearly ready for erection in the ship. Nearest is X, then A, Y, B. *(Harland & Wolff)*

*Vanguard*'s Y turret being erected at Clydebank in October 1945, with gun slides and cradles and 15in front armour plates in place. X turret needs only the last roof plates fitted. Across the basin is the 'Battle' class destroyer *Barrosa. (NRS)*

### *VANGUARD*

With the 16in *Lion* class suspended, Churchill was keen to get another modern battleship built. He was advised that by taking the four mountings removed from *Courageous* and *Glorious,* it would be possible to complete a new battleship by 1944. John Brown had been earmarked to build the *Lion* class *Conqueror* of the 1939 Programme, so was selected in 1940 to build *Vanguard* although she was not formally ordered until March 1941.

The mountings had been stored at Rosyth, but required to be converted to Mark I/N standard with 30° elevation, which was approved in May 1940. The redesign and most of the modifications were done by Vickers-Armstrong at Elswick and Barrow, but the actual re-erection was undertaken by Harland & Wolff at Scotstoun, where four of the original COW pits had been re-instated. Estimated cost was £380,000, about one-third of what four brand new mountings would have cost. A major change was to place the magazines below the shell rooms for greater protection. But the main hoist cage still had the cordite charges carried above the shell, so a separate cordite hoist was required to bring the charges up two deck levels to feed into the cage via flashtight hoppers.[13]

Many smaller changes were made which included a thicker gunhouse roof, pneumatic run-out, improved flashtightness round the gun loading cage, bevel instead of worm drive training motors, improved shell bogies at the bottom of the main hoist, better ventilation, mantlet plates over the gun ports instead of blast bags and pneumatic weather seals between gunhouse floor and top of barbette armour. A major change was to fit RP.10 remote control gear but only for training not elevating. The revolving weight was increased to 850 tons. Ammunition stowage was 95 APC, 5 HE and 9 practice. The mountings were given unusual Reg N^os^ rather than the usual two digits for left and right: A 10/9744. B 10/9772, both from *Glorious*, X 503/7679, Y 443/13 both from *Courageous*. 503 was *Renown*'s Fairfield yard number whose other three mountings came from COW and 443 *Repulse*'s John Brown yard number (Elswick).

[13] *15-inch B.L. Mark I Gun on Twin Marks I and I/N Mountings for H.M.S. Abercrombie and Vanguard, 1952.* ADM 224/169

*Vanguard* was broken up without ever firing her guns in anger. By September 1960, demolition at Shipbreaking Industries' Faslane yard had reached A and B turrets with roof plates and projecting gun barrels removed. *(Author)*

Given other shipbuilding priorities, *Vanguard* was only launched on 30 November 1944 but it was still hoped to complete her in time for the war against Japan. Scotstoun finally completed the mountings in mid-1945. Most of the components were shipped the 2½ miles downstream to Clydebank in the coaster *Sea Fisher* from July. Much of the erection was done by floating crane, using the 150-ton crane transferred from Southampton to the Clyde in about 1941. X mounting was the first to be shipped, starting with a 20-ton trunk on 25 August. Lifts were made each day until the two 100-ton guns were shipped on 31 August with the roof plates fitted on 3 September. Y mounting was next in September, followed by A in November and B in December 1945.[14]

*Vanguard* departed Clydebank on 2 May 1946 for speed and gunnery trials. She never fired her guns in anger, being broken up in 1960.

### THE 15-INCH TWIN MOUNTING MARK II

The Admiralty planned a new class of capital ship in late 1915, with a 30-knot speed and 15in guns. Both triple and twin mountings were considered, and even 18in, finally deciding on four twin 15in rather than the three in *Repulse.* Initially three ships were ordered in April 1916 with a fourth in July, but then Jutland revealed shortcomings in British battlecruiser design. Better protection was required and longer range guns with better flashtightness in the gun mountings. Rather than adopt the existing 15in Mark I mounting, they tasked Vickers with designing and building four new Mark II mountings for the first ship, using the existing 15in gun Mark I.

Elevation of 30° was required to add 24% to the Mark I's range with 20°. A deeper gunhouse and turntable was needed to accommodate the breech in recoil at maximum elevation. While the basic design was on the same lines as the Mark I, there were many detailed changes. All-angle loading was retained but only up to 20°. Flashtightness was improved with a new door at the top of the central hoist and by fitting revolving cordite hoppers between the handing room and hoist. The gun run-out mechanism was changed from hydraulic to pneumatic at 1000lb/sq in. Inside the gunhouse the sights were changed from vulnerable hoods in the roof to ports in the front face. Stops were fitted in the loading trays, cages and bogies to stop shells surging in a seaway. Magazines were retained above the shell rooms, regarded as less vulnerable to a torpedo hit. Gun shield armour was thickened to reduce the risk of shell penetration setting off cordite charges; front 600lb Krupp cemented, side front 480lb KC, rear 440lb KC. The flat roof was formed of three bolted Krupp non cemented 200lb plates. The floor was 120lb high tensile steel, needed more for structural support than protection. A 30ft rangefinder was fitted at the rear of each turret. Inside the gunhouse, the control position was enclosed by 120lb armour. The usual four steam driven hydraulic pumps were similar to those that had been used since the *King Edward VII* class of 1905. The result was a mounting about 100 tons heavier than the *Queen Elizabeths* at 860 tons revolving weight.

The ship design was revised after Jutland to incorporate better protection especially over the magazines, with 'load' displacement 41,200 tons and 31 knots. The four keels were laid in the autumn of 1916: *Hood* at John Brown, *Howe* at Cammell Laird with Vickers mountings, *Rodney* at Fairfield with Coventry Ordnance Works mountings, and *Anson* at Armstrong Whitworth with Elswick mountings. But with priority given to merchant ships and convoy escorts, construction of all but *Hood* was suspended in March 1917, then cancelled in March 1919.

Later than expected, *Hood's* mountings were shipped from Barrow in the 2677 grt coaster *Horden* in four voyages, one mounting at a time between July and November 1919, X and Y first.[15] The vessel would have been time-chartered by Vickers from her owners Burnett

[14] Log Book of Southampton 150-ton Floating Crane, 1943–46. Marine Technology Special Collection, Newcastle University.

[15] John Brown Shipyard Diary. Glasgow University Scottish Business Archive.

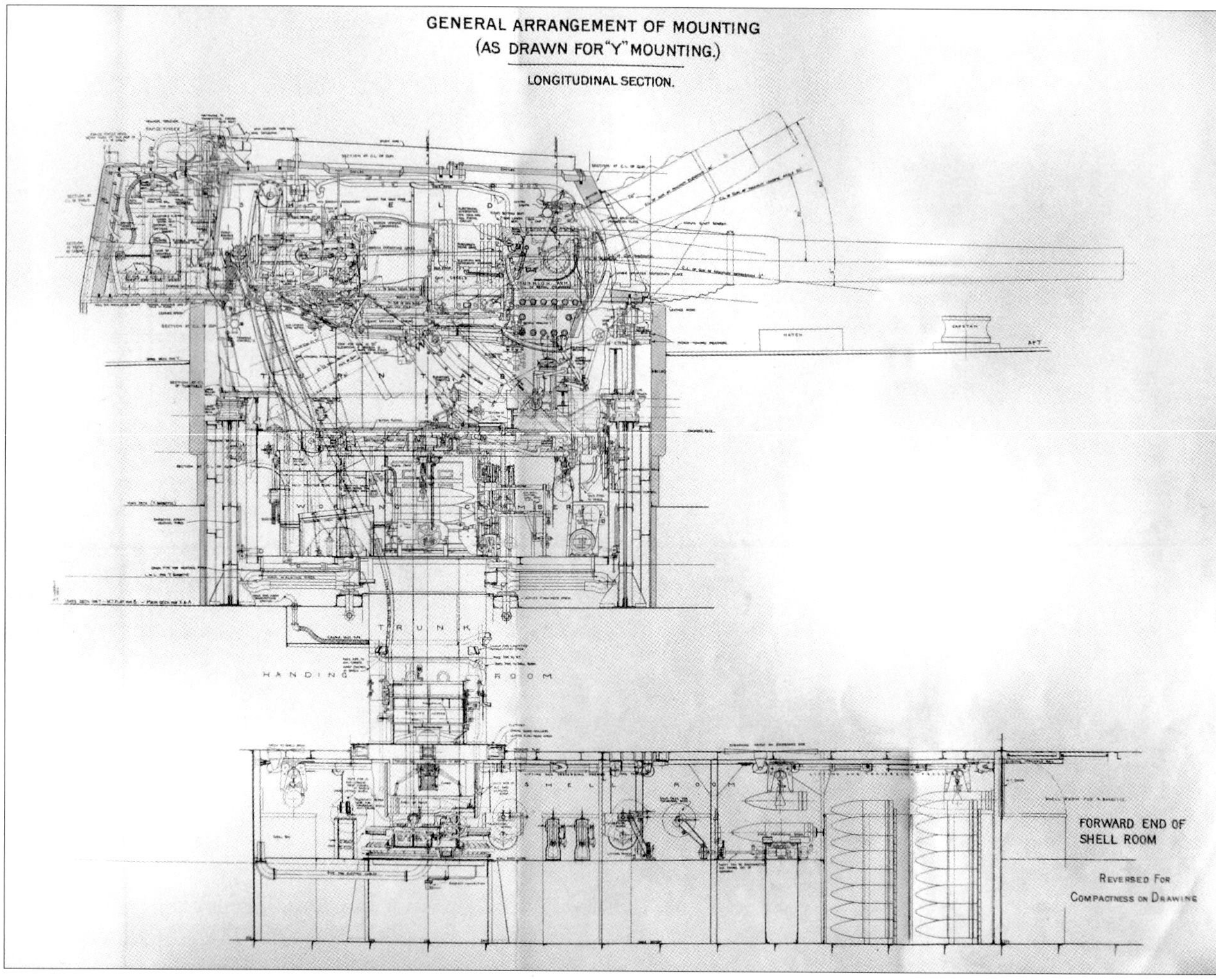


*Hood* was the only vessel fitted with the Vickers designed 15in Mark II, although her three cancelled sisters would have had the same. The increased elevation of 30° required a deeper gunhouse and working chamber, and flashtightness was improved. The shortest trunk Y mounting is illustrated, constrained by the low quarterdeck and fine lines aft, with shells stacked eight high in bins. *(ADM 186/249)*

**15in Mounting Twin Mark II**

| | |
|---|---|
| Rotating weight, tons | 860 |
| Max elevation, degrees | 30 |
| Max depression, degrees | -5 |
| Training speed, deg/sec | 2 |
| Elevating speed, deg/sec | 5 |
| Recoil, in | 47.25 |
| Max range at 30°, yards | 30,200 |

Steamship Co, who would have offered her at a daily hire rate while Vickers paid the voyage costs – fuel, loading and discharge and port dues. *Hood* departed from Clydebank on 9 January 1920 for drydocking at Rosyth, with trials in March. The Admiralty paid Vickers £816,534 for her mountings, double the price of the Mark I, partly due to their increased complexity but mostly due to wartime inflation which had also trebled merchant shipbuilding costs. Profit was recorded as £130,824 or 19% of cost.[16]

*Hood's* 15in Mark II X mounting receives its first 100-ton gun at Clydebank on 7 August 1919. *(NRS)*

Minor changes were made to the mountings over the years. Ammunition outfit was reduced in 1931 from 120 rounds per gun to 100, 80 APC and 20 CPC, which saved about 200 tons in a ship now badly overweight. Although the Admiralty was well aware of her shortcomings, there were not enough resources pre-war to modernise her along the lines of *Queen Elizabeth* with lighter machinery, better deck protection and improved anti-aircraft armament. The first time she fired her guns in anger was on 3 July 1940 at the French battleships at Mers el-Kebir. When she and *Prince of Wales* encountered *Bismarck* in the Denmark Strait on 24 May 1941, *Hood* opened fire on the cruiser *Prinz Eugen* rather than *Bismarck* which was fired on by *Prince of Wales.* Garzke estimates that *Hood* fired nine salvoes from A and B turrets and one from X until the hits from *Bismarck* caused first a fire then one which set off a massive magazine explosion which destroyed the battlecruiser.[17]

[16] Vickers Gun Mounting Ledger 170G. BAE Systems, Barrow shipyard archives.

[17] Garzke, Willian H, Dulin, Robert, Jurens, William. *Battleship Bismarck*. Naval Institute Press, 2019

## CHAPTER TWELVE

# The 18-inch Mounting

FISHER WAS KEEN to mount the largest possible guns in one of his 'large light cruisers'. Elswick advised that they could build an 18in gun but that it would take eighteen months. They would need a new gun shop to handle the 62ft long 152.6-ton guns, for which the Admiralty would have to pay; estimated cost was £111,000. *Furious* was ordered from Armstrong, Whitworth in March 1915 to carry two single 18in guns, known for security purposes as the 15in B, sometimes as B CD (coast defence). The company requested a cost-plus contract given all the uncertainties over costs and with wages rising. The barbette diameter of the single mountings was the same as the twin 15in, so that the latter could be mounted should the 18in prove unsuccessful. Two such mountings were also ordered, eventually being fitted in monitors.

The gun was essentially a scaled up 15in of 40 calibres length. It was designed to take the largest propellant then manufactured, 630lb of MD45 cordite. The shell itself weighed 3320lb, 73% more than the 15in. With a muzzle velocity of 2270ft/sec it was expected to range 29,500 yards at 30° elevation. Three guns were ordered, one spare. The first gun was tested at Ridsdale in September 1916. The mounting itself was based on the 15in twin but with fixed angle loading as in *Agincourt*, and with thinner gunhouse armour (9in front, 5in roof, 263 tons) and a revolving weight of 826 tons. The two mountings plus equipment on the fixed structure weighed 1840 tons in total, 10% of her displacement. Maximum ammunition stowage was to be 120 rounds per gun.

Meanwhile, the hull was under construction at Naval Yard Berth Nº 1. Most of her 18 boilers were installed before launch on 15 August 1916. The newly completed 250-ton floating crane that Armstrong was building for the Admiralty was used to install her Y mounting at the end of 1916. Contemporary records are unclear as to whether A mounting was installed early in 1917 but it was probably not. In March the Admiralty approved for *Furious* to be converted to a seaplane carrier, initially fitting a hangar and flying off deck over the forecastle, and plating over A barbette. She left the Tyne for trials to start on 3 July. Her aft 18in was fired, shaking up the lightly built hull.

In September the Admiralty decided to convert her fully to a seaplane carrier, removing her after 18in and fitting a flying-on deck aft. Vice Admiral Bacon of the Dover Command was advised of the availability of the three guns. Three 12in gun monitors were adapted to carry a single 18in aft, on a coast defence mounting capable of 45° elevation but only limited training abeam. The first mounting was tested at Silloth on the Cumbrian coast in May 1918. When supplied with 8 crh shells and supercharges, a range of 40,500 yards could be achieved. *General Wolfe* was ready for operations off the Belgian coast in August 1918, followed by *Lord Clive* in October. A full account of their conversion and operation is given in *Big Gun Monitors*.

Consideration was given to fitting 18in guns in post-war battlecruisers, discussed by Friedman in *Naval Weapons of World War One*. But none were built, while the three existing 18in guns were used for trials purposes post-war, the last being scrapped in 1947.

*Furious* alongside Naval Yard shortly after commissioning on 26 June 1917, her single 18in aft evident, plus seaplane flying off deck forward. *(Author's collection)*

CHAPTER THIRTEEN

# 16-inch Triple Mountings

## THE 16-INCH TRIPLE MOUNTING MARK I

The Admiralty planned four new 'G3' battlecruisers of 48,000 tons in 1921 mounting nine 16in guns in triple mountings. These would incorporate all the experiences from WW1. On the basis that Armstrong Whitworth had the most recent experience of designing the heaviest gun mountings – 18in for *Furious* – they were given a provisional order in August 1921 to design the 16in mountings. The ships were ordered in October 1921. But only a few weeks later the Washington Treaty banned ships of that size so in February 1922 they were cancelled. However, Britain was allowed to build two new 16in battleships but only of the new limit of 35,000 tons standard displacement. While these could still mount nine triple turrets with good protection, speed would have to be sacrificed to a modest 23 knots. *(See colour plates on pages 161, 162)*

On 11 December 1922 *Nelson* was ordered from Armstrong's Naval Yard on the Tyne and *Rodney* from Cammell Laird at Birkenhead, both desperate for warship orders so quoting loss-making prices for hull and machinery. On the next day, the Admiralty's Director of Contracts accepted the tender and prices from Armstrong Whitworth for six triple 16in mounting plus three spare mountings. This was only a provisional order, but enough for Elswick to start work. Vickers were to supply the gunshield armour and director firing gear; the guns were ordered separately.[1] As a consolation prize, Vickers were awarded the contract for the twelve twin 6in mountings for their secondary armament. [see page 126] Work proceeded at a slow pace, limited by the provisional sums granted in the annual Navy Estimates, with an advance payment of

[1] *H.M. Ships Nelson and Rodney: Gun Mountings, 1922–30.* ADM 116/3426

A triple 16in in the easternmost pit in Nº 24 Shop at Elswick with guns at low elevation. Given the date of 28 September 1926, it is probably one of *Rodney*'s. *(Tyne & Wear Archives)*

This 1942 aerial view of *Nelson* well illustrates the layout of her triple 16in turrets. *(Author's collection)*

£263,000 in March 1923. The formal contract was not signed until 27 July 1926, and then on the basis of reduced 1924 prices as in the table below. The Admiralty stated that the prices allowed for costs incurred before 1922; it also specified the maximum degree of interchangeability of mechanical components.

The triple mountings were the first for a British battleship, although Vickers had supplied 12in triples to Italy before WW1. All three mountings were grouped forward of the superstructure with machinery aft, which helped reduce armour belt and deck protection weight to keep within the 35,000-ton limit. The mountings differed appreciably from the previous twin 15in in *Hood*, particularly in terms of improved flashtightness requiring modified loading arrangements and with elevation increased to 40°. To reduce vulnerability to long range plunging fire, the cordite magazines were now placed below the shell rooms. Although the middle deck protection had been increased to 250lb (6¼in), it was also reckoned that the bins full of relatively inert shells would provide a degree of extra protection to the magazines below. The loading arrangement were changed to include handing rooms between magazine and shell rooms and the hoists attached to the mountings, requiring flashtight doors between ammunition stowage and hoists. Shells and charges now had to be transferred from the fixed structure to bogies in the handing rooms which rotated with the main mounting to whatever bearing the turret was trained on. Shells were moved horizontally by overhead rails then turned vertically onto the bogies then moved to scuttles on the rotating part to one of the three shell pusher hoists. Each hoist delivered a shell in line with the gun breeches, where they were tilted horizontal before ramming into the breech at the fixed loading angle of 3°. Instead of the usual working chamber transfer, the space was used for training engines. Meanwhile the charges in six 83lb parts were moved on runways to the cordite handing room where each was transferred manually to the two-level loading 'canisters'. Three hoists, one per gun, then delivered them vertically to the rear of the gunhouse. They were tilted horizontally in a bucket then a traverser moved them across to the loading tray in line with the breech for ramming. The designed loading cycle was 30 seconds *ie* two rounds per gun per minute, although in service it was over 40 seconds.[2]

Other changes included using an oil based hydraulic fluid which was less prone to leakages and corrosion than the previous water. This fluid was delivered to a central pivot at the base of the central trunk rather than through walking pipes. Water storage was now needed on the mounting for gun washout and cordite drenching purposes. The run-out cylinder was operated by high pressure air, putting less demand on the hydraulic system. Three 750hp steam driven hydraulic pumps were fitted alongside the magazines, one per turret, delivering to the various hydraulic engines on the mounting at 1000lb/sq in. Instead of blast bags, each gun port was sealed by a curved blast-excluding mantlet plate fitted to the gun cradle. A revised turntable design reduced the roller path diameter to 33ft which saved weight, also affecting the barbette armour. Two powerful swashplate 400hp training engines were fitted, although only one could drive at a time. More extensive interlocks were required to avoid more than one flashtight opening being open at a time during loading operations. Although a 41ft rangefinder was fitted in each turret, there was only one sight fitted, a periscope in the silent cabinet fitted on the right side of the gunhouse which also housed the control officers. The left side

[2] *Handbook for 16-inch B.L. Mark I Gun on Mark I Triple Mounting, 1928.* ADM 186/285-287

Demolition of *Rodney*'s 16in A and B gun mountings on 19 August 1948 reveals the thickness of the gunhouse armour: side rear 9in, barbette side 15in. *(Marine Technology Special Collection)*

The internals of *Rodney*'s X turret at Inverkeithing on 19 August 1948, showing the great thickness of a 16in gun body (52in, breech ring 65in), some of the wire winding, the run-out cylinder on the gun slide and the control cabinet at right into which eight numbers squeezed. *(Marine Technology Special Collection)*

was taken up with three 6pdr sub-calibre guns for practice firing. Ready-use shells were stowed at the rear of the gunhouse.

The gunhouse was heavily armoured, with front plates of 640lb, side front 440lb, side rear 360lb and rear 360lb, all priced at £240 per ton, with five 290lb bolted removable roof plates, priced at £205 per ton. Payment was paid for each plate based on its actual weight, which because of rolling tolerances of 2% under and 0% over were slightly lighter than designed, a hidden weight 'saving'. 45% of the price was paid when the plate was rolled to thickness and its face treated, 45% when bent, machined and completed at the maker's works and 10% when received at the gun mounting manufacturer's works. Inside the rear plate was fitted another curved plate weighing 80 tons to balance the gunhouse. Both plates were pierced by 72 4.5in diameter holes to vent gases from any burning cordite charge. Barbette armour varied from 600lb on the more vulnerable sides to 480lb nearer the centre line.

The base of each barbette plate was now seated on a cast steel armour chock bolted to the deck which improved the rigidity of the structure.

While the nominal shell stowage was 80 rounds per gun (plus 10 practice), the actual capacity was 381 in A shell room, 387 in B and 303 in X, an average of 119 per gun, all APC as built. This was a device to reduce weight as ammunition was included in the standard displacement – about 700 tons was 'saved' in this way. Interestingly as all the ammunition spaces were adjacent, watertight doors were fitted to allow transfer between the mountings. [The US *Iowa* class could transfer between forward and after turrets using an overhead rail along their third deck.]

A and X were so-called short trunk mountings with a revolving weight of 1464 tons with guns, while superfiring B with its longer hoists was called long trunk, weighing 1483 tons. The three together weighed 4411 tons compared with 3520 tons for *Hood*'s four Mark II 15in, a much higher proportion of displacement. Associated equipment of magazine and shell room gear, hydraulic machinery and water added another 250 tons per mounting. The crew for B comprised 24 in the gunhouse and training engine space, 36 in the shell and handing rooms and 40 in the magazine and handing rooms, 100 in all; A and X had 12 fewer in ammunition supply.[3] B had the widest training arc at 330°, with A at 295° and X at 125° on either side although giving severe blast to the superstructure.

A new Mark I gun was designed, 45 calibres bore and wire-wound. Campbell gives 29 made (*ie* 60% spare) 10 from Elswick, 8 from Vickers at Sheffield, 5 from Beardmore and 6 from RGF Woolwich. Elswick carried out proofs of their guns at Ridsdale with their first (S.13) on 20 March 1925, with a 2048lb projectile at 2800ft/sec with 545lb MD45 propellant.[4] *Furious*' Nº 1 18in gun had been linered down to 16in by Elswick in 1921 for cordite trials purposes.[5] Changes were made to the rifling after excessive bore wear was found.

As Naval Yard's heaviest fitting out crane was only 30 tons, the 150-ton *Titan II* floating crane was used to lift most of the components of both *Nelson*'s 16in and 6in mountings. Starting shortly after her launch on 3 September 1925, *Titan II* made about 100 lifts over a period of twelve months ranging from a 10-ton hoist component to a 165-ton turntable – 10% over the crane's rated capacity! The first mounting to be completed was B whose last 108-ton gun was lifted in on 3 September 1926, then A on 18 September and finally X on 1 October.[6]

The contract for 'Gun Machinery and Shields for 16-inch 45-calibre Guns' included the price per ship (see table, left).[7]

The Elswick accounts at 31 December 1927 record the sales value for both ships including three spare mountings and extras at £3,266,324, cost £2,812,066, profit £454,258 (16% on cost), say £1.64M per ship.[8] Over the next six years the Admiralty placed fourteen contracts with Vickers-Armstrongs Elswick worth £70,000 for modifications to the main roller paths, adding vertical rollers, pivot trays and additional safety gear.

After the mountings had been delivered, negotiations were entered by Armstrong (rather than the new V-A company) about claims for an extra £259,000 for modifications and the £30,000 spent on a third large pit – despite the large profit. The Admiralty probably argued that the latter cost should have been included in the establishment charges built into the tender prices, but Armstrong probably countered that delivering six mountings from only two pits would have delayed the ships (each mounting spent about nine months in a pit) and that introducing modifications to the shell room equipment disrupted produc-

### 16in BL Mark I Gun

| | |
|---|---|
| Calibres, bore | 45 |
| Length overall, in | 742.2 |
| Weight incl BM, tons | 108 |
| Muzzle velocity nominal ft/sec | 2614 |
| Shell, lb | 2048 |
| Charge SC280, lb | 498 |
| Life, efc | 250 |

### 16in Triple Mounting Mark I

| | |
|---|---|
| Rotating weight A and X, tons | 1464 |
| Max elevation, degrees | 40 |
| Max depression, degrees | -3 |
| Loading angle, degrees | 3 |
| Training speed, deg/sec | 4 |
| Elevating speed, deg/sec | 10 |
| Recoil, in | 47 |
| Rounds per gun per minute | 1.5 |
| Max range at 40°, yards | 38,400 |

### Gun Machinery and Shields for 16in 45-calibre Guns

| | Item | £ |
|---|---|---|
| 1 | Turntables, roller path, chambers | 170,383 |
| 2 | Turning gear, engines, training rack | 105,570 |
| 3 | Gun slides, carriages, cradles, rammers, run-out & recoil cylinders | 254,393 |
| 4 | Pumping engines | 57,619 |
| 5 | Pipes, valves etc on turntables | 56,773 |
| 6 | Shell room machinery, lifting presses, | 112,937 |
| 7 | Magazine machinery, flashtight doors, hoist cages | 51,365 |
| 8 | Transporting gear, cordite hoist containers, tilting trays, bogies | 143,978 |
| 9 | Pipes outside revolving structure | 26,442 |
| 10 | Brackets, rails etc | 18,049 |
| 11 | Gun shield. Fixed price | 61,401 |
| 12 | Gun shield armour, priced per ton. Estimated 850 tons | 200,000 |
| 13 | Miscellaneous | 28,933 |
| 14 | 2 spare mountings | 65,240 |
| | Erection in pits | 87,392 |
| | Delivery & erection of mountings (*Nelson*) * | 128,480 |
| | Total | 1,568,955 |

* *Rodney* at Birkenhead £168,285. Excludes guns and director firing gear. Approx £450/ton excluding guns.

[3] Hodges, Peter. *The Big Gun*, p139
[4] Tyne & Wear Archives. DA.VA/6/35/10
[5] *Big Gun Monitors*, p228
[6] *Titan II* Crane Log 1926. Marine Technology Special Collection. 01250/8
[7] *Nelson* and *Rodney* Ships Covers. NMM
[8] Tyne & Wear Archives. DS.VA/3/3/3

*Nelson* on 15 March 1949 shortly after arrival for breaking up at Inverkeithing. Ward's have already cut off part of B mounting's barrels to reduce the weight of the remaining portions to be lifted off. *(Marine Technology Special Collection)*

tion. While some of these extras were agreed, the other claims were settled by the Treasury in March 1930 for £55,000.[9]

Not surprisingly, the novelty of so many new mechanisms resulted in teething problems. Campbell notes: 'The turrets gave much trouble initially and several modifications had to be made in the shell handing rooms and elsewhere, while a vertical guide roller race was added just below the main horizontal one. The interconnected shell bogies made it necessary to load all three guns together.'[10] Hodges gives details of the various changes.[11] The guns needed to be fired in sequence to reduce blast: centre gun separate from the wing guns.

During the final action against *Bismarck* on 27 May 1941, *Rodney* fired 375 rounds of 16in APC shells, an average of 42 per gun. While there were no major breakdowns, the right gun of A missed 11 salvoes due to the gun slide not locking for loading, while its shell hoist jammed later. Other salvoes were missed due to mechanical or drill failures but overall output was 77% of possible at 1.6 salvoes per minute (salvoes were fired to help spotting using four or five shells from the three turrets).[12] *Nelson* and *Rodney* both carried out shore bombardments off Sword Beach at Normandy, the latter firing over 500 rounds, mostly HE, stocks of which were limited, the former firing 224.

Overall the 16in was regarded as less successful than the well tried and consistent 15in, despite its heavier punch. With its higher rate of fire in practice, eight 15in could deliver a heavier weight of shell per minute than nine 16in.

[9] ADM 116/3426.
[10] Campbell, John. *Naval Weapons of World War Two*, p22
[11] Hodges, Peter. *The Big Gun*, p94
[12] Roberts, John. 'The Sinking of Bismarck'. *Warship* 28

## 16-INCH TRIPLE MOUNTINGS FOR THE *LION* CLASS

For the two 1938 Programme battleships, the displacement could be increased to 40,000 tons, which would allow in a balanced design nine 16in guns, the same as the US Navy and as was believed at the time the Japanese (although they were actually 18in). The ship design of this new *Lion* class was essentially a scaled-up version of the *King George V* class with three triple turrets. Vickers-Armstrongs at Elswick was tasked with the design, which owed more to the 14in than the 16in in *Nelson* and *Rodney*. A new 16in gun was also needed, both the Mark II and III being essentially similar. This 45-calibre gun was an enlarged 14in weighing without the 12-ton balance weight 118.7 tons. Performance with a 2375lb shell and 2400ft/sec muzzle velocity was estimated as 38,200 yards at 40° elevation with a 6 crh shell. Eighteen production guns (12 right guns, 6 left) were ordered from Elswick in the third quarter of 1938 at a price of £30,000 each plus £2970 for each breech mechanism, enough for the first two ships. The Mark II mountings were ordered on 30 July 1938, three from Elswick, three from Barrow. However, erection in the pits could not start until the last 14in mountings had been finished. Based on the planned timetable for the latter, the first 16in mounting could be completed early in 1942, with the first ship completed in July 1942. The mountings had the same 40° maximum elevation as the 14in and would weigh almost the same as the quad 14in, namely 1560 tons. Armour would be slightly heavier with 15in front plates, 10in side front, 7in side rear and rear, 6in roof.

Meantime tenders had been invited on 15 December 1938 from six shipbuilders, including for the first time for a battleship Harland & Wolff. Two ships were ordered on 15 March 1939, *Lion* from Vickers-Armstrongs Tyne shipyard and *Temeraire* from Cammell Laird. The 1939 Programme was to include two more, one to be named *Conqueror* from John Brown and *Thunderer* from Fairfield. But war had broken out before they could be formally ordered, although John Brown had already allocated Yard Number 567 to their ship, later used for *Vanguard*. *Lion*'s keel was laid on Berth Nº 2 at Naval Yard on 4 July 1939 and *Temeraire*'s on 1 June, but work was suspended in October, by which time 218 tons of keel material had been erected for *Lion*.

Little progress had been made on *Lion*'s mountings by the end of 1939 with priority given to completing the 14in mountings. Notional delivery of Barrow's mountings was put back to mid-1943, although preliminary work had started there on jigs and tools. Work was formally suspended in June 1940. At the end of 1941, instructions were given to proceed with structural work on the 166-ton turntables, albeit at low priority. Elswick reported 'Nos. 1 and 2 turntables are being assembled on the shop floor. Work of assembly of the first two cordite platforms, central ammunition hoists and training engine space, training gear space and shell platforms is well advanced.'[13] Work even started on assembly in one of the pits. Then a letter was received from the Admiralty on 16 October 1943 to suspend work at both Elswick and Barrow. By that time about £390,000 had been spent, over half the cost of one mounting. Campbell notes only four guns made, which roughly equates with the £170,000 spent by Elswick making 16in guns by 1944,

[13] Cambridge University Library, Manuscripts. Vickers Archive 207

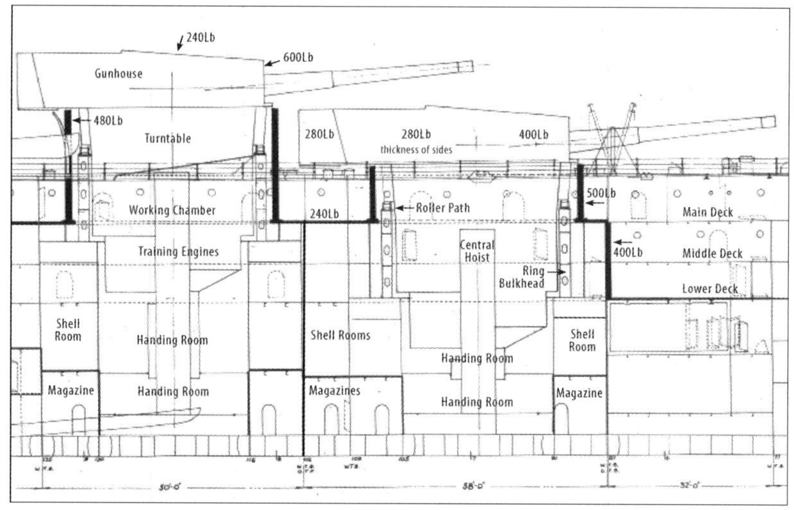


The triple 16in Mark II mounting for the *Lion* class battleships was based on the 14in Mark II mounting rather than the 16in Mark I. The arrangement of A and B mountings is shown, Y was similar. Construction of the ships was suspended in 1939, but work on the first gun mountings was restarted at Elswick in 1942, but suspended again in 1943 and finally cancelled. *(Ian Johnston)*

although there could have been more. Trials using them were made at Isle of Grain and Shoeburyness 1945 to 1947, including flashless propellant. Formal cancellation of the other guns only came in early 1946.

Construction of *Lion* and *Temeraire* remained suspended throughout the war, although from time to time the Admiralty Board considered resuming construction. The design was revised in 1942 to take account of wartime experience, although Director of Naval Construction Sir Stanley Goodall regarded all such activity as a waste of his staff's time, rightly identifying aircraft carriers as more important.[14] Indeed ordering three light fleet carriers from Naval Yard in 1942 precluded them from restarting *Lion* until at least 1944. There were many factors affecting their suspension: the shortage of heavy gun mounting capacity (the War Office was demanding more heavy coast defence mountings as well as large numbers of anti-aircraft mountings including 5.25in), the lack of armour making capacity which had been largely diverted to tank manufacture and as ever shortage of skilled labour, quite apart from the lack of a strategic need. Formal cancellation only came at the end of 1945. However, *Vanguard* did proceed but only because her four twin 15in were existing ones removed from *Courageous* and *Glorious*. Astonishingly, as late as March 1944 the Admiralty asked Elswick to prepare a new Mark III design of triple 16in mounting with RP.40 remote power control, lighter with thinner armour, a faster firing cycle and training speed, and a new cordite hoist.[15] A new design of Mark IV gun was even put in hand.[16] More surprisingly, the Admiralty was still discussing the possible use of 16in guns as late as 1948.[17] The exigencies of post-war finances finally put an end to such unrealistic discussions.

[14] Buxton, Ian. *Diary of a Wartime Naval Constructor: Sir Stanley Goodall*. Seaforth Publishing, 2022

[15] Tyne & Wear Archives. DS.VA/6/29/6

[16] Campbell, John. *Naval Weapons of World War Two*, p24

[17] Admiralty: Guns and Mountings. ADM 1/20970

## CHAPTER FOURTEEN

# 8-inch Twin Mountings

FOLLOWING THE WASHINGTON TREATY, the Admiralty decided to build its next class of cruisers right up to the limit of 10,000 tons with 8in guns. It found that a well-balanced trade protection design with good speed and endurance could mount only four twin 8in guns. The mounting design was a complete departure from previous heavy cruiser mountings with many innovations. A high elevation of 70° was required for anti-aircraft fire which not only needed a more spacious recoil area in the gun well but made all-angle loading impractical. So a fixed angle loading of 10° was adopted, giving a wedge shape to the bottom rear of the gunhouse for the rammer. Magazines and shell rooms needed to be below the waterline and a protective deck, so both were fitted at the same hold level, protected by armoured boxes. One pusher hoist serving each gun, with shells vertical rather than in wire-lifted cages with shells horizontal, could more easily be made flashtight. They also permitted a smaller ammunition trunk diameter and reduced roller path diameter, saving weight – important with a limit on displacement. The speed and acceleration of cages would have meant a slower rate of fire. However, it was thought that pusher hoists could only be angled in one plane not two, so requiring a break in shell supply from shell room to gunhouse. This was achieved by using a fixed vertical transporter hoist to move shells to a handing room one deck above, which were then transferred to the bottom of the pusher hoist to deliver straight to the gunhouse. In the handing room each shell was moved via a bogie to a ring around the bottom of the trunk holding 30 shells, which could be rotated to bring each shell in line with a door at the bottom of the pusher hoist on the revolving trunk. A hoist was installed for each gun, delivered to the rear of the gun where it was swung horizontal for loading and ramming.

The result was a mounting steadily increasing in weight as the design progressed, ending up weighing over 180 tons, over double the weight of the twin 6in in *Nelson* and *Rodney.* In the superimposed long trunk B and X mountings, whose revolving weight was 187 tons, it was 58ft from magazine floor to gunhouse roof. The whole mounting rotated within the 18ft diameter ring bulkhead which supported the roller path. Gunhouse protection was only 1in all round; supplied for the *London* group by Vickers and Armstrong Whitworth from Openshaw at £160/ton.[1]

While the design was being worked up, probably by Elswick, a new 8in BL gun was being developed. The Mark VIII was a wire-wound gun 50 calibres long, weighing 17.2 tons with breech mechanism. It was composed of five layers: the rifled inner A tube, the main A tube, wire winding, outer B tube and a jacket each about 60% of the gun length. At the rear end was the usual breech ring screwed to the A tube to hold the breech mechanism. A 4.2-ton balance weight was needed to bring

[1] Armour Plate. 1925–6. ADM 116/3351

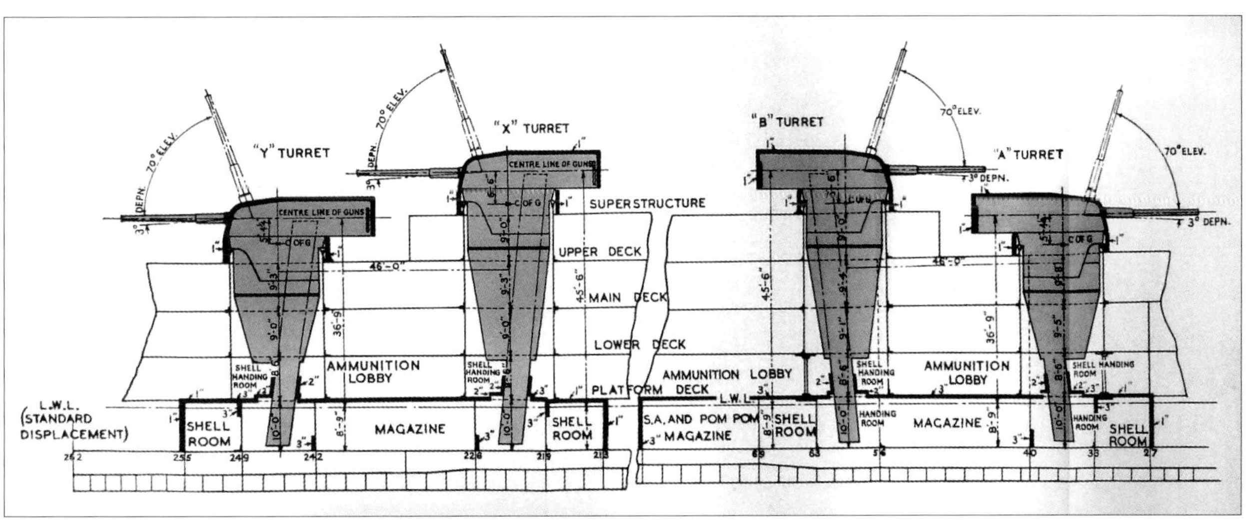


The 'County' class cruisers had magazines and shell rooms at hold level, protected by armoured boxes. Fixed hoists lifted shells to a handing room where they were transferred to the revolving structure. *Kent* class illustrated. *(ADM 234/190)*

its centre of gravity back to the trunnions. The Mark VIII* used the autofrettage technique to make a gun with only inner A tube, A tube and jacket, all designed to withstand 20.5 tons/sq in pressure. Instead of using the wire to induce a compression on the inner tubes to resist the pressure of firing, autofrettage applied internal pressure during forming, which when released put the outer part of the tube in compression. Only three layers were necessary instead of five as illustrated by the Mark VIII*. *(See colour plate on page 163)*

Left and right guns were interchangeable, each capable of being elevated independently, allowing each gun's performance to be adjusted for differing bore wear. In the Mark I mounting, elevation was by pinion driven by a 13hp hydraulic motor engaging with a toothed arc fitted to the underside of the gun cradle. With the gun, cradle, recoil and run-out cylinders and elevation arc, the whole elevating mass was 29.2 tons. The planned elevation speed was 10°/sec but such a high speed was found to be impossible to control accurately so was reduced to 4°/sec. Two 40hp swash plate training motors were fitted beneath the gunhouse, in theory capable of 8°/sec but in practice only 5°/sec. In the conical trunk, the upper compartment housed the 120hp electrically powered pump with oil as the pressure medium, while the lower compartment held air blast equipment, wash out and drain tanks. Twin cordite hoists were fitted in a rectangular trunk on the turret centre line with two 33lb half charges carried horizontally. A conventional cage lifted by hydraulic cylinder and wires delivered them into the gunhouse without an intermediate handing room.[2] *(See colour plate on page 164)*

Five 'County' class cruisers were ordered under the 1924 Programme at a cost of £2M each, one from each main dockyard, *Cumberland* from Vickers and *Berwick* from Fairfield. Vickers at Barrow received the order for four twin 8in mountings each for *Kent* and *Cumberland*. The other three sets were ordered from Armstrong Whitworth for *Berwick*

[2] *Handbook for 8-inch Marks VIII, VIII*, VIII** Guns on Twin Marks I, I* Mountings 1944.* ADM 234/190

A view looking into Elswick's N° 24 erecting shop in 1926 showing two twin 8in Mark I mountings with a cone-shaped ammunition hoist, probably for *Berwick* (Reg N°s 1–4). In the background is a triple 16in, probably for *Rodney*. *(Tyne & Wear Museums)*

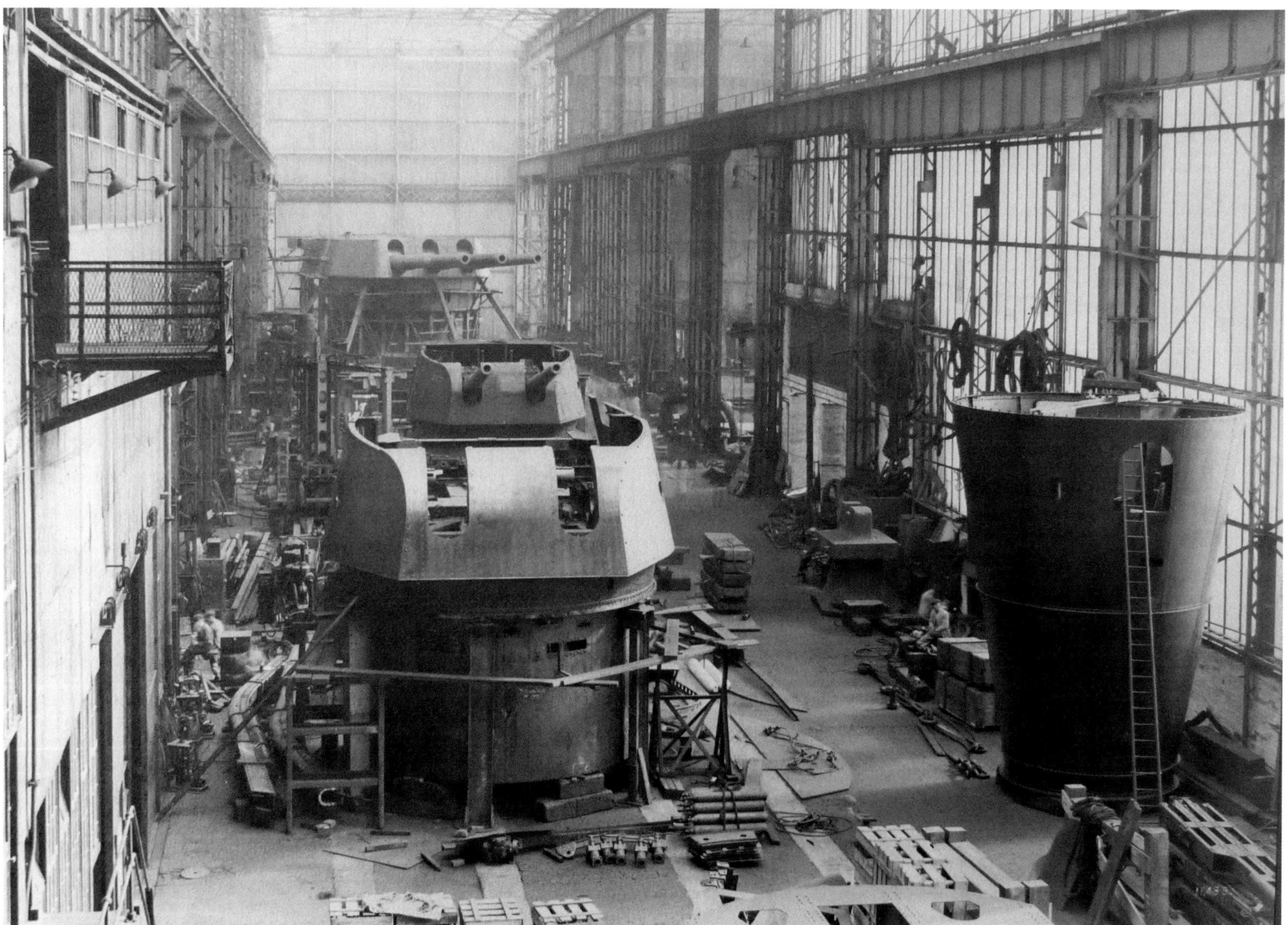

Three 8in Mark I mountings in Elswick's No 7 Shop probably for their first contract *Berwick* and *Suffolk*. *(Vickers)*

(Reg Nos 1-4) and *Suffolk* from Elswick and *Cornwall* from Openshaw. Two spare mountings were also ordered from Elswick. Openshaw also received orders for the gunshield armour. In the Vickers contract the price was broken down as follows:[3]

**Four Twin 8in Mark I Mountings for HMS *Cumberland*.**
**Registered Nos A: 16, B: 18, X: 15, Y: 17.**
Vickers Barrow contract 205G ordered 14 June 1924.
Final claim 10 February 1928.

| *Item* | | £ |
|---|---|---|
| 1 | Turntable | 57,210 |
| 2 | Training gear | 25,165 |
| 3 | Cradles and slides | 99,160 |
| 4 | Pumping plant | 13,995 |
| 5 | Pressure and exhaust piping | 13,110 |
| 6 | Shell room gear and bogies | 81,072 |
| 7 | Gun loading gear and hoists | 67,370 |
| 8 | Brackets and rails | 10,605 |
| 9 | Gun shield structure | 25,220 |
| 10 | Gun shield armour | 18,165 |
| 11 | Hoist armour | 3,658 |
| 12 | Lighting and turret danger signals | 14,055 |
| 13 | Director firing gear* | 40,542 |
| **Sub-total** | | **469,337** |
| Delivery of mountings (Barrow) | | 850 |
| Erection on board | | 34,000 |
| Erection of director gear | | 1,010 |
| Painting turrets | | 400 |
| Alterations to gun pits | | 19,600 |
| One set of jigs, tools and gauges | | 48,750 |
| Price adjustments and miscellaneous | | 15,119 |
| **Total price** | | **589,066** |
| Costs | | 441,910 |
| Profit | | 147,156 |
| Price each mounting | | 147,266 |

* Barrow £16788, Erith £23754

[3] Vickers Gun Mounting Contract Ledger 205G. BAE Systems Barrow shipyard archive.

An 8in gunhouse with turntable and working chamber below weighing over 90 tons being loaded on to a cargo ship at Elswick, probably destined for *Berwick* at Fairfield. The brackets are probably for lashing it to the hatch coaming. *(Tyne & Wear Museums)*

The price corresponded to about £800 per ton and profit to 33% of cost – generous but useful in enabling the works to survive a lean period. All eight guns were Mark VIII* made by Vickers at Sheffield in 1925.[4] *(See colour plates on pages 163-165)*

The building programme envisaged two months between delivery of the mounting components at the shipyard and completion of erection, then four months to gunnery trials, with final delivery two months later, although this date usually slipped. *Kent* building at Chatham was the last of the first batch to be completed, but the Captain of *Excellent* responsible for her gun trials early in 1928 reported that the state of her Barrow 8in turrets was 'below the standard expected', and that more care should be taken to have skilled supervision in erection. But things must have improved as *Canberra*'s trials a few months later were 'very satisfactory', also a Barrow set.[5]

The guns were ordered separately under a different Vote in the Navy Estimates. Campbell has 168 guns made over the three Marks, but this was only enough for the fifteen ships built and their spares, so probably excludes the 37 new guns that Elswick alone built 1940–43. Those from Elswick cost the Admiralty around £6000 each plus £950 for the breech mechanism.

Two similar cruisers were ordered in 1925 for Australia, both from John Brown, with four Mark I mountings for *Australia* from Armstrong and four for *Canberra* from Vickers – the companies had not yet merged. From the movements of cargo vessel *Brompton Manor,* two of *Australia*'s mountings arrived at Clydebank on 29 July 1927 from Manchester (*ie* made at Openshaw), and two arrived on 11 August from the Tyne (*ie* Elswick). She also brought *Canberra*'s from Barrow on 16 January and 17 February 1928. The same 1735 grt ship had carried

[4] Ships Book: HMS *Cumberland*. ADM 136/26

[5] Cambridge University Library, Manuscripts. Vickers Archive 149.

*Berwick*'s two forward 8in mountings with the ship nearly complete at Fairfield early in 1928. *(Ian Johnston)*

*Barham*'s 15in mountings to Clydebank from February to June 1915 under her original name *Aydon*.

The Mark I* mounting was then developed with the lower elevation speed of 5½° per sec. It was fitted in the next four ships, with *Devonshire*'s and *Shropshire*'s coming from Barrow, *Sussex*'s from Elswick while *London* got hers from Openshaw, the latter the last heavy gun mountings built at the former Whitworth works. Two spares were also ordered from Elswick. These four also got larger shell rooms, stowing 360 SAPC shells per turret compared with 305 for the Mark I, plus 40 HE.[6] The nominal capacity was only 100 rpg but this was a 'weight saving' device which could be ignored in wartime. One magazine served each pair of turrets, each with 240 cases each holding five half charges, *ie* 600 full charges, implying that some rounds would be fired with half charges. *(See colour plate on page 164)*

Six months after launch at Hebburn, *Sussex*'s mountings were lifted from Elswick's barge by the *Titan II* floating crane, starting with Y on 26 July 1928, in four main lifts: 27-ton ammunition trunk, 91-ton turntable and gunhouse, then two 27-ton guns with slides. Then each Thursday another mounting until the last, A, was shipped on 16 August, seven months before completion.[7]

The Mark II mounting was fitted in all the later 8in gun ships with the Mark VIII or VIII* gun, which had the loading angle reduced to 6° so enabling the rammer to be sited on a level gunhouse floor. The elevating gear was changed to cylinder and ram enabling more precise control. It was found possible to orient the pusher hoists in two planes, doing away with the shell transfer compartment, so that there was one common cordite and shell handing room at hold level for each pair of mountings. In the handing room were two shell loading levels, the upper feeding the left gun, the lower the right gun. The bottom of the hoist was right on the mounting centre line fed by two tilting buckets from the trays. The twin hoist then reached right up to the gunhouse, each angled out to deliver in line with the breech of each gun and its

6 *Handbook for 8-inch BL Mark VIII Gun on Twin Marks II and II* Mountings, 1931.* ADM 186/307 & 308

7 *Titan II* Log 1928. Marine Technology Special Collection 01250/10

rammer. Those same rammers and loading trays could be traversed across the gunhouse to load the cordite charges delivered abreast the breeches by a vertical twin hoist. In the gunhouse was also a local sight between the guns which could be used in conjunction with the control officer's periscope in the silent cabinet on the right side of the gunhouse; on the left was a 6pdr sub-calibre gun. A large 23ft FM.3 rangefinder spanned the gunhouse rear. Gunhouse protection was splinter-proof 1in armour, which was also used to protect the training rack in a ring fixed to the upper deck and reaching up to the gunhouse floor, where a leather weatherproof apron was fitted. The magazine crown was 2½in NC, sides 4in C.[8] *(See colour plate on page 165)*

The Mark II was fitted in *Dorsetshire* and *York* (which had only three turrets) made by Elswick and *Norfolk* by Barrow, while *Exeter* had three Mark II* mountings with 50° elevation to save weight, made by Elswick. Two spares were also ordered from Barrow, one long trunk, one short. Eight Mark II mountings were ordered for the two 1928 Programme cruisers in February 1929, but the ships were cancelled early in 1930; Barrow claimed £120,000 for work done on the £430,000 contract for *Northumberland* and Elswick £100,000 for *Surrey*.[9]

More 8in mountings continued to be ordered from Elswick and Barrow after the Vickers-Armstrongs merger. By 1930, Elswick was making the mountings very efficiently despite receiving prices from the Admiralty about 25% lower than the earlier mountings.[10] Their last three 8in contracts showed profits around 50% of costs, a margin that enabled the works to survive the depths of the depression of the 1930s and retain enough skilled men for when re-armament started.

| *Ship* | *Mountings* | *Sales value £* | *Costs £* | *Profit £* | *Price per mtg* |
|---|---|---|---|---|---|
| *Dorsetshire* | 4 Mark II | 450,239 | 303,403 | 146,835 | 112,560 |
| *York* | 3 Mark II | 338,559 | 233,040 | 105,519 | 112,853 |
| *Exeter* | 3 Mark II* | 329,103 | 208,610 | 120,493 | 109,701 |

Service experience soon showed that these complex high elevation mountings suffered from many minor faults, mostly in the ammunition supply and interlocks – there were at least 26 of the latter. These were gradually ironed out but resulted in a reduction in rate of fire from a planned 5.5 rpg to only 3 to 4:

> The operation of the main armament was not found to be entirely satisfactory and all of the class had considerable attention devoted to this problem, especially during 1927 and 1928 before they were accepted into service in the latter year. … As the mountings were so complicated, many years elapsed before all of the problems had been solved – as late as 1938 they were giving cause for concern: from *London* (16 June 1938) 'One would wish that the 8in mountings and torpedoes would behave as they should. The prolonged firing of 20 rounds per gun from *London* was a disappointment from the material side'.[11]

The conical ammunition trunk of an 8in mounting loaded onto a barge, so probably a Mark I* for *Sussex* building on the Tyne, as shells are fed in at a higher level than the four scuttles for charges. *(HMS* Excellent*)*

An addendum to the Mark I handbook had to be issued in 1935 detailing all the changes.[12]

Barrow had delivered *Devonshire*'s four mountings to Devonport in the third quarter of 1928. Four months after completion, when in the Aegean on 26 July 1929 *Devonshire* experienced an explosion in X turret killing 17 men. The whole ship could have been lost had the fire penetrated to the magazine. Lt Cdr Gilbert Roberts was gunnery officer at the time and his biography described what happened:

> The result of the enquiry into the accident in X turret gunhouse was found to be an incorrectly assembled interlock. This interlock was to prevent the breech of a loaded gun being opened unless the gun had fired and had recoiled. In this case the salvo fired and the left hand gun did so, but the right hand gun had not fired for some reason. The breech was opened and this should not have been

[8] ADM 186/307 & 308.

[9] Vickers Gun Mounting Contract Ledger 223G.

[10] Tyne & Wear Archives. DS.VA/3/5/3 & 4

[11] Raven, Alan & Roberts, John. *British Cruisers of World War Two*. Arms & Armour Press, 1980, p125

[12] ADM 234/549

'County' class cruiser *Devonshire* at Malta in the early 1930s, probably after her X mounting had been replaced following an explosion in July 1929. *(World Ship Society)*

> possible. With the firing electric tube being faulty, still smouldering, the fresh air caused flame to shoot into the cartridge already in the gun. The cartridge ignited, driving flame into the next cartridge to be loaded. This in turn ignited the left hand gun cartridge and the whole issue went up with a terrific bang. The one thing that really held up was the flash-proofing which prevented the flash going down to the magazine.[13]

The gun trials a few months earlier should have picked up that the interlock was not working correctly. The damaged turret was removed at Devonport and sent back to Barrow for repair at the end of 1929, and re-erected on a pit in the second quarter of 1930. One of the spare mountings was quickly fitted so she was back in the Mediterranean by January 1930. Repair work which included new roof plates was completed in March 1931 at a price of £30,432 and the turret sent to Portsmouth as a spare.[14]

Designed ammunition stowage was 150 rounds per gun. Between 1926 and 1928 the Admiralty ordered 17,400 semi armour piercing capped shells from Hadfield and 10,600 from Firth, both of Sheffield, at an average price of £35 each, excluding filling and fuze, 23 ship outfits.[15]

In 1928 Spain ordered two 8in cruisers designed by Sir Philip Watts, retired DNC, but based on the *Kents*. Their main armament was four twin 8in similar to the British Mark I. In February 1929 Vickers-Armstrongs received the order for one twin 8in mounting for the first ship, *Canarias* building at Ferrol, to be made at Barrow for delivery in 22 months. With two guns made by Elswick and design drawings at £25,000, the total price came to £180,786 – which included commission of 6% – whose pockets did that end up in? The remaining guns and mountings were made in Spain, requiring £47,000 worth of jigs and gauges to be supplied by V-A for manufacture at the San Carlos works of Sociedad Española de Construcción Naval in Cadiz, 40% owned by V-A (book value £395k) and John Brown.[16]

During the war the Army took over six Mark VIII*** guns for coast defence near Dover on 70° mountings[17] Elswick making the six mount-

13 Williams, Mark. *Captain Gilbert Roberts RN*. Cassell, 1979, p66
14 Vickers Gun Mounting Contract Ledger 228G.

15 Corker. *op cit*. Appendix C
16 Vickers Gun Mounting Contract Ledger 222G.
17 Campbell, John. 'British Heavy Coast Defence Guns in World War II'. *Warship* 1995

*Australia* was the last cruiser to be scrapped still with her 8in mountings, although X had been removed in 1945. She had just been towed to Ward's shipbreaking yard at Barrow when Ken Royall took this photo on 8 July 1955.

ings at a cost of £208,000.[18] Unusually it was Harland & Wolff rather than Vickers-Armstrongs who undertook the repair of *Sussex*'s mountings after her bombing and fire in the Clyde in 1940. Elswick repaired *Exeter*'s A and B mountings in 1940 after the River Plate battle for £12,000. In the case of *Norfolk*'s 8in damaged in the fight against *Scharnhorst,* X mounting was removed from her at Palmers Hebburn yard by *Titan II* in January 1944 and sent to Elswick for repair, being re-installed in August although other repairs were not complete until November. *Norfolk*'s report of her part in the *Bismarck* action on 27 May 1941 noted that 527 rounds of 8in were fired over about 90 minutes, varying from 103 from B turret's right gun to only 8 from X turret's left gun (probably due to double ramming of cordite charges). Her CO Captain A J L Phillips concluded: 'For 8in Mark II mountings, the final state of affairs and the percentage output are considered most satisfactory.'[19]

**8in BL Mark VIII Gun**

| | |
|---|---|
| Calibres, bore | 50 |
| Length incl BM, in | 431.1 |
| Weight incl BM, tons | 17.2 |
| Nominal muzzle velocity, ft/sec | 2800 |
| HE shell, lb | 256 |
| Charge SC205, lb | 66 |
| Life, efc | 600 |

**8in Mark I Twin Mounting**

| | |
|---|---|
| Rotating weight A and Y, tons | 183 |
| Total incl on fixed structure, tons | 208 |
| Max elevation, degrees | 70 |
| Max depression, degrees | -3 |
| Rounds per gun per minute | 4 |
| Max range, yards | 30,650 |

In 1941, the Admiralty flirted with a new class of 8in cruiser requiring new guns (Marks X and XI) and a new triple Mark V mounting, even placing an order for three mountings with Barrow. V-A optimistically claimed they could complete fifteen such mountings by 1944. But as so often, the resources to build the hulls and mountings were simply not available so they were suspended, later cancelled, while the need for such large cruisers disappeared as the war progressed.

Post-war the 8in cruisers with their large crews were soon paid off in the RN, only *Devonshire* surviving as a training cruiser just with A mounting until 1954 and *Cumberland* losing all her 8in when she became a trials cruiser in 1951. In the RAN, *Australia* and *Shropshire* survived until 1955, both being towed to the UK for badly needed scrap steel.

[18] Tyne & Wear Archives. DS.VA/3/5/16

[19] ADM 199/1188

CHAPTER FIFTEEN

# 6-inch Gun Mountings after the First World War

### THE CHANGE TO TWIN MOUNTINGS IN BATTLESHIPS

The 6in guns mounted in cruisers in WW1 proved powerful weapons, so were considered the right calibre for trade protection cruisers post-war. However, DNO considered that such single mountings had drawbacks. They only had partial shields so loading crews were subject to enemy shell splinters as well as heavy spray breaking over the deck and gun blast from nearby mountings. While some mountings were fitted on the centre line, others were at the sides with limited arcs of fire and with awkward magazines close to machinery spaces. With hand loaded 100lb shells, crews were soon fatigued with prolonged firing. Future engagements were likely to be at longer ranges requiring greater elevation and full director control (rather than individual gunlayer firing).

Thus in 1920 DNO wanted the guns to be power worked and in fully enclosed gunhouses, which would provide protection from nearby gun blast, splinters and spray. The hoists for shell and cartridge could be suspended under the gunhouse and rotate with it, similar to larger calibre guns. Fitting two guns in one mounting would need smaller crews than two singles. One gun from each mounting could be used in salvo firing, then the other, simplifying shell splash spotting. Triple mountings were considered but ammunition supply would be more difficult than a twin, and only of advantage if twelve guns rather than eight were required for larger ships. While such a mounting would be heavier and more costly than two singles, the advantages far outweighed the disadvantages. Such a twin 6in mounting could also be used as secondary armament in battleships, rather than the single 6in in casemates in pre-war ships. The design of the 1921 cancelled 'G3' battlecruisers would have included eight such mountings.

Completed in 1926, *Enterprise* was the first cruiser to mount a twin 6in mounting, the Vickers Mark XVII, her other five being singles. *(World Ship Society)*

Both Vickers and Armstrong Whitworth were asked to prepare designs of twin 6in mountings – Coventry Ordnance Works was no longer in the business. Mock ups were trialled at Whale Island which confirmed the feasibility of the proposed loading arrangements. An order was placed with Vickers in September 1921 for one mounting to be designated Mark XVII. This was planned to replace the two forward single 6in aboard the cruiser *Enterprise*, then fitting out. She had been launched by John Brown in December 1919 but was transferred to Devonport Dockyard for completion, being towed from Clydebank on 9 April 1920. With priority given to other ships completing in the early 1920s, lack of funding slowed work on her, so she was not finally completed until March 1926. *Diomede* had been completed in 1922 with a single 6in in an enclosed Mark XVI mounting as proof of the weatherproof concept.

With 40° elevation, the guns could range to over 20,000 yards, compared with about 14,000 of a casemated single 15° mounting. Hydraulic power from a 50hp electrically driven VSG pump below the turntable powered training, elevating and hoists. The mounting was a so-called 'long trunk' *ie* the hoists extended down four decks to the magazine and shell room in the hold, and rotated with the mounting. Both shell and cartridge hoists were endless chain hoists, so-called dredger hoists somewhat akin to the bucket ladder on a harbour dredger. The guns themselves were the Mark XII, as fitted in battleships and cruisers, so as to be the same as on the ship's five other CP.XIV single mountings, with the same 100lb shell.[1]

[1] *Handbook for 6-inch BL Mark XXII Gun on Mark XVIII Twin Mounting and 6-inch Mark XII Gun on Mark XVII Twin Mounting, 1927.* ADM 186/280

*Rodney* in 1937 showing her three starboard 6in mountings. *(Author's collection)*

The original contract was for £65,000, with payments to be made:

1/5 When 50% of the material required for the construction of the mounting has been delivered at the contractor's works, satisfactorily tested and stamped by the Overseer.
1/5 When 95% of the material required for the construction of the mounting has been delivered at the contractor's works, satisfactorily tested and stamped by the Overseer.
1/5 When the pedestals, recoil presses and pistons have been machined and the cradles and carriages have been rough machined.
1/5 When 95% of the material for the mounting has been delivered [to the ship]
1/5 When the mounting has been erected on board and finally accepted by the Admiralty after examination and proof, in satisfaction of the contract ...

This last payment was claimed on 22 January 1926. The final figure including modifications and extras was £73,579 which included design and erection, resulting in a loss to Vickers of £1872.[2]

A similar turret was designed by Vickers for *Nelson* and *Rodney*, twelve being ordered from Barrow in December 1922 as Contracts 194G and 195G with a final value of £460,000. Two spares were also ordered at the same time. Twelve sets of pusher hoists from the magazines were supplied from Elswick at £70,000.[3] The twin Mark XVIII mountings were sited abreast the mainmast, the lower ones commanding arcs of 150° (forward mounting) and 145° (aft), the superfiring two 180°. The three 16in triples were all forward of the superstructure. Elswick received the more valuable order for their six triple 16in mountings, although DNC Berry in his 1929 INA paper on the ships wrongly credits Elswick with *Nelson*'s 6in mountings as well. As well as firing against surface ships, they could be used to put up an anti-aircraft barrage with time-fuzed shells by elevating up to 60°. These were of the 'short trunk' two stage hoist arrangement with the magazines below the shell rooms on two levels on the centreline aft of the boiler rooms. Angled pusher endless chain hoists delivered to flashtight working spaces at the sides of the middle deck, where the 100lb shells and the 30lb charges in leather Clarkson cases were manually transferred into the hoists which revolved with each gunhouse. The cases were not expendable so had a separate chute to return them to the magazine. Loading was at the fixed angle of 5°, with a telescopic chain rammer. Under favourable conditions 5 rounds per gun per minute could be achieved. While this was slower than the single 6in on pedestal mountings in the *Queen Elizabeth*s, it was more accurate. Hydraulic power was used for training and elevating (in separate cradles), from a VSG oil pump driven by a 50hp electric motor under the gunhouse. The 9-ton 50-calibre guns were the new Vickers designed wire-wound 6in Mark XXII, fitted in just these two ships; only 40 were made including the similar XXII*.[4] 3pdr sub-calibre guns were provided.

[2] Vickers Gun Mounting Contract Ledger 197G. BAE Systems Barrow shipyard archive.
[3] Armstrong Whitworth Elswick accounts 1927. Tyne & Wear Archives. DS.VA/3/3/3
[4] Campbell, John. *Naval Weapons of World War Two*, p37

| **6in BL Mark XXII* Gun** | |
|---|---|
| Calibres, bore | 50 |
| Length overall, in | 309.7 |
| Weight incl breech mechanism, tons | 9.01 |
| Shell weight, lb | 100 |
| Charge, cordite MC26, lb | 30 |
| Muzzle velocity, new gun, ft/sec | 2,945 |
| Life, efc | 700 |

| **6in Mark XVIII Twin Mounting** | |
|---|---|
| Revolving weight, tons | 85 |
| Max elevation, degrees | 60 |
| Max depression, degrees | -5 |
| Elevating, deg/sec | 8 |
| Training, deg/sec | 5 |
| Roller path diameter, ft | 14 |
| Rounds per gun per minute, max | 5 |
| Magazine capacity, rounds per gun | 150 |
| Range at 45°, yards | 25,000 |

Four director control towers were fitted for low angle fire with 12ft rangefinders, two of which could also control high angle fire via a HACS Mark III system.[5] *(See colour plates on page 166)*

The gunshields had 1½in front D quality steel, 1in the rest.

*Nelson*'s mountings were delivered from Barrow to the Tyne in Robertson's 712 grt coaster *Sagenite.* The log of *Titan II* floating crane noted: 'March 19 1926: Lifted from steamer *Sagenite* onto jetty at Naval yard 3 x 33-ton barbettes (*sic*) and 1 x 8-ton ammunition trunk. April 9: Lifted from Naval yard jetty 3 x 33-ton barbettes and 2 x 8-ton ammunition trunks and placed aboard *Nelson*.'[6] As the revolving weight of the complete mounting was about 85 tons, this suggests that the two guns with cradles weighing 13 tons each were shipped separately. Armstrong's Naval Yard's 250-ton cantilever crane was not built until 1931, so *Titan II* had to be hired. As the crane was also used for *Nelson* on at least 58 other days lifting 16in components up to 165 tons, boilers, armour plates and smaller fittings, no doubt Craven (the managing director) pointed to the savings that could be made with their own crane in building future battleships.

With Armstrong Whitworth receiving the contract for *Nelson* and *Rodney*'s triple 16in mountings, the Admiralty had wisely allocated their twelve twin 6in mountings to Vickers. Barrow's contracts 194G and 195G were dated 12 December 1922, the day after the ships were ordered. The contract price was £28,000 each, with erection at £3,600 each.

The actual costs for *Nelson* for six mountings claimed from the Admiralty were:

| *Item* | £ |
|---|---|
| Turntables | 34,811 |
| Training gear | 21,630 |
| Elevating gear | 14,490 |
| Cradles | 20,760 |
| Loading gear | 31,242 |
| Gun shields | 21,042 |
| Fittings | 9,780 |
| Electrical pumping plant | 14,370 |
| Design | 14,758 |
| Delivery | 3,540 |
| Erection | 21,600 |
| **Total** | **208,023** |

The last instalment was claimed on 13 October 1927. Extras and later price adjustments brought the final total to £237,465, yielding a profit of £16,432 if half the design costs were allocated to *Rodney,* about 7.5% on cost. *Rodney*'s total price to the Admiralty came to £222,093 without design costs. Her last claim was made on 5 December 1927, resulting in a profit of £35,583, thus benefiting from the learning curve. Average price over twelve mountings was £38,296, or about £570 per ton of revolving weight excluding guns.[7]

*Rodney* used her 6in to good effect in the sinking of *Bismarck*, firing 706 rounds, and at Normandy in shore bombardment with 454.[8] Both *Rodney*'s and *Nelson*'s mountings were removed at Rosyth to reduce their draft aft before they were towed to the berth at Ward's shipbreaking yard at nearby Inverkeithing in 1948–9.

## CRUISER TWIN MOUNTINGS

The 8in cruisers were expensive, with two of the 1928 Programme cancelled to 'save' money; 6in were regarded as better value, and more

*Nelson* and *Rodney* each carried six twin 6in Mark XVIII mountings, Vickers having received the contract to build them in December 1922. This view in Bay 13 at Barrow shows several in course of erection, the nearest at 60° elevation. *(Author's collection)*

[5] Jordan, John,'The 6in Turrets of *Nelson* and *Rodney*'. *Warship 2020*, p184

[6] *Titan II* Crane Log, 1926. Marine Technology Special Collection 01250/8

[7] Vickers Gun Mounting Contract Ledger 194G

[8] Ballantyne, Ian. *HMS Rodney*. Pen & Sword Books, 2008, p145

could be built within any treaty tonnage or budgetary constraints. Work was put in hand to design a new 6in gun, the Mark XXIII. It was of 50 calibres bore length, comprised of just an A tube, jacket and breech ring, *ie* no wire winding, weighing 6.9 tons with an Asbury breech mechanism. A rear balance weight brought the centre of gravity in line with the trunnions, which also housed the recoil and recuperator (run-out) cylinders. Since manual loading was no longer required, it could fire a heavier 112lb shell (100lb only in *Leander*) with a lower muzzle velocity with improved ballistics and a better armour penetration and longer life. Guns were interchangeable left and right, although the breech mechanism needed adjusting.

A new twin mounting would be needed. *Enterprise*'s twin 6in was reported on favourably as were *Nelson* and *Rodney*'s, so formed the basis of the Mark XXI designed by Elswick in 1929 which armed the *Leander* class cruisers, intended for trade protection and scouting for the Fleet. This would be a 'short trunk' design, *ie* the upper hoists would rotate with the gunhouse, but the lower hoists from the magazine would be fixed and deliver to a flashtight ammunition lobby to be transferred to the upper hoists. They were designed to deliver eight rounds per gun per minute. Although the general layout of the gunhouse was similar to the Mark XVIII, significant differences included: hand ramming which allowed loading at angles between 12½° and maximum depression -5°, cordite in cardboard containers which were disposed via chutes in the rear of the gunhouse. A 65hp electric motor with hydraulic oil pump powered the elevating (up to 60°) and training mechanisms and endless chain hoists. Gunshield thickness was 1in splinter-proof steel.[9]

A prototype was ordered from Elswick (now part of Vickers-Armstrongs) followed by an order on 12 May 1930 for four mountings for *Leander*, which was the only 6in cruiser ordered under the 1929–30 Programme. The contract signed by Director of Contracts W St D Jenkins specified 'the detailed design, manufacture, trials, delivery and erection on board at Devonport Yard of the complete 6-inch gun machinery installation for a cruiser, together with the supply and delivery to a Home Yard of one Reserve Mounting…' The guns were ordered separately. As the specification had not yet been finalised, no prices were listed. Stage delivery dates included:

| | | |
|---|---|---|
| Preliminary pit trial of first (advance) mounting | | 1.9.31 |
| Pit trials of component mountings: | Nº 1 | 1.3.32 |
| | Nº 4 | 15.5.32 |
| Delivery of last revolving structure | | 15.6.32 |
| Ready for gun trials | | 1.10.32 |
| Completion in all respects in readiness for taking over | | By end Dec 1932 |

V-A had to report when the first cradle was ready for proof (firing at Ridsdale) and the earliest dates by which the first two guns were required for proof of mounting and erection purposes. Any material prepared for the 8in mountings of *Surrey* and *Northumberland* (now

The *Leander* class cruisers each mounted four twin 6in Mark XXI. Those for *Ajax* (Vickers contract 245G) are shown being erected in Bay 13 at Barrow in 1934. They cost £58,500 each. *(Dock Museum 4675)*

cancelled) should be utilised for this order where possible – wishful thinking as there was little in common between an 8in and a 6in mounting.[10] *(See colour plates on page 167)*

Barrow received its first order on 15 December 1930 for four mountings for *Achilles* (230G) building at Cammell Laird and two spares (231G). Four for *Ajax* building at Barrow were ordered on 7 May 1932 (245G) for completion in March 1935. The final price agreed with the Admiralty included £16,500 for erection, £18,178 for 79 tons of gunhouse armour and £22,232 for director firing gear of which £8907 was for Crayford who made some of the instruments. The three mountings for *Arethusa*, also for the 1931 Programme, (246G) were ordered from Barrow the same day (7 May 1932), four months before the hull from Chatham Dockyard. Average price for each mounting for both ships was about £59,000 or about £740 per ton. That price excluded the guns, which were ordered separately by the Admiralty and supplied to V-A as a free issue for fitting to each mounting.

The revolving parts of the four mountings were the same, but the fixed hoists were varied in length to suit the height of the turret. The weight including hoists varied from 85.9 tons for Y, the shortest, to 94.2 tons for B, the longest, plus 13.8 tons for the guns.[11] The four other *Leander*s and three *Amphion*s (all sold to Australia) all had four Mark XXIs. The *Arethusa* class were smaller, to increase the number of cruisers within Treaty tonnage limits, so had only three 6in mountings each. The total crew including ammunition handling for each mounting was about 44, about 16 of whom were in the gunhouse. Nominal magazine capacity was 200 rounds per gun.

[9] *Handbook for 6-inch BL Mark XXIII Gun on Twin Mark XXI Mounting, 1932.* ADM 186/310 & 311

[10] *Leander* class Ships Cover. NMM

[11] Correspondence with Peter Cannon 2020–23

After *Achilles* was broken up in India as *Delhi*, her Y turret was installed at the Royal New Zealand Navy's Devonport base near Auckland. *(Author)*

The breech of *Achilles*' preserved left-hand gun. Engraved at the top is LEFT TOP POSITION B.L. 6-INCH STEEL MARK XXIII V.A. 1939 N. NO. S/4917. The gun was made at Elswick in 1939, the S signifying Ship, as opposed to L for Land artillery. The breech mechanism has R.G.F. 1944, *ie* Woolwich. *(Pete Cannon)*

Unfortunately, no Admiralty gun mounting registers have been found, so details are not available for the allocation or disposal of all 49 mountings. But those for the Australian ships are known as are some Barrow ones, so the table opposite shows the most likely.

*Orion*'s A mounting received a direct hit off Crete in May 1941 and was presumably replaced by one of the spares at her repair at Mare Island Navy Yard, completed in February 1942. Barrow completed reconditioning one mounting from *Ajax* in the third quarter of 1942, presumably removed after battle damage in the Mediterranean in 1941 and replaced by one of the spares. One mounting from *Aurora* was reported by V-A as being reconditioned at Barrow in 1943, presumably A or B which were damaged when she was mined in the Mediterranean in December 1941. *Achilles*' Y mounting is preserved at Devonport Dockyard, Auckland.

After in-depth research, Peter Cannon concluded that the mountings were quite reliable and well liked in service. Intermittent faults included

**Twin 6in Mark XXI Mountings**

| Reg No | Maker | Ship | Ordered | Delivered |
|---|---|---|---|---|
| 1 | Elswick | *Leander* X | May 1930 | 4Q 1932 |
| 2 | Elswick | *Leander* | May 1930 | 4Q 1932 |
| 3 | Elswick | *Leander* | May 1930 | 4Q 1932 |
| 4 | Elswick | *Leander* | May 1930 | 4Q 1932 |
| 5 | Elswick | Spare? | May 1930 | 1933 |
| 6 | Elswick | *Neptune*? | Dec 1930 | 3Q 1933 |
| 7 | Elswick | *Neptune*? | Dec 1930 | 3Q 1933 |
| 8 | Elswick | *Neptune*? | Dec 1930 | 3Q 1933 |
| 9 | Elswick | *Neptune*? | Dec 1930 | 3Q 1933 |
| 10 | Elswick | *Orion*? | Dec 1930 | 2Q 1933 |
| 11 | Elswick | *Orion*? | Dec 1930 | 2Q 1933 |
| 12 | Elswick | *Orion*? | Dec 1930 | 2Q 1933 |
| 13 | Elswick | *Orion* A or B | Dec 1930 | 2Q 1933 |
| 14 | Barrow | *Achilles*? | Dec 1930 | 3Q 1933 |
| 15 | Barrow | *Achilles* Y | Dec 1930 | 3Q 1933 |
| 16 | Barrow | *Achilles*? | Dec 1930 | 3Q 1933 |
| 17 | Barrow | *Achilles*? | Dec 1930 | 3Q 1933 |
| 18 | Barrow | Spare? | Dec 1930 | 1933 |
| 19 | Barrow | Spare? | Dec 1930 | 1933 |
| 20 | Elswick | *Amphion/Perth* X | May 1932 | End 1934 |
| 21 | Elswick | *Amphion/Perth* Y | May 1932 | End 1934 |
| 22 | Elswick | *Amphion/Perth* A | May 1932 | End 1934 |
| 23 | Elswick | *Amphion/Perth* B | May 1932 | End 1934 |
| 24 | Barrow | *Ajax*? | May 1932 | End 1934 |
| 25 | Barrow | *Ajax*? | May 1932 | End 1934 |
| 26 | Barrow | *Ajax*? | May 1932 | End 1934 |
| 27 | Barrow | *Ajax*? | May 1932 | End 1934 |
| 28 | Barrow | *Arethusa*? | May 1932 | End 1934 |
| 29 | Barrow | *Arethusa*? | May 1932 | End 1934 |
| 30 | Barrow | *Arethusa*? | May 1932 | End 1934 |
| 31 | Elswick | Spare | 1932 | 1934 |
| 32 | Elswick | Spare | 1932 | 1934 |
| 33 | Elswick | *Phaeton/Sydney* X | Dec 1932 | 2Q 1935 |
| 34 | Elswick | *Phaeton/Sydney* Y | Dec 1932 | 2Q 1935 |
| 35 | Elswick | *Phaeton/Sydney* A | Dec 1932 | 2Q 1935 |
| 36 | Elswick | *Phaeton/Sydney* B | Dec 1932 | 2Q 1935 |
| 37 | Barrow | *Galatea* | Dec 1932 | Early 1935 |
| 38 | Barrow | *Galatea* | Dec 1932 | Early 1935 |
| 39 | Barrow | *Galatea* | Dec 1932 | Early 1935 |
| 40 | Barrow | *Apollo/Hobart* A | Dec 1932 | Mid 1935 |
| 41 | Barrow | *Apollo/Hobart* Y | Dec 1932 | Mid 1935 |
| 42 | Elswick | *Apollo/Hobart* X | Dec 1932 | Mid 1935 |
| 43 | Elswick | *Apollo/Hobart* B | Dec 1932 | Mid 1935 |
| 44 | Barrow | *Penelope* | Dec 1933 | Mid 1936 |
| 45 | Barrow | *Penelope* | Dec 1933 | Mid 1936 |
| 46 | Barrow | *Penelope* | Dec 1933 | Mid 1936 |
| 47 | Barrow | *Aurora* | Dec 1934 | Mid 1937 |
| 48 | Barrow | *Aurora* | Dec 1934 | Mid 1937 |
| 49 | Barrow | *Aurora* | Dec 1934 | Mid 1937 |

| **6in BL Mark XXIII Gun** | |
|---|---|
| Calibres, bore | 50 |
| Length overall, in | 309.8 |
| Weight incl breech mechanism, tons | 6.9 |
| Shell weight, lb | 112 |
| Charge, cordite SC150, lb | 30 |
| Muzzle velocity, new gun, ft/sec | 2750 |
| Life, efc | 1020 |

| **6in Mark XXI Twin Mounting** | |
|---|---|
| Revolving weight, tons | 93 |
| Max elevation, degrees | 60 |
| Max depression, degrees | -5 |
| Roller path diameter, ft | 14 |
| Gunshield thickness, in | 1 |
| Rounds per gun per minute, max | 8 |
| Maximum range, yards | 24,800 |

sluggish run-outs, sticky obturator pads and the like after prolonged firings, which had to be managed by the gun crews and ordnance maintenance staff. With continuous firing, there could be occasional hoist breakdowns but the remaining hoists and/or auxiliary hoisting arrangements managed to keep up satisfactory rates of supply.[12]

### TRIPLE 6-INCH GUN MOUNTINGS

With the 7000-ton *Leander*s and 5200-ton *Arethusa*s under way, the Admiralty addressed the issue of some other navies, including the Japanese, having much larger cruisers with fifteen 6in guns. It reckoned that a ship of about 9000 tons could mount twelve 6in, be better protected and be primarily used for trade protection. A triple mounting would be feasible in a larger ship, with the ammunition supply using the previous short trunk system. By positioning the centre gun 2ft 6in behind the wing guns, there would be less congestion between the three loading crews and less interference in flight between three shells, as well as moving the centre of gravity nearer the centre of rotation.

In 1933, design studies were put in hand by Vickers-Armstrongs, with most of work being done at Elswick, as Barrow was more focused on anti-aircraft mountings. Meanwhile, various ship options were undertaken by DNC department, but it was not until 1 May 1934 that orders were placed for the first two of what were initially called the *Minotaur* class, later *Southampton* but more generally known as the 'Towns'. V-A had already been awarded the contract on 19 December 1933 for eight triple mountings, designated Mark XXII, plus one spare. Those for *Newcastle* building on the Tyne were allocated to Elswick, while those for *Southampton* building at Clydebank were allocated to Barrow (Contract 265G), continuing the general practice of allocating mountings for east coast shipyards to Elswick and west coast to Barrow. Prices including director firing gear were yet to be agreed; there was now no competition or tenders, V-A being the only supplier. Delivery was required in 2¾ years.[13] *(See colour plates on page 168)*

The guns were to be Mark XXIII, already fitted in the *Leander*s, capable of ranging to 24,800 yards at 44° elevation. Maximum elevation was reduced to 45°, depression -5°, with hand loading between 12½° and -5°. Designed rate of fire was 8 rounds per gun per minute (compared with 4 for the 8in), mostly firing CPBC (common pointed ballistic cap) but also high explosive (HE) and semi-armour piercing (SAP), with 200 rounds per gun stowed, plus some practice. The

[12] *Ibid*

[13] *Southampton* class Ships Cover. NMM

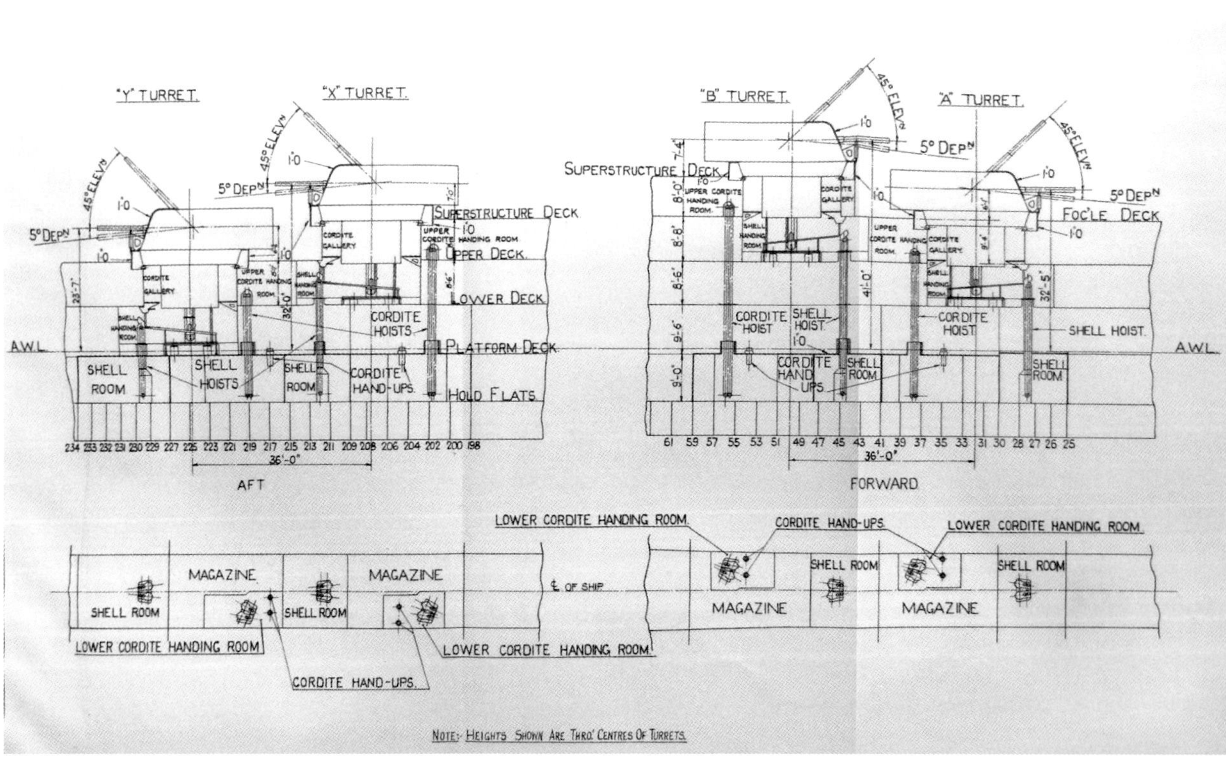


Four triple 6in Mark XXII mountings were fitted in the *Southampton* class cruisers. The layout plan shows the short trunk design with the magazines in armoured boxes. The fixed shell hoists delivered to a handing room one deck below the gunhouse, the cordite hoists two decks below, thence to the rotating mounting. *(IWM LBY 30581-1)*

Army as well as the Navy used the gun, Campbell giving 469 made.[14]

Both magazines and shell rooms were at hold level in armoured boxes with 80lb non cemented on the roof and on the sides 180lb (4.5in) cemented upper strake and 120lb NC lower strake. From there fixed endless chain hoists delivered to ammunition lobbies two decks below the gunhouse floor (a double hoist for shells) and one deck below

[14] 'British Naval Guns 1880–1945' Nº 13. *Warship* 30

*Southampton*'s triple 6in Mark XXII Y mounting being installed at John Brown's yard on 3 July 1936. *(UCS 118-542-28)*

*Belfast* and *Edinburgh* were the first cruisers to complete with four triple 6in Mark XXIII. *Belfast's* original Y was taken to replace *Trinidad's* A lost in transit to Devonport on 23 February 1941. She is seen firing all guns after her 1959 modernisation. She fired 1996 6in rounds at Normandy in June 1944. *(Author's collection)*

(charges). The hoists were enclosed in 1in D.1 steel (2in NC in last three ships). In these lobbies both shells and charges were transferred manually to the hoists on the revolving part of the mounting. Gunhouse protection was a meagre 40lb (1in) NC on front, sides and roof (the same as the 'Counties' 8in), but protection below was 80lb D.1 to the side and 40lb D.1 to the less vulnerable forward and aft segments. NC armour was much more expensive than high tensile D.1 plating, which cost about £26 per ton.

Like the earlier 6in mountings, Mark XXII had a 103hp electric motor driving a hydraulic pump using mineral oil fitted beneath the turntable. This powered the training and elevating gears and upper shell hoist. The cordite hand-up tube from the working chamber into the gunhouse was short with charges light enough to be manually handled. Air bottles for the gun run-out mechanisms and clearing the bores at 1250lb/sq in were stored at shell transfer level. In front of the gunhouse was safety depression gear which would prevent any gun being depressed to levels which endangered the structure. Training buffer stops prevented the mounting training more than 145° either side of the centreline.[15]

Orders for the next three followed in December 1934, the same month that the hulls were ordered. Barrow received the order for *Glasgow* (276G) at the same time an order for one spare (277G), while Elswick got *Sheffield* and *Birmingham*. A provisional price for mounting components of £227,000 was used for calculating progress payments. Included in the *Glasgow* contract was £3200 for delivery to the Clyde (March 1937), £18,000 for erection on board and £18,150 for director firing gear. Non cemented armour on the gunhouse and around the revolving hoists was priced at £192.8 per ton, 110 tons from English Steel Corporation at Sheffield amounting to £21,300. The final price including extras, spares, as fitted drawings and adjustments came to £278,521 was not settled until August 1942, five years after ship completion.[16]

The next three from the 1935 Programme included thicker gunhouse armour, as well as a second director control tower aft. With a more angular gunhouse shape, the front was 160lb (4in) cemented, with 80lb NC on sides and roof. Including a thicker gunhouse floor (1½in vs ¾in), 35 tons was added to the revolving weight, now about 165 tons. On 4 November 1935, *Liverpool's* four were ordered from Barrow (299G), with *Manchester* and *Gloucester* from Elswick, the same month as the hulls. Assuming three spare mountings, that made a total of 35 Mark

[15] *Handbook for 6-inch BL Mark XXIII Gun on Triple Mark XXII Mounting, 1937.* Imperial War Museum LBY 30581

[16] Vickers Gun Mounting Contract Ledger 276G

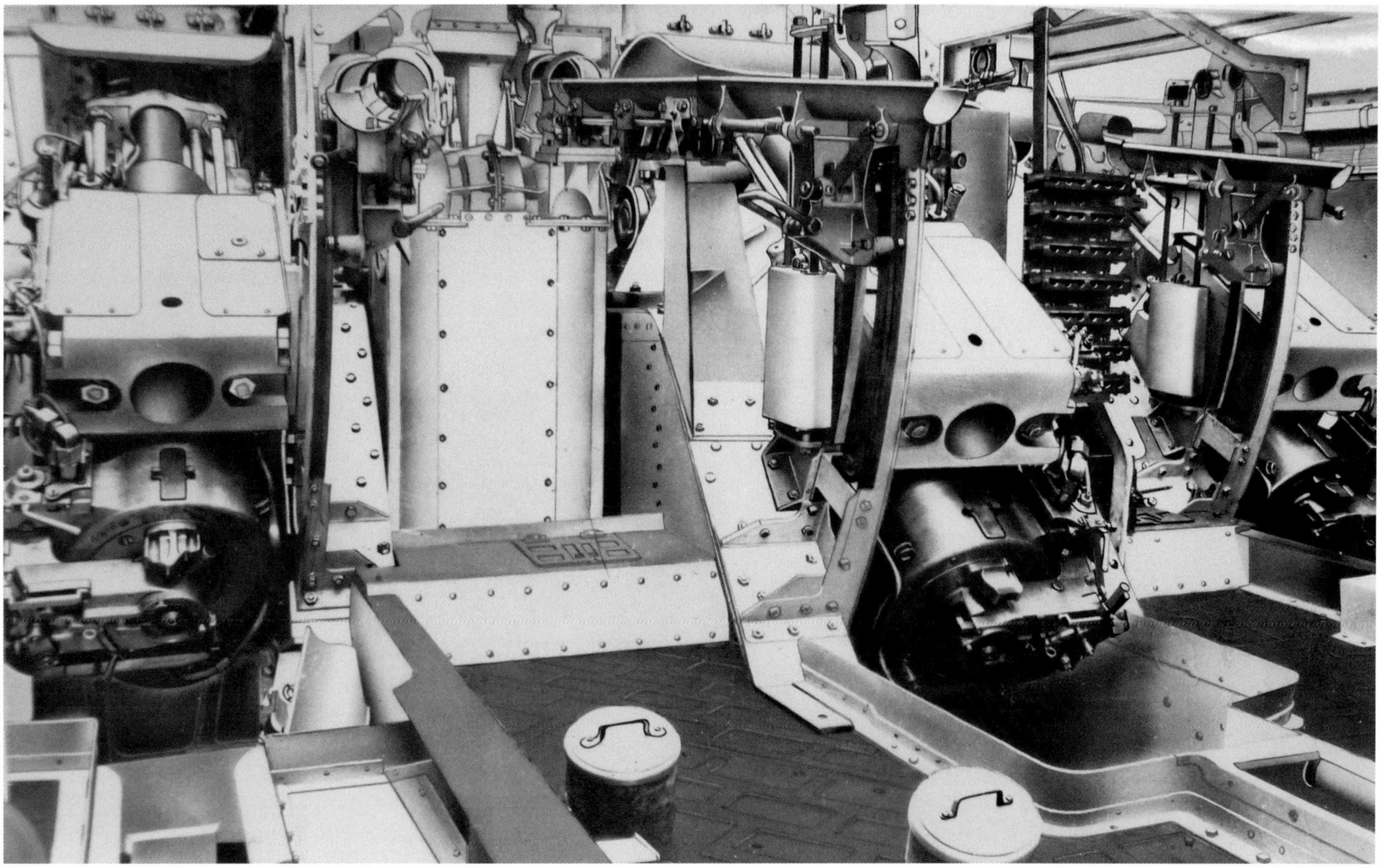

**Above:** Inside a new triple 6in Mark XXIII gunhouse at Elswick in October 1939. Cordite hand-ups at the bottom, shell hoists between the guns, balance weight and recuperator above gun. *(Tyne & Wear Archives).*

**Right:** Each 6-inch mounting carried a tally plate with its Registered Number and weight, this one from *Mauritius'* A mounting. Each cradle also had a plate, this one from her left gun. *(Author)*

XXII altogether. Completion of the mountings took three years but did not delay ship delivery, although some directors were not fitted at completion. The 1939 edition of *Progress in Naval Gunnery* reported that the 'Mark XXII mountings have proved to be extremely reliable in service'.[17]

When *Liverpool*'s A mounting was badly damaged and she lost her bow after being torpedoed in the Mediterranean in October 1940, she was repaired at Mare Island Navy Yard. Her whole fore end was rebuilt and her A mounting was replaced, probably by one of the spares in 1941. One of *Sheffield*'s mountings, probably Y damaged by a mine in March 1942, was under repair at Barrow early in 1943.

Barrow received an order from Argentina in September 1935 for a 6500-ton cruiser/training ship, with three triple 6in mountings broadly similar to the British ones. The mountings were ordered from Barrow in November 1935 under contract 293G, valued at £184,000 excluding

[17] *Progress in Naval Gunnery, 1939*. ADM 239/137

Three views of installing *Bermuda*'s 6in mountings at Clydebank. **Above:** One mounting about to be lifted out of its support frame by John Brown's 150-ton derrick crane. It had been delivered from Barrow by coaster *Empire Jack* early in 1942. The two Scotch boilers may be for refrigerated cargo vessel *Hororata. (UCS118-568-126)*

*Bermuda*'s A mounting (Reg Nº 37) being lifted into its barbette on 11 May 1942, without guns and roof plates to keep below the crane's 150 tons capacity. *(NRS)*

### 6in 50-calibre QF Guns for Argentina

Bore length 300in. Total 314.55in.
Weight of gun and breech mechanism 6 tons 10 cwt 2 qtr. Replaceable barrel 2t 0c 1q.
Muzzle velocity 2953ft/s. Working pressure 19 tons/sq in.
100lb projectile SAP 4lb burster TNT. HE 10.75lb. Cartridge 69lb.
45° elevation, -7° depression.
Estimated range 25,100 yards at 45°.
Penetration with SAP 3in NC 25° to normal at 1350ft/sec.
3 47mm sub-calibre guns.

guns, with Barrow, Elswick and Crayford supplying the rest of the armament. *La Argentina* was delivered in January 1939, a year late for which V-A paid a penalty of £82,780 reducing the overall value of the contract to £1,365,776. [18]

For the 1936 Programme, it was planned to go right up to the 10,000-ton standard displacement Treaty limit. Initially quadruple 6in mountings were considered, but a better balanced ship design could be achieved by retaining four triple 6in. Designed as 'long trunk' with the rotating hoists going right down to magazine level, revolving mountings would be heavier than short trunks but require smaller crews by eliminating the ammunition lobby transfer crew. The new mounting was designated Mark XXIII with Mark XXIII guns, retaining the thicker gunhouse protection of the last three 'Towns'. With machinery spaces further aft compared with the 'Towns' and the 4in magazine

[18] Vickers Gun Mounting Contract Ledger 293G

*Bermuda*'s X and Y mountings installed by mid-1942, the former with control position and rangefinder on top. *(UCS 118-568-144)*

forward, X and Y mountings had to be raised one deck (effectively the same level as the flush deck 'Counties') to provide space for shell rooms which were above the magazines. The 180lb C side belt was extended to cover the magazine spaces, with 80lb NC deck over. Hoist trunks were protected by 80lb NC (sides) and 40lb (fore and aft sectors).

*Belfast*'s mountings (321G) were ordered from Barrow on 10 June 1936 with Registered Numbers 1-4, with *Edinburgh*'s four from Elswick. A replacement Y mounting for *Belfast* was ordered from Barrow (391G) in May 1941 after her original Y was removed for fitting in place of *Trinidad*'s A lost when *Shoal Fisher* was mined en route from Barrow to Devonport; delivered in July 1942. Crew averaged 46 per mounting.

*Belfast* has been a museum ship in the Pool of London since 1971, where her gun mountings can be seen. Wingate's book has a good working description and drawings.[19] Johnstone-Bryden's has excellent photos of her gunhouse and shell rooms.[20] *(See colour plates on page 169)*

## THE 'COLONIES'

The 1936 London Treaty removed limits on the number of cruisers that could be built but set an upper limit of 8000 tons standard displacement. During 1936, the Admiralty explored a number of designs of 8000 tons, some with 6in and some with 5.25in guns. The design adopted had four triple 6in Mark XXIII mountings, whose long trunk design had the advantages of removing the transfer space, giving a reduced crew and a faster rate of ammunition supply.[21] Twenty mount-

[19] Wingate, John. *HMS Belfast 1939–72*. Profile Publications, 1972

[20] Johnstone-Bryden, Richard. *HMS Belfast Cruiser 1939*. Seaforth Publishing, 2013

[21] *Handbook for 6-inch BL Mark XXIII Guns on Triple Mark XXIII Mounting, 1939*. ADM 234/191

ings were ordered from V-A at the end of 1936 before the ship design had been finalised, with a provisional price of £203,000 per ship – with 6in mountings plus guns that made up about 12% of a cruiser's cost. Eight progress instalments of 10% each were to be paid when specified amounts of material had been received and when it had been machined. The ninth instalment was paid when 95% had been delivered and the final 10% after satisfactory trials.[22] The five ships were ordered in December 1937: *Fiji, Nigeria, Mauritius, Kenya* and *Trinidad.* Differences with *Belfast* included thinner gunhouse armour to save weight and an armoured control position fitted on top of X mounting but retaining the same nominal 200 rounds per gun, a mix of SAP and HE plus practice. Front and roof was 2in NC, sides and rear 1in NC, about 37 tons weight each for A, B and Y and 43 tons for X, usually from Colvilles or English Steel Corporation. Hoist protection was reduced to 1in NC sides, ¾in NC fore and aft.

The 1938 Programme ordered four more ships, eight mountings ordered from Barrow and eight from Elswick in September 1938, six months before the hulls were ordered, which not only reserved a place in V-A's production programme but enabled the company to recruit

[22] Contract and purchase papers: Gun mounting gear. ADM 116/5310

### 6in Mark XXIII Triple Mounting

| | |
|---|---|
| Revolving weight, tons | 155 |
| Max elevation, degrees | 45 |
| Max depression, degrees | -5 |
| Roller path diameter, ft | 19 |
| Recoil, in | 16.5 |
| Rounds per gun per minute, av | 6 |
| Maximum range at 45°, yards | 24,800 |

### Triple 6in Mark XXIII Mountings

| Reg Nº | Maker | Ship | Ordered | Delivered | Notes |
|---|---|---|---|---|---|
| 1 | Barrow | *Belfast* | Jun 1936 | 4Q 1938 | |
| 2 | Barrow | *Belfast* | Jun 1936 | 1Q 1939 | |
| 3 | Barrow | *Belfast* | Jun 1936 | 1Q 1939 | |
| 4 | Barrow | *Belfast.* Y? | Jun 1936 | 1Q 1939 | To *Trinidad* 1941 |
| 5 | Elswick | *Edinburgh* | Jun 1936 | 4Q 1938 | |
| 6 | Elswick | *Edinburgh* | Jun 1936 | 4Q 1938 | |
| 7 | Elswick | *Edinburgh* | Jun 1936 | 4Q1938 | |
| 8 | Elswick | *Edinburgh* | Jun 1936 | 1Q 1939 | |
| 9 | Elswick | *Fiji.* B? | 3Q 1936 | Sep 1939 | Delivered in *Shoal Fisher* |
| 10 | Elswick | *Fiji* | 3Q 1936 | 3Q 1939 | |
| 11 | Elswick | *Fiji* | 3Q 1936 | 3Q 1939 | |
| 12 | Elswick | *Fiji* | 3Q 1936 | Dec 1939 | Delivered in *Shoal Fisher* |
| 13 | Elswick | *Nigeria* | Dec 1936 | Jan 1940 | |
| 14 | Elswick | *Nigeria* | Dec 1936 | Feb 1940 | |
| 15 | Elswick | *Nigeria* | Dec 1936 | May 1940 | |
| 16 | Elswick | *Nigeria* X | Dec 1936 | Jun 1940 | Removed *c*1955 before becoming *Mysore* |
| 17 | Elswick | *Mauritius* A | Dec 1936 | Oct 1940 | |
| 18 | Elswick | *Mauritius* Y | Dec 1936 | Jul 1940 | |
| 19 | Elswick | *Mauritius* B | Dec 1936 | Aug 1940 | |
| 20 | Elswick | *Trinidad* X | Dec 1936 | 2Q 1941 | Ordered for *Mauritius* |
| 21 | Barrow | *Kenya.* Y? | Dec 1936 | 4Q 1939 | |
| 22 | Barrow | *Kenya.* B? | Dec 1936 | 4Q 1939 | |
| 23 | Barrow | *Kenya* X | Dec 1936 | 1Q 1940 | Removed at Chatham 1945 |
| 24 | Barrow | *Kenya* A | Dec 1936 | 1Q 1940 | |
| 25 | Barrow | *Trinidad* Y | Dec 1936 | 1Q 1941 | |
| 26 | Barrow | *Trinidad* B | Dec 1936 | 1Q 1941 | |
| 27 | Barrow | *Mauritius* X | Dec 1936 | Aug 1940 | Ordered for *Trinidad.* Removed at Birkenhead 1945 |
| 28 | Barrow | *Trinidad* A | Dec 1936 | Feb 1941 | Lost in *Sea Fisher* 23.2.41 |
| 29 | Elswick | *Gambia* A | Sep 1938 | Jun 1941 | |
| 30 | Elswick | *Gambia* Y | Sep 1938 | Aug 1941 | |
| 31 | Elswick | *Gambia* B | Sep 1938 | Oct 1941 | |
| 32 | Elswick | *Gambia* X | Sep 1938 | Nov 1941 | Removed at Devonport 1949 |
| 33 | Barrow | *Jamaica* A | Sep 1938 | Oct 1941 | |
| 34 | Barrow | *Jamaica* B | Sep 1938 | 4Q 1941 | |
| 35 | Barrow | *Jamaica* X | Sep 1938 | 4Q 1941 | Removed at Portsmouth 1945 |
| 36 | Barrow | *Jamaica* Y | Sep 1938 | 1Q 1942 | |
| 37 | Barrow | *Bermuda* A | Sep 1938 | May 1942 | Planned for *Ceylon.* Delivered in *Empire Jill* |
| 38 | Barrow | *Bermuda* B | Sep 1938 | Mar 1942 | Planned for *Ceylon.* Delivered in *Empire Jack* |
| 39 | Barrow | *Bermuda* X | Sep 1938 | Mar 1942 | Planned for *Ceylon.* Delivered in *Empire Jack.* Removed at Clydebank 1944. |
| 40 | Barrow | *Bermuda* Y | Sep 1938 | Mar 1942 | Planned for *Ceylon.* Delivered in *Empire Jack* |
| 41 | Elswick | *Uganda* Y | Sep 1938 | 2Q 1942 | Suspended 1940 |
| 42 | Elswick | *Uganda* A | Sep 1938 | 2Q 1942 | Suspended 1940 |
| 43 | Elswick | *Uganda* B | Sep 1938 | 3Q 1942 | Suspended 1940 |
| 44 | Barrow | *Ceylon* A | Sep 1938 | Dec 1942 | Planned for *Uganda.* Suspended 1940 |
| 45 | Barrow | *Ceylon* B | May 1939 | Jan 1943 | Planned for *Bermuda.* Suspended 1940 |
| 46 | Barrow | *Ceylon* Y | May 1939 | Jan 1943 | Planned for *Bermuda.* Suspended 1940 |
| 47 | Elswick | *Newfoundland* Y | May 1939 | Aug 1942 | Planned for *Bermuda.* Suspended 1940 |
| 48 | Elswick | *Newfoundland* B | May 1939 | 3Q 1942 | Planned for *Bermuda.* Suspended 1940 |
| 49 | Elswick | *Newfoundland* A | May 1939 | 3Q 1942 | |
| 50 | Elswick | *Swiftsure* | Apr 1941 | Oct 1943 | Planned for *Newfoundland* |
| 51 | Elswick | *Swiftsure* | Apr 1941 | Jan 1944 | Planned for *Newfoundland* |
| 52 | Elswick | *Swiftsure* Y | Apr 1941 | Feb 1944 | Planned for *Newfoundland* |
| 53 | Barrow | *Ontario* A | Apr 1941 | Aug 1944 | Planned for *Tiger.* All guns RGF |
| 54 | Barrow | *Ontario* B | Apr 1941 | Sep 1944 | Planned for *Tiger.* All guns RGF |
| 55 | Barrow | *Ontario* Y | Apr 1941 | Jun 1944 | Planned for *Tiger.* All guns RGF |
| 56 | Elswick | *Superb* | 4Q 1941 | Nov 1944 | |
| 57 | Elswick | *Superb* | 4Q 1941 | Dec 1944 | |
| 58 | Elswick | *Superb* | 4Q 1941 | Jan 1945 | |
| 59 | Barrow | *Belfast* Y | May 1941 | Jul 1942 | Replacement for Nº 4 |

Derived from gun trial reports and Vickers gun mounting ledgers.

more men with a secure order book. In May 1939 sixteen more were ordered with Registered N^os^ 45–60, but in the event only two ships were actually ordered (*Bermuda* and *Newfoundland*) so the later production slots were re-allocated in 1941. With priority in 1940 being given to ships which could be completed quickly, mountings for *Uganda, Ceylon* and *Newfoundland* were suspended. When the ships were resumed they had only three mountings each, resulting in re-allocations as shown in the table. The contracts were also modified, with payments revised, re-timed and split between Barrow and Elswick. With mountings being switched between ships and the usual long delays in final settlement of claims, the resulting prices recorded at Barrow and Elswick are confusing. The best that can be estimated is that a typical price for a war-built mounting was about £80,000, of which armour made up about 10%, plus per ship a further £10,000 for Barrow making one director tower and Crayford £14,000 for the fittings. So with a revolving weight of about 155 tons less 21 tons for guns, say £600 per ton.

Elswick delivered the mounting components by barge to the nearby Tyne shipyards, but Barrow used coasters to deliver them to the Clyde yards. Coaster *Shoal Fisher* delivered *Fiji*'s B mounting on 11 September 1939 from Barrow, leaving Clydebank the next day.

One of *Uganda*'s three triple 6in Mark XXIII erected in a pit in Nº 7 Shop at Elswick on 23 February 1942. The roof plates are only tack bolted as they and the guns will be removed for separate transport and erection at the ship at Swan Hunter's shipyard. *(Vickers)*

To reduce top weight and increase anti-aircraft armament, during 1944–45 or during post-war refits, X mounting was removed from all the cruisers with triples (at a cost of about £3200) except *Belfast,* whose stability had been improved by bulging. Post-war, *Belfast's* 1959 modernisation included fitting RP.40 remote control.

For Mark XXIII gun see page 130.

## TRIPLE 6-INCH GUN MOUNTINGS MARK XXIV

The 1941 Programme saw six 6in cruisers ordered all with Mark XXIII mountings, but they had a somewhat chequered history as the war progressed and priorities changed. Removing X turret from *Uganda, Ceylon* and *Newfoundland* as built had in theory enabled their production slots, usually the bottleneck in building ships with heavy gun mountings, to be allocated to a new three-mounting ship, *Swiftsure,* ordered in May 1941 at V-A's Walker yard, with an Elswick mounting. Two similar were ordered in May 1941, *Bellerophon* (later *Tiger*) at Clydebank and *Minotaur* (later *Ontario*) at Belfast, both with Barrow mountings. In December three more were ordered: *Superb* at Swan Hunter, *Tiger* at V-A Walker and *Defence* at Scotts of Greenock.

In 1942 a modified design of triple 6in mounting was designed, broadly similar to the Mark XXIII but with 60° elevation, power loading and remote power control (RP.10) to improve anti-aircraft performance. *Tiger, Bellerophon* and *Defence* were changed to this Mark XXIV, with *Tiger's* from Elswick, the others from Barrow. In the 1942 Programme, two more were ordered, *Blake* from Fairfield, with mountings from Elswick despite being for a west coast yard, *Hawke* from Portsmouth (Barrow), plus three spares from Barrow.

*Superb's* mountings were ready by early 1944, but with priority given to light fleet carriers, the ship was not ready to accept them until fifteen months after launch at Wallsend. It had been planned to lay down *Tiger* after *Swiftsure's* launch from Berth N^o^ 4 at Walker in February 1943, but possibly due to slow progress with the Mark XXIV, two destroyers were laid down instead, followed by a tank landing ship. So construction of what had now been renamed *Bellerophon* never started and she was cancelled early in 1946. Her first mounting was doing pit trials early in 1946 but was re-allocated as a spare. The mountings for *Bellerophon* at Clydebank, now renamed *Tiger,* were 75% complete by the end of 1945, but construction was suspended a few months after launch. The mountings for *Defence* were in the pits 85% complete at Barrow early in 1946, when she too was suspended.[23]

Of the 1942 Programme, *Blake* was also suspended though with mounting completion planned for 1949, while *Hawke* was cancelled, although her machinery had already been built by V-A Barrow. The Admiralty arranged for some Mark XXIV to be completed and stored at Rosyth and others to be cancelled. Precise details of the eighteen ordered are not known, but it appears that six from Barrow were moved to Rosyth in 1950 and six cancelled. Five from Elswick went to Rosyth with one being retained for research at Elswick. Friedman[24] reported a Vickers-Armstrongs cruiser design for Venezuela in the 1950s which might have used some of the stored Mark XXIVs, three of which were only 20% complete. The three suspended *Tigers* were briefly considered to be fitted with updated Mark XXIVs but eventually completed from 1959 each with two twin Mark 26 mountings as described below.

[23] Cambridge University Library, Manuscripts. Barrow and Elswick Works Quarterly Reports, Vickers Archive 216–224

[24] Friedman, Norman. *British Cruisers: Two World Wars and After.* Seaforth Publishing, 2010, p306

## SCRAPPING HEAVY GUN MOUNTINGS

In June 1944, DNO, Rear Admiral C H Woodhouse, started circulating a discussion paper ('docket' in Admiralty parlance) on the proposed scrapping of heavy gun mountings after the war, at that time assumed to be end 1946, and the necessary number of spare mountings to retain. The paper did the rounds including Director of Dockyards, Director of Plans, Director of Armament Supply and Deputy Controller, each adding their comments.[25] A major unknown was which battleships and cruisers would be retained in the post-war fleet or in reserve. The initial assumption was that the battleships would be *Vanguard,* the four *King George V* class, *Queen Elizabeth, Valiant, Renown, Nelson and Rodney,* although in the event all but the first five had gone for scrap by 1948. The plan was to retain one spare mounting for every six remaining in service, which would mean seven 14in mountings, as 'mounting' meant only the elevating portion. A complete mounting could weigh over 1000 tons making storage difficult so only the more valuable portions such as cradle and slide (21 tons for the 14in) were kept. Storage of guns was less of a problem as they were transportable by rail; there were reckoned to be 135 15in in 1945. But for cruisers, storing the entire mounting was practicable, albeit at a waterside facility as transport by sea would be required. That limited the choice to one of the dockyards or taking over part of the Royal Ordnance Factory at Dalmuir. But DoD was not keen to tie up valuable real estate in this way, either by storing mountings within the working radius of a large crane or inside one of the gun mounting workshops, which he wished to keep for refitting work. The issue was

[25] Cruiser and battleship turret gun mountings: Post-war policy for scrapping. ADM 1/16670

Post-war several triple 6in mountings were stored at Rosyth. Of this group in July 1957, one is cocooned, one is unprotected (a Mark XXIV?) and one twin 5.25in cocooned, possibly *Vanguard's* spare. Alongside is the Danish *Rolf Krake,* formerly the 'Hunt' class *Calpe. (Author)*

*Jamaica*'s A and B turrets being dismantled at Dalmuir in August 1961. Between the guns are shell hoists and loading trays. A steam turbine condenser lies on the quayside full of valuable non-ferrous metals. *(Author)*

complicated by having to consult the War Office, which used some naval guns for its coast defence sites.

The outcome was that all gun mountings would be scrapped with the ship when it went for breaking, as long as the one in six ratio of reserves was achieved. The Ministry of Supply still required Dalmuir, but Harland & Wolff's nearby Scotstoun works could be used as the Admiralty had paid for major extensions pre-war, although only suitable for medium calibre mountings like spares of the new 4.5in Mark VI. Smaller items such as 18-ton 15in slides could be stored at the Coventry gun mounting store (which was on part of the old Coventry Ordnance Works site). All the RN 8in cruisers had been scrapped by 1950, apart from training ship *Devonshire* and trials ship *Cumberland*, which would not need spares. Only the RAN retained operational 8in cruisers, for which the existing spare mounting at Sydney would be sufficient. But DoD had to accept a number of 6in and 5.25in mountings stored under the 250-ton cantilever crane at Rosyth in the 1950s. An aerial photo in May 1960 shows five triple 6in mountings, possibly X mountings removed in 1944–45 with one cocooned. Three twin 5.25in are also visible, one cocooned so possibly *Vanguard*'s spare.

All the cruisers with triple 6in mountings that were broken up in the UK still had their mountings aboard when disposed of for scrapping, although *Sheffield* had had her guns removed. Thus their scrap arisings and value became part of the overall ship demolition process. But there were the 6in triples stored at Rosyth within the working radius of the 250-ton crane on the west wall in 1960. Some Mark XXII and Mark XXIII mountings had been retained as spares while there were still ships in service with such, but the Mark XXIVs which were never installed in any ships were probably scrapped first. Rosyth Dockyard had a process for scrapping redundant equipment such as rail wagons, which were usually sold to Shipbreaking Industries which had a shipbreaking yard within the Dockyard from 1926 to 1963. In 1962 the Dockyard sold two triple 6in mountings to SI for £2675 – 2% of what they cost to build – which was little more than their mild steel scrap value. The Dockyard staff clearly did not realise the great value of the non-ferrous metals and the nickel bearing gunhouse armour, but SI certainly did, having broken up cruisers at their Faslane yard. While the guns and slides (12 tons each) were not included in the sale, SI extracted from the two mountings 256 tons of saleable material worth £8669 giving a healthy profit of £4826 before overheads. Even deducting say £500 for overheads left as much as they could expect from breaking up a destroyer of around 2500 tons displacement. It took only two months from March 1962 for the main demolition and a further two to send all the sorted material to buyers and receive the proceeds. The wage cost equates to about eight men for eight weeks.[26]

| *Value* | *Tons* | *Sales £* | *£/ton* |
|---|---|---|---|
| Steel scrap | 185 | 2,036 | 11 |
| Armour | 46 | 1,733 | 37 |
| Non ferrous metal | 25 | 4,800 | 196 |
| **Total** | **256** | **8,669** | |

| *Costs* | *£* |
|---|---|
| Wages | 754 |
| Cutting gases | 156 |
| Carriage | 232 |
| Metal analysis | 26 |
| Purchase price | 2,675 |
| Total costs | 3,843 |
| Sales value | 8,669 |
| Profit | 4,826 |

### THE 6-INCH TWIN MOUNTING MARK 26

Post-war the Admiralty made many studies into the role of the cruiser, in particular the need for better anti-aircraft capability. A design of a new triple 6in 80° Mark XXV mounting was proposed in 1945 to arm a new 15,350-ton *Neptune* class cruiser. Designs using 5.25in and 8in mountings were also prepared, all discussed by Friedman, although none were ordered.[27] The US already had the *Worcester* class cruisers under construction with a new rapid firing Mark 16 twin 6in 47-calibre dual-purpose mounting capable of firing 12 rounds per minute per gun, so something similar was proposed. This new British mounting was designated Mark XXVI, although Roman numerals were dropped in 1950 when it became Mark 26. Consideration was given to using the US 6in/47 gun and ammunition in the new mounting but too many objections were raised.

This was to be a dual-purpose twin mounting capable of firing 20 rpg at elevations up to 78½° (14° depression). It was to be automatic *ie* no manual handling, with ammunition stowed vertically on circular rings around the hoists, mechanically loaded from there to the breech. There was an upper and lower indexing ring each holding 40 shells (AP or HE) fed from stowage in the two shell rooms, which could be moved to line up with the hoists.[28] A new quick firing gun Mark N5 was developed from 1944, as QF was better for fast power loading than BL used in the previous 6in turret guns, *ie* the charge was in a brass cartridge case which needed to be ejected through the front of the gunhouse after firing. The cartridges were stowed in the magazine below the lower shell room. Each gun would be in a separate cradle with a loose barrel that could be replaced on board by withdrawing through the rear of the gunhouse. The heat from the rapid firing rate would need to be dissipated by water cooling. The mountings were to be interchangeable *ie* all the same hoist length suitable for A and Y positions. The gunhouse roof had a distinctive upward slope to accommodate the breech end at maximum depression. The specification assumed a crossing target at 500 knots to be engaged at a minimum range of 2000 yards, with enough power on the mounting to cope with 15° roll and 5° pitch.[29] No sub-calibre gun was provided.

A prototype was ordered from Vickers-Armstrongs at Barrow late in 1947, who also collaborated with the Admiralty in new cruiser designs, one of which was the *Minotaur* class with five twin Mark 26, but never built. By this time the British shipbuilding industry was focused on merchant ship construction, especially for export, and both defence budgets and RN manpower were being cut back. Studies were also made in the early 1950s to re-arm some of the 'Towns' and 'Colonies' and *Superb* with Mark 26s, but there was neither the money nor the gun mounting nor dockyard capacity to do so, and all those ships went for scrap with their original 6in mountings. The most that could be afforded was to complete the three *Tiger* class cruisers which had been laid up incomplete soon after launch. They already had machinery installed and superstructure erected but no armament fitted (three 6in Mark XXIV). The superstructure would need to be completely rebuilt and the ships would need to be refitted with 440V AC electrics (previously 220V DC) with about treble the generating capacity of the 'Colonies' or 4500kW. A 6in twin mounting could be fitted in A and Y positions, each needing about 500kW – to train at 40° per second required a motor of about 200hp (150kW). The three new twin 3in Mark 6 dual-purpose mountings to be fitted in the *Tiger*s would also require a lot of power. The 6in magazine areas would need to be rebuilt to hold about 400 rounds per gun, *ie* some 800 rounds compared with 600 for each Mark XXIII but *Blake*'s A magazines held 1090 rounds and Y 798. With little manual handling, the mounting crew could be reduced to about 26 compared with 46 in the Mark XXIII. It was originally planned to have seven in the gunhouse: officer of the turret, captain of the turret, gunlayer, trainer, sightsetter, communications number and artificer. Later with full RPC the gunlayer, trainer, sightsetter were replaced by a single aimer for local control. The revolving weight in operation (with hoists loaded) would be about 163 tons, which included 34 tons of gunhouse NC armour: 3in front, 2in sides, rear and roof 1.5in, barbette 2in. In 1949 this was deemed sufficient protection against atomic blast. Including the fixed parts of the hoists and off-mounting training gear, the total installation came to about 192 tons. It was reckoned that four of these new 6in would be equivalent to the twelve 6in in the Soviet *Sverdlov* class cruisers.

A variety of cruiser designs with two to five Mark 26s were drawn up in the early 1950s. But the least expensive solution remained a revised design of the *Tiger*s with two each. Six new Mark 26 had already been ordered at the end of 1949, three from Barrow, three from Elswick for notional delivery by 1955. Four of the mountings would have the

[26] Marine Technology Special Collection. 12074

[27] Friedman, Norman. *British Cruisers: Two World Wars and After*, p266

[28] *Handbook for 6-inch Mark N5 Gun on 6-inch DP Twin Mark 26 Mounting, 1963.* ADM 234/870

[29] 'Technical History of Development of 6in Dual Purpose'. Tyne & Wear Archives. DS.VA 6/40/1-2

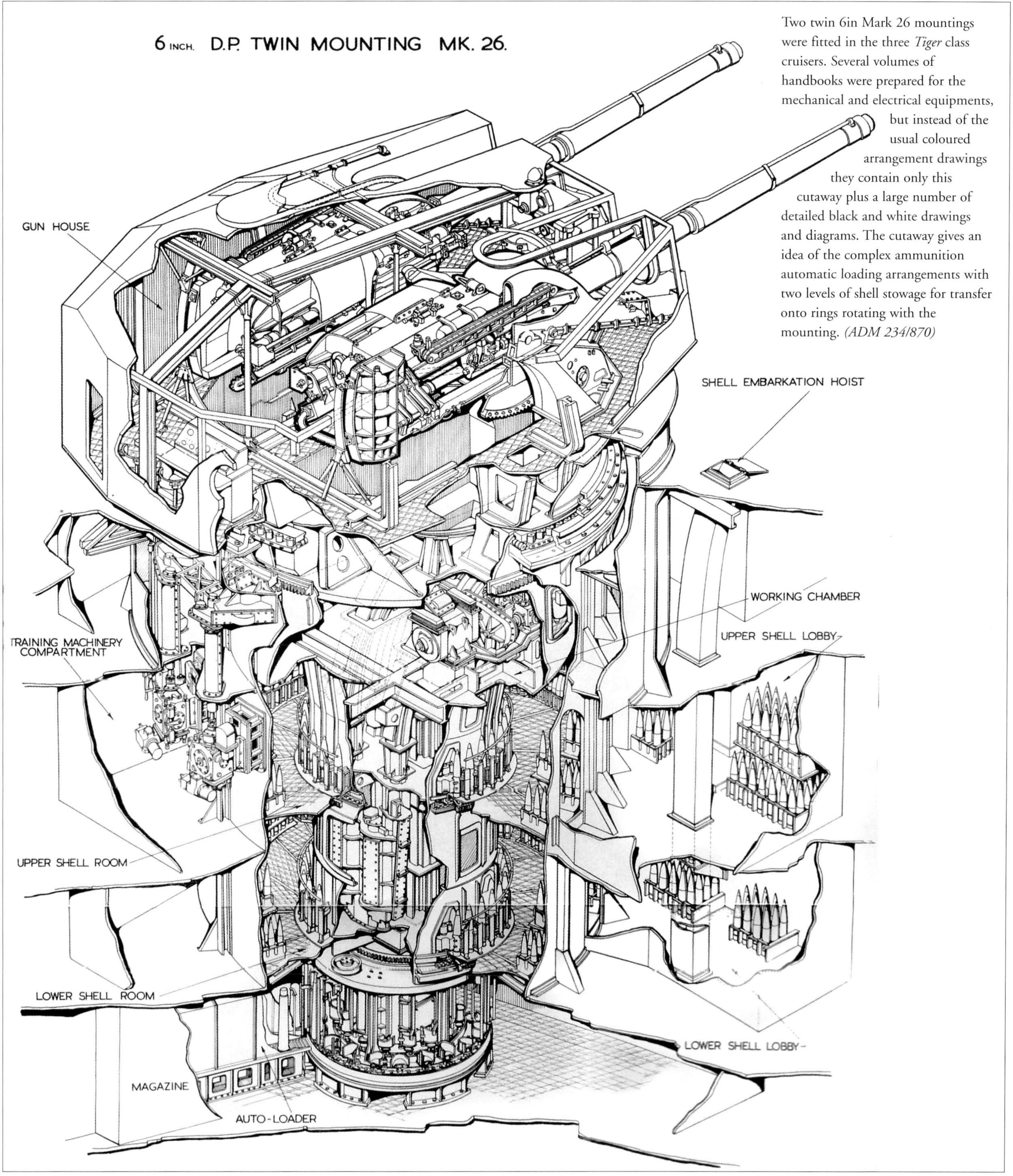


Two twin 6in Mark 26 mountings were fitted in the three *Tiger* class cruisers. Several volumes of handbooks were prepared for the mechanical and electrical equipments, but instead of the usual coloured arrangement drawings they contain only this cutaway plus a large number of detailed black and white drawings and diagrams. The cutaway gives an idea of the complex ammunition automatic loading arrangements with two levels of shell stowage for transfer onto rings rotating with the mounting. *(ADM 234/870)*

*Cumberland* as a trials cruiser mounted the first twin 6in Mark 26 forward and a twin 3in Mark 6 aft in May 1956. *(Author's collection)*

electro-hydraulic RP.15 remote power system (*Blake* and *Lion)* and two an all-electric system, RP.50 for elevating and RP.53 for training (*Tiger*).[30] It was found that with the training engine off mounting, the 6in gun pits were not large enough to take the new mountings, so both at Barrow and Elswick three 14in pits were adapted. Unlike previous 6in, the training rack was fixed to the mounting (rather than the ring bulkhead) with the motor and pinion drive in a compartment on the lower deck; this reduced the revolving weight and inertia. Many discussions and studies about details of the design were made between 1948 which were carefully recorded.[31]

With the onset of the Korean War in 1950, some consideration was given to completing the *Tigers* earlier by using the existing triple 6in Mark XXIV, two per ship, plus four twin 4.5in Mark 6 and three twin 40mm STAAGs. That would mean stopping work on the Mark 26 and reverting to DC and the idea was dropped.

V-A were also working on the 4.5in and 3in mountings, so progress on the 6in was slow but it was planned to have the 6in mountings ready in 1957. Eventually orders to complete the three *Tiger* hulls were placed in October 1954, to be as much command ships as gunnery cruisers. As well as workshop capacity, V-A was short of gun mounting installation and setting-to-work teams who were also involved in the 4.5in Mark 6 programme. Warship building priorities in the 1950s had been focused on completing the carriers ordered during WW2, and building new fast frigates and submarines, all factors which delayed completing the cruisers until 1959–61.

A prototype 3in Mark 6 mounting had also been ordered in 1947 from Barrow. After land trials, the plan was to install both prototypes in the trials cruiser *Cumberland* at Devonport from December 1954 to October 1955. When DNO pointed out this would mean abandoning the 1955 summer trials season (usually in the Mediterranean) only the 3in was installed (aft) and trialled in 1955, while the 6in was installed in B position and trialled in 1956, together with their associated MRS3 Mod.1 fire control systems and Type 903 radar. After the trials, *Cumberland's* mounting was planned to become a spare. Shop trials on the prototype mounting with five-round bursts had been carried out at Barrow in May 1953, then the elevating portion trialled at V-A's Eskmeals range further up the Cumberland coast.

The first Mark 26 installed was *Tiger's* Y mounting at Clydebank in December 1957 followed by A in March 1958, both from Barrow. Gunnery trials were carried out in February 1959, during which A shot off 67 rounds and Y 60.[32] *Lion* was next with two Elswick mountings fitted in late 1959, then *Blake* with A from Barrow, followed by Y from Elswick with trials in May 1961. Both 6in and 3in contracts were 'cost plus' with an agreed profit rate between 7 and 10% on cost. Each production mounting cost the Admiralty about £800,000, ten times the cost of a triple 6in, although say five after inflation. Therefore, two

[30] ADM 234/870, *op cit*

[31] DS.VA 6/40/1, *op cit*

[32] Gunnery trial reports *Tiger*, *Lion* and *Blake*. *Tiger* class Ships Covers, NMM

*Tiger*'s forward twin 6in Mark 26 at Portsmouth in August 1959. The circular door in the rear was for withdrawing a 6in barrel. The water-cooled barrels of her twin 3in are visible. *(Author)*

constituted 12% of the final ship cost of about £14M, a percentage which would be doubled when the guns and the 3in mountings (about £400,000 each) and fire control systems were included.

When *Blake* and *Tiger* were converted to helicopter carriers in 1969 and 1972, Y turret was removed to make way for the hangar and flight-deck; the two midship 3in mountings were also removed. Only HE (£160 each) and starshell were then carried.

The gun had an external jacket over 60% of the gun length, and a loose replaceable barrel, between which was an annular cooling space (coolant 70% water, 30% ethylene glycol). Jacket and barrel each had a separate registered number series.

The relatively low muzzle velocity gave a good accuracy, but the gun could jam if the steel cartridge cases were out of tolerance. But in-service problems arose with the mounting, mostly resulting from the very high rate of fire which could not be sustained for long; 17 rpg was about the fastest obtainable. Marland concluded:

> The 6in Mk 26 was a disaster: a large complex turret that rarely achieved the gun functioning burst-fire requirement. … The 6in weapon rarely achieved its designed rate of fire of 20 rounds per minute per gun due to mechanical failures in its highly stressed feed system. Initial firings were successful, but by later life the 6in system had become an embarrassment. … In contrast to the 3in Mk 6, the 6in Mk 26 was problematic, due to the high torques in the feed system; particular problems were the scuttles and tilting buckets, the three-part feed tray, and complexity caused by a Recoil Energy Power System [used to supplement the regular pumps] used on the elevating mass of the gun.[33]

**6in QF Mark N5 gun**

| | |
|---|---|
| Calibres, bore | 50 |
| Length, in | 315 |
| Weight incl BM, tons | 6.85 |
| Shell AP, lb* | 130 |
| Cartridge, lb | 77 |
| Charge, lb | 34.5 |
| Muzzle velocity, ft/sec | 2500 |
| Max range, yards | 24,500 |
| AA ceiling, ft | 49,500 |
| Time of flight, sec | 51 |

* HE rounds 135lb

Over the same period, Vickers-Armstrongs also produced 25 twin 3in Mark 6 for the *Tiger*s and Canadian destroyer escorts which Marland describes as 'In contrast, the 3in Mk 6 was an outstandingly good AA weapon'. The lighter fixed ammunition (38lb) was less demanding mechanically with a more compact feed system despite a much higher rate of fire (90 rpg).

[33] Marland, Peter. 'Postwar Weapons in the Royal Navy'. *Warship* 2015.

*Bermuda* in October 1943, the last 'Colony' to be completed with four triple 6in turrets. *(Admiralty)*

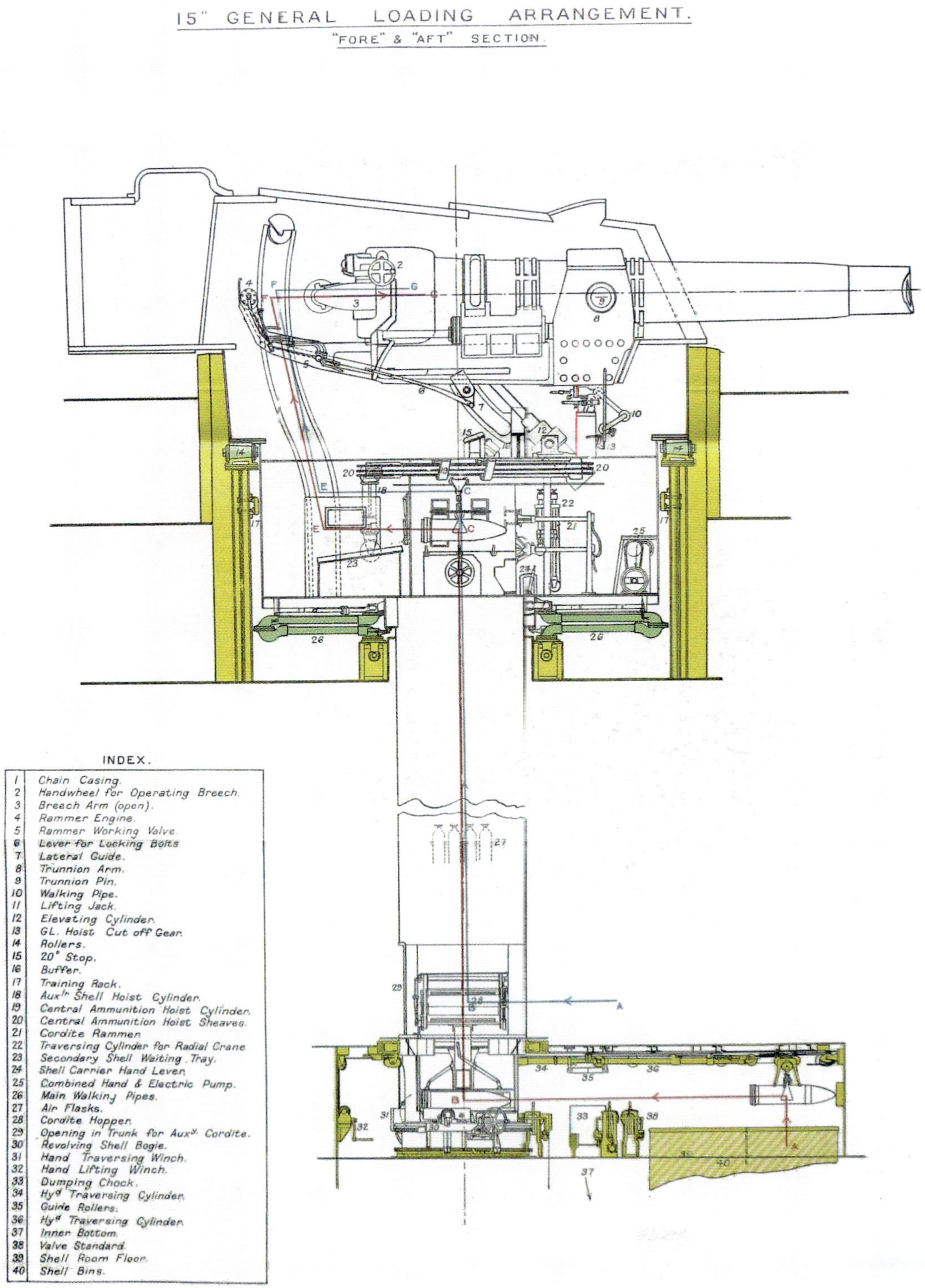


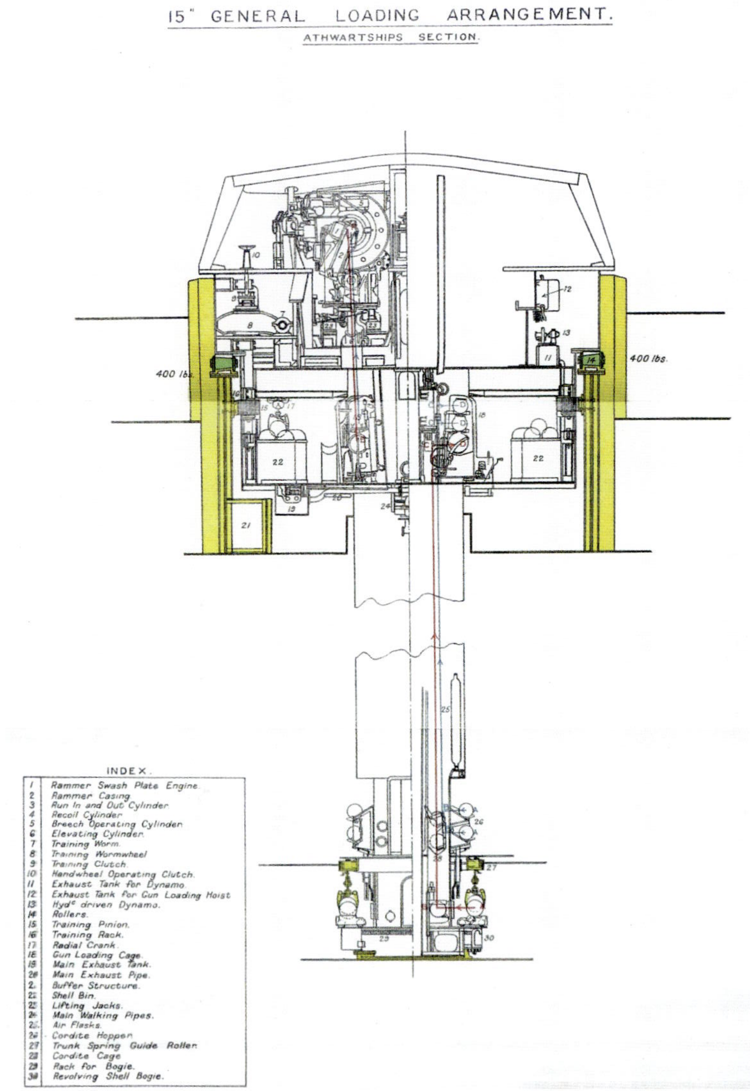


**Twin 15in general loading arrangement**
These drawings from the 1921 *Manual for Power Worked Mountings* shows the ma.n operating components of a 15in mounting, with the loading path of the shells shown in red and cordite charges in blue. *(ADM 186/256)*

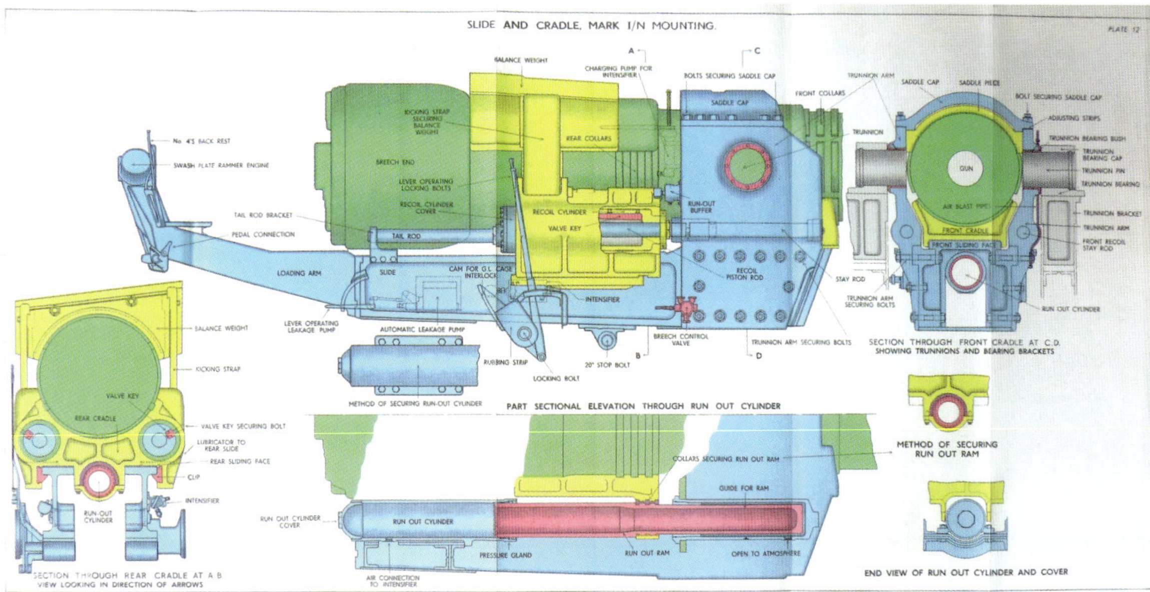


## Gun slide and cradle

The drawing of the gun slide and cradle for the 15in Mark I/N mounting illustrates well the core parts of a heavy gun mounting. The surface of the forged steel slide (blue) was machined to allow the gun cradle (yellow) holding the gun (green) to recoil on firing and then run-out for reloading. The slide was pivoted on trunnions which allowed the whole mechanism to be elevated. At the rear of the slide was the loading arm containing a chain rammer which pushed the shell and charges into the breech once the gun loading cage was in position. One pneumatic run-out cylinder (pink) was fitted within the slide, while the two hydraulic recoil cylinders (blue) were fitted within the cradle. The yellow balance weight on top made it possible to elevate to 30° within the existing turret dimensions. *(ADM 234/168)*

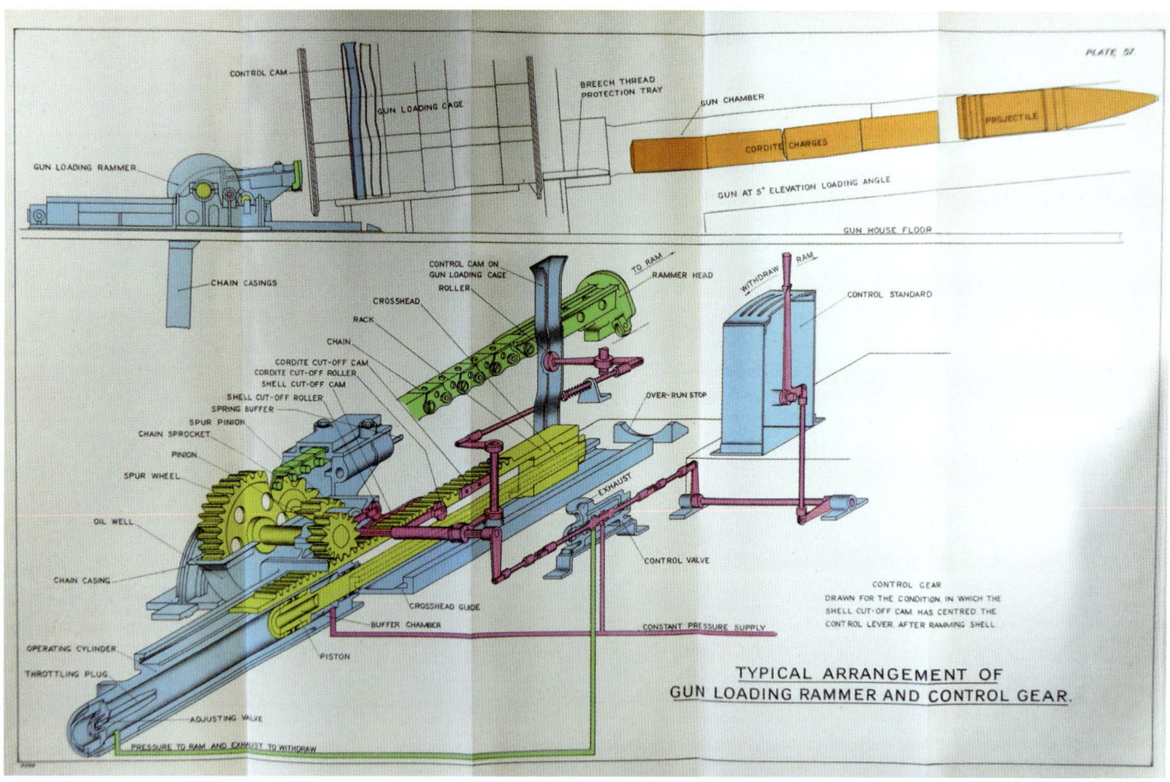


## Chain rammer

The chain rammer for a 14in Mark VII gun was fixed at a 5° loading angle, ramming first the shell then in two stages the four quarter charges. The chain was moved by hydraulically driven spur gears and could flex only one way. *(ADM 234/271)*

**METHODS OF CARRYING PRESSURE INTO REVOLVING STRUCTURE**

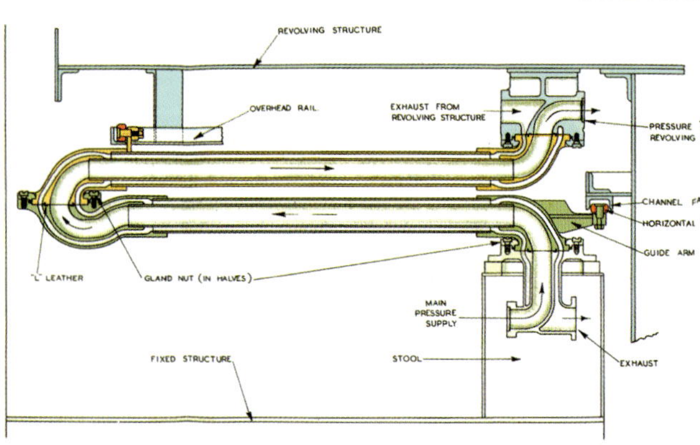


**WALKING PIPES (15 INCH MOUNTING)**

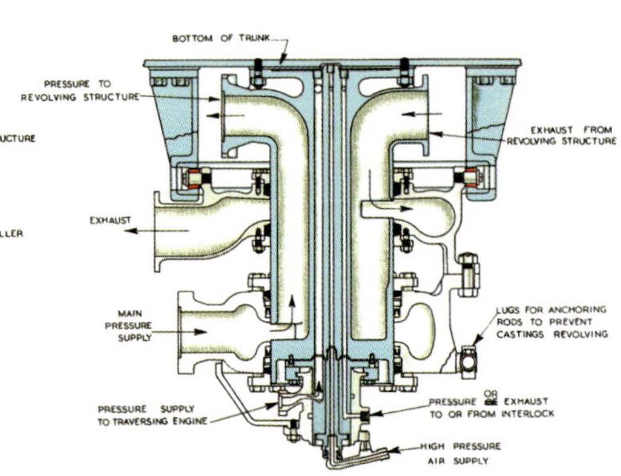


**CENTRAL PIVOT (14 INCH MOUNTING)**

## Methods of carrying pressure into revolving structure

Hydraulic fluid under pressure had to be transferred from the pumps and piping on the hull to the revolving structure. Walking pipes fitted below the working chamber allowing the turret to rotate about 150° either way were fitted in the WW1 era and earlier mountings. *Nelson, Rodney* and the *King George Vs* had a central pivot which delivered the fluid to an annular space at the bottom of the trunk.

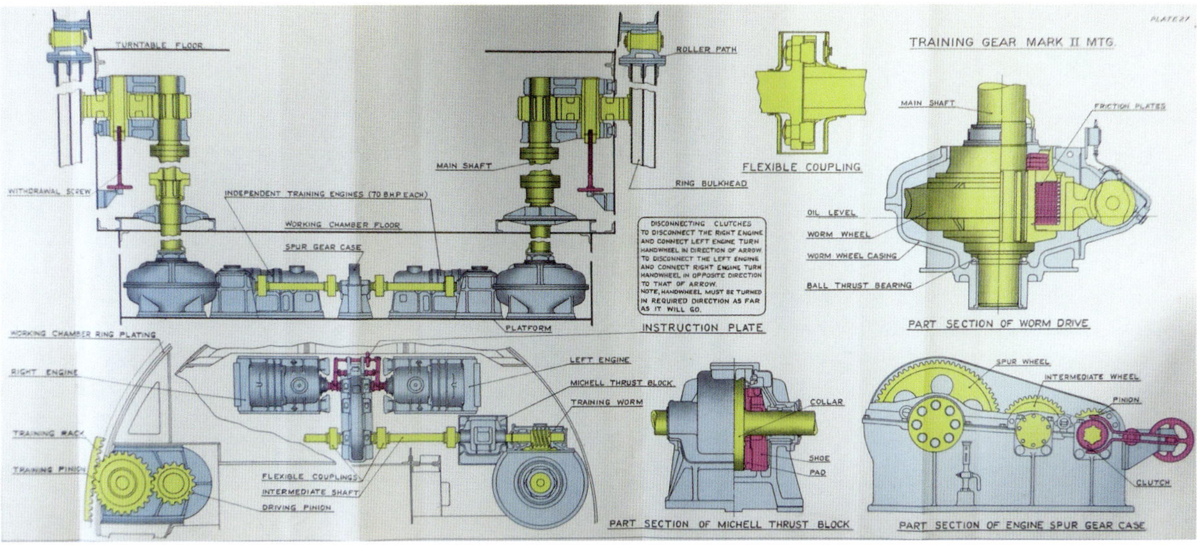


## Training drive

The training drive for the twin 14in Mark II mounting with two hydraulically driven 70 hp motors, only one clutched in at a time. The worm drive to the main shaft which engaged the training rack prevented any reversal from turret motion. *(ADM 234/271)*

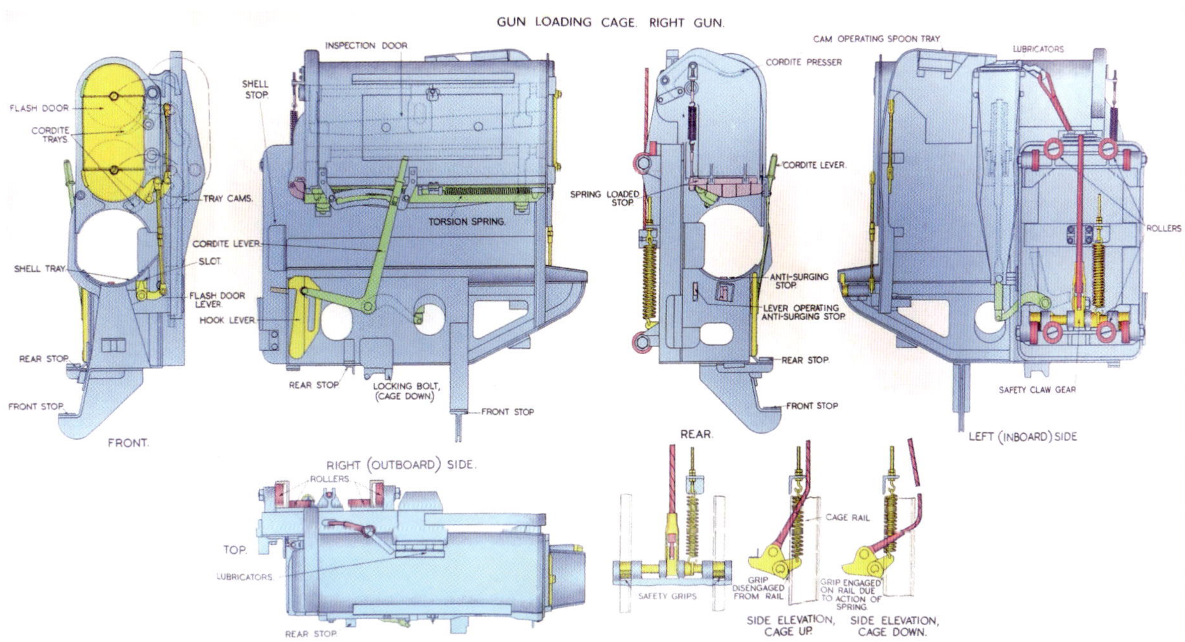


## Loading cage

The 15in gun loading cage lifted shell and cordite charges from the working chamber to the gunhouse for ramming into the breech, with the shell at the lower level and two cordite trays each holding two quarter charges at the upper level. *(ADM 234/168)*

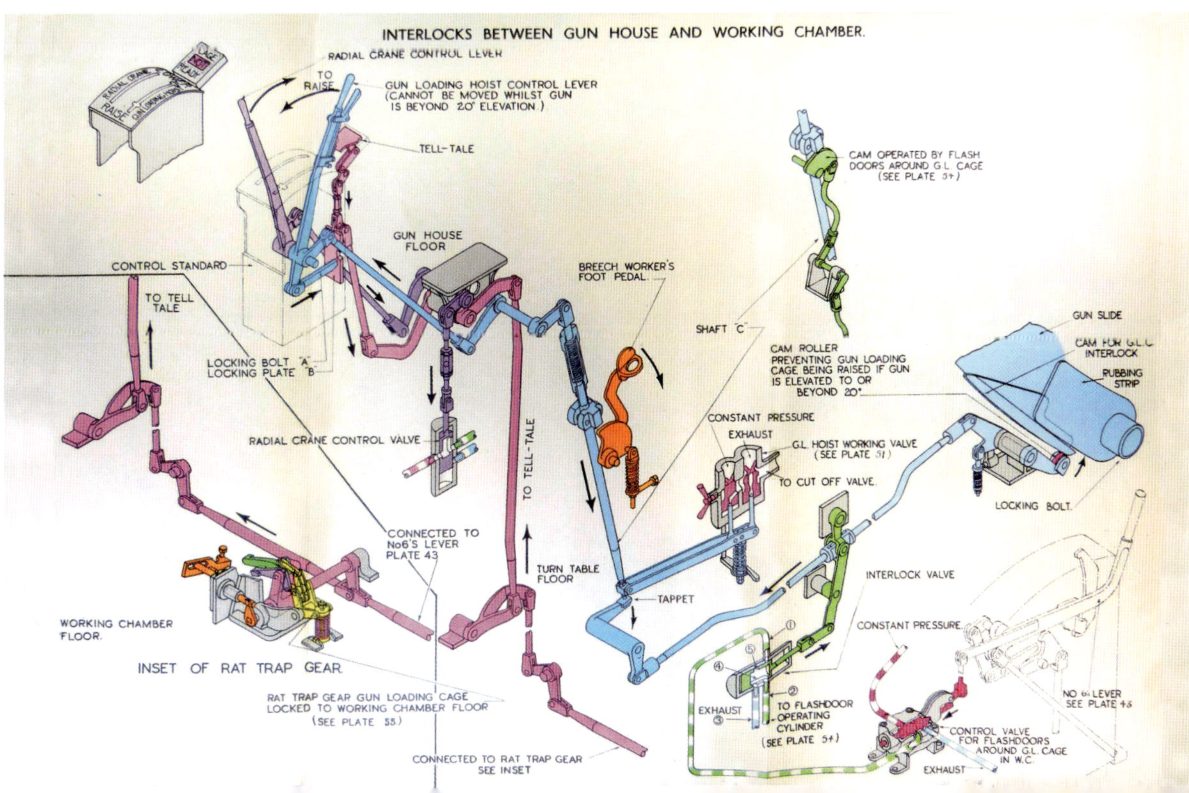


## Interlocks

Interlocks were fitted to prevent dangerous movements in loading gear, using rod linkages such as illustrated in the 15in Mark I/N between gunhouse control standard and working chamber. *(ADM 234/168)*

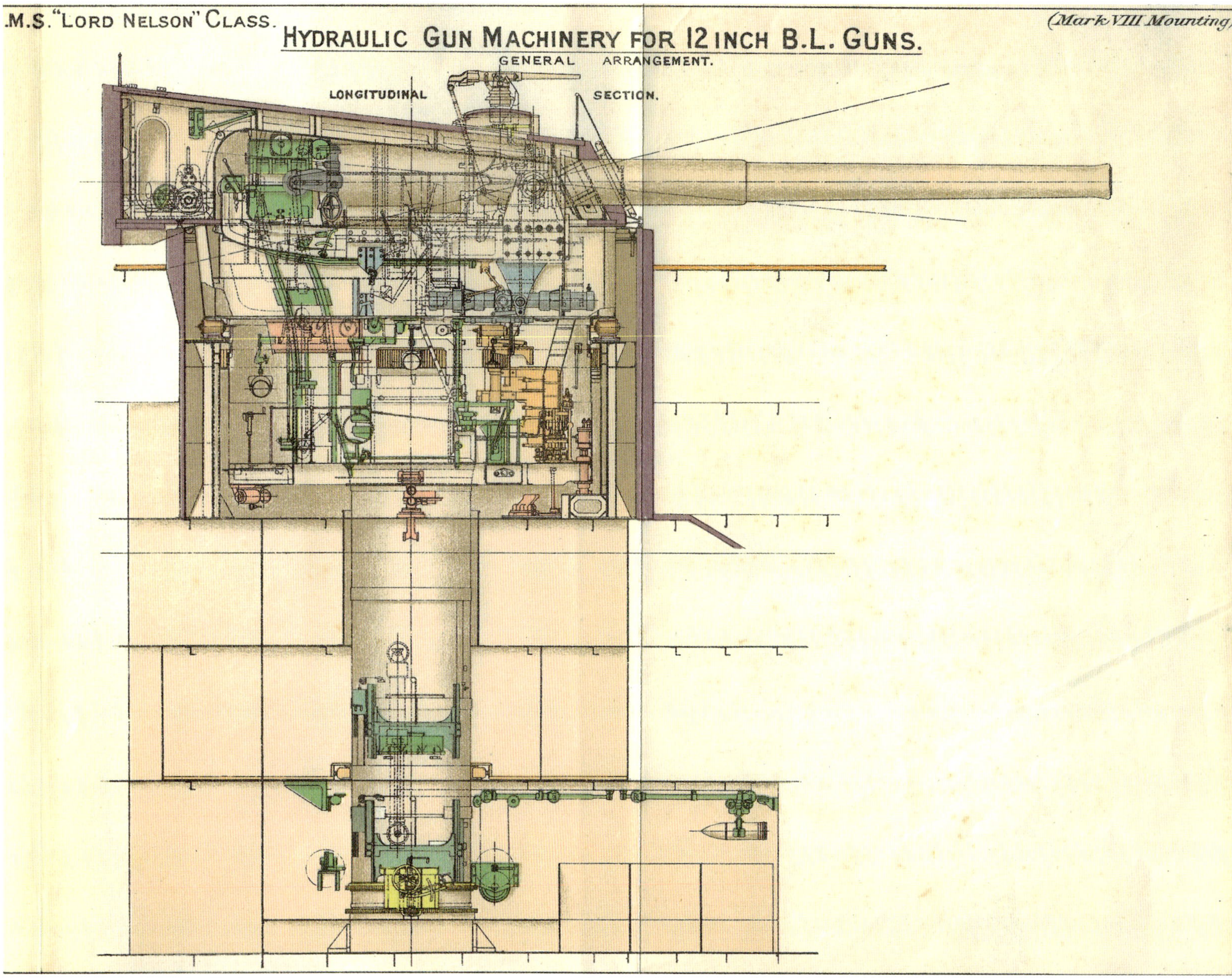


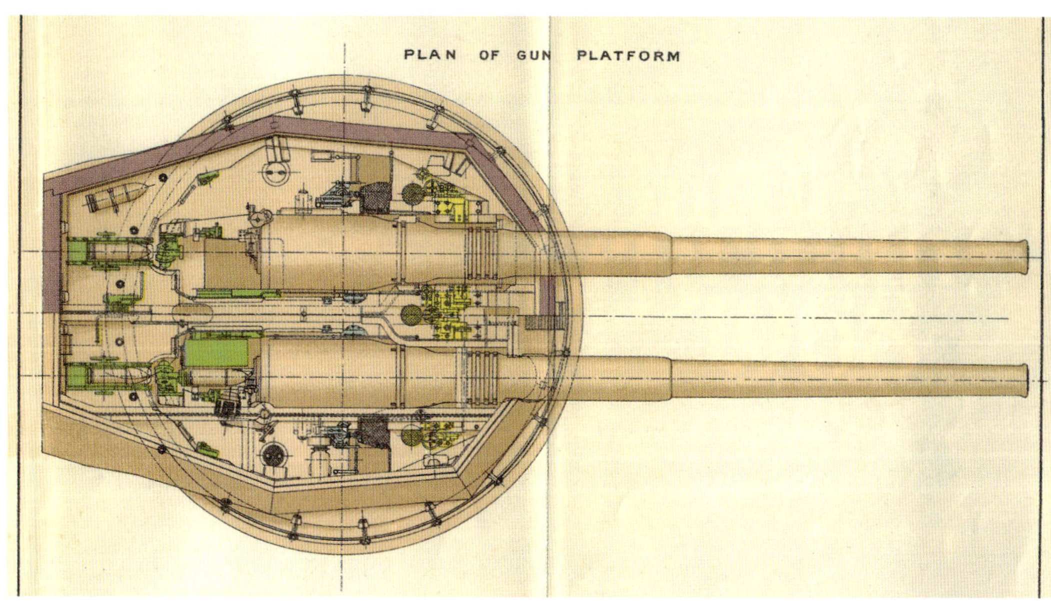


**Hydraulic gun machinery for twin 12in B.VIII**
The Vickers designed twin 12in B.VIII carried the new 45-calibre 12in Mark X gun and was fitted in the two *Lord Nelsons* (two each), *Dreadnought* (five), *Indomitable* and *Inflexible* (four each). It could be loaded at any angle between its maximum of 13½° and -5°. It had the more compact chain rammer on the gun loading arm, which reduced the diameter (and weight) of the barbette. Gunhouse side armour plates were flat, so were slightly cheaper than curved. *(Addenda to Manual of Hydraulics 12-inch BL Mountings B.VIII (Vickers))*

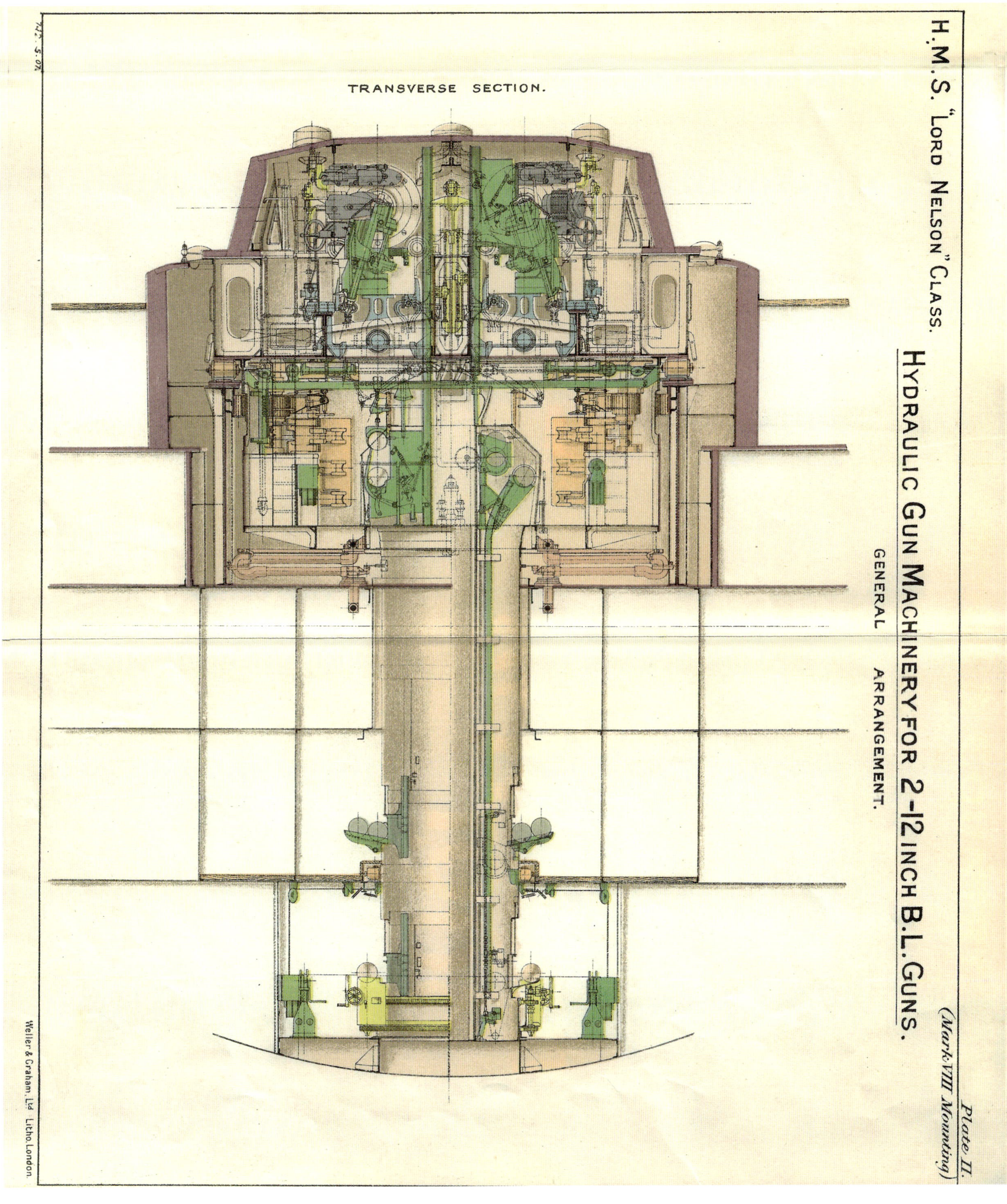
H.M.S. "Lord Nelson" Class.
Hydraulic Gun Machinery for 2-12 inch B.L. Guns.
General Arrangement.
Plate II.
(Mark VIII Mounting)
TRANSVERSE SECTION.
Weller & Graham, Ltd Litho. London.

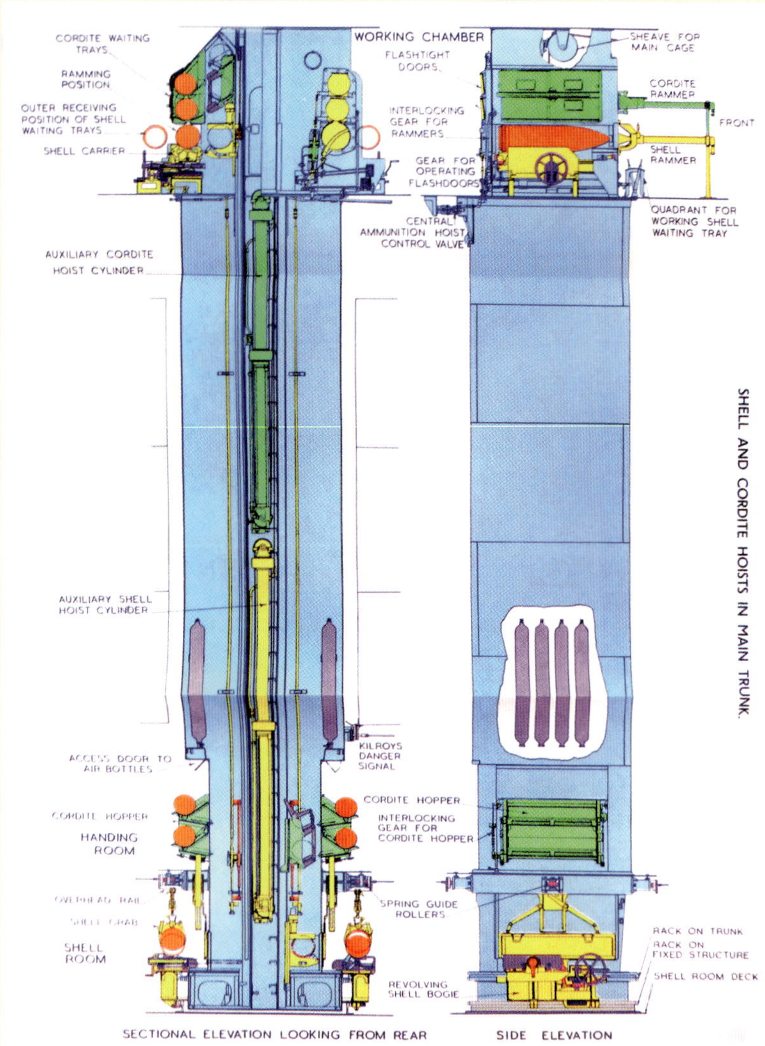


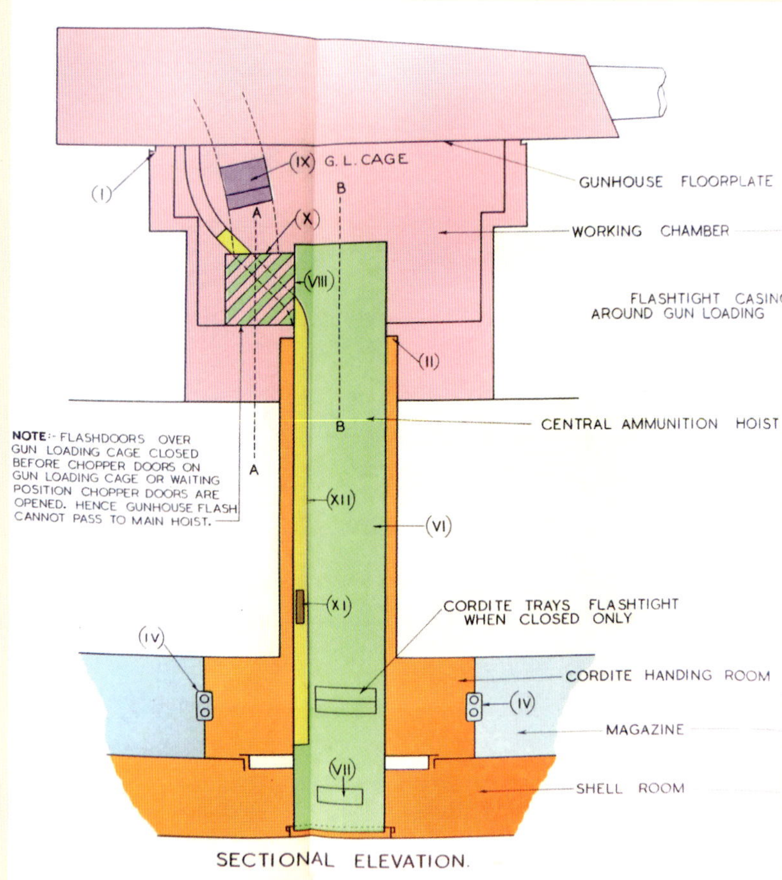


**Flashtight zones**

The main flashtight zones of the 15in Mark I/N are shown, with the gun loading cage and auxiliary cordite hoist having their own enclosures. *(ADM 234/168)*

**Main hoist trunk components**

The main hoist trunk components of the 15in Mark I/N mounting are clearly shown linking the shell and cordite handing rooms and working chamber. Air bottles in purple for gun air blast. *(ADM 234/168)*

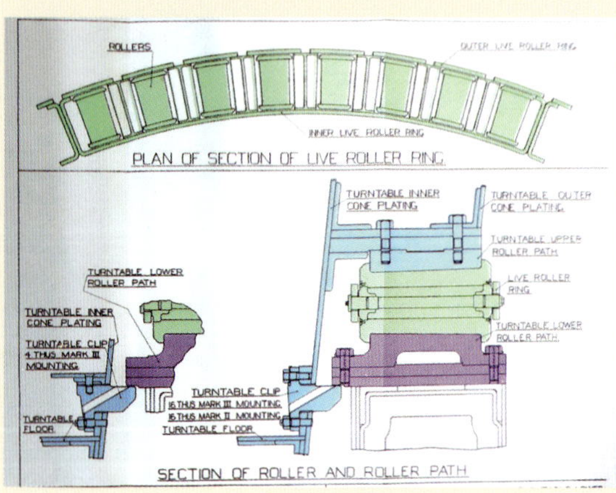


**Roller path**

A section of the roller path of the twin 14in mounting with 48 rollers (green) secured inside a live ring with all rotating together on top of the lower roller path (purple) secured to the ring bulkhead. The clips (blue) prevented the whole mounting lifting off its support, *eg* in a seaway. *(ADM 234/271)*

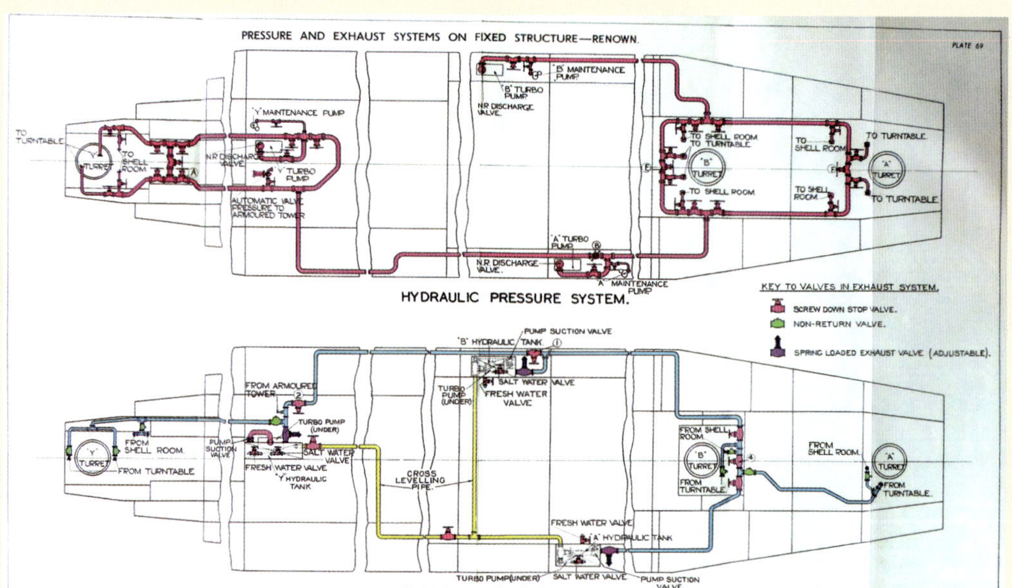


**Hydraulic pressure and exhaust systems**

Each capital ship had two to four steam driven pumps which delivered hydraulic fluid at pressure of about 1650lb/sq in though pressure and exhaust piping. *Renown*'s is shown after her conversion to Mark I/N now with three turbine-driven pumps. *(ADM 234/168)*

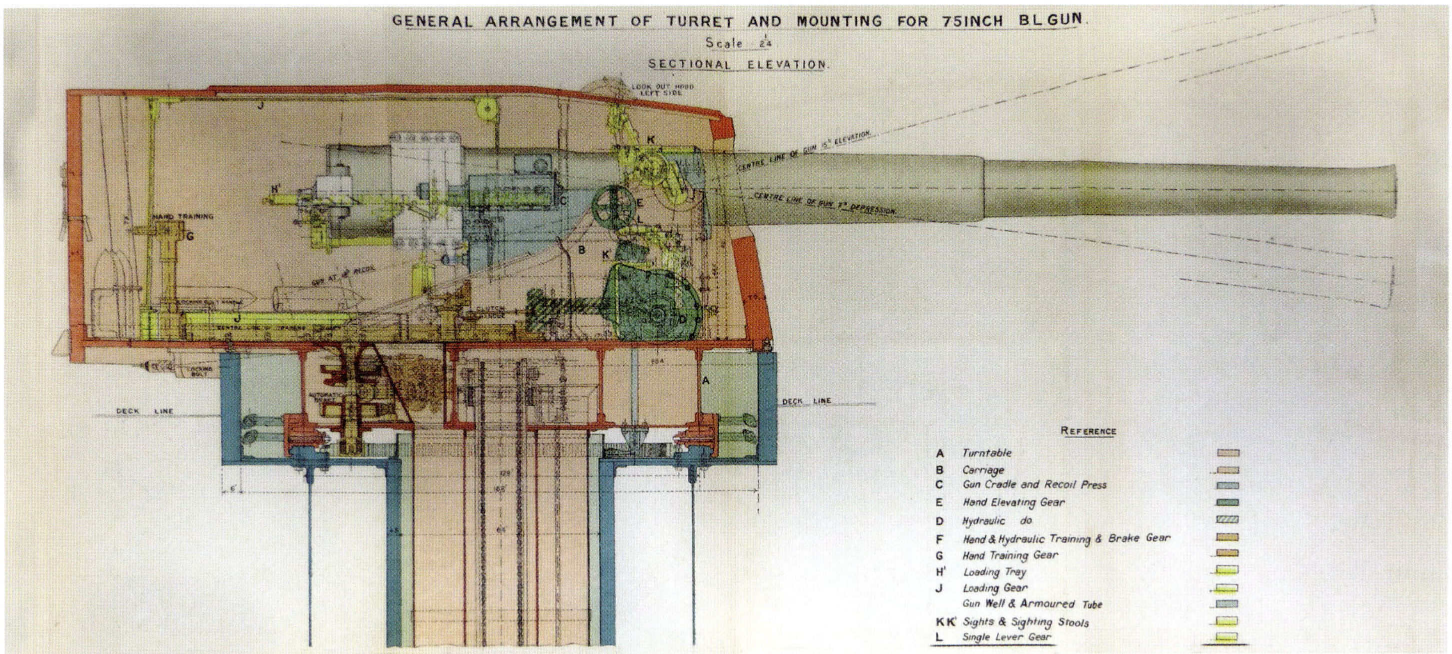


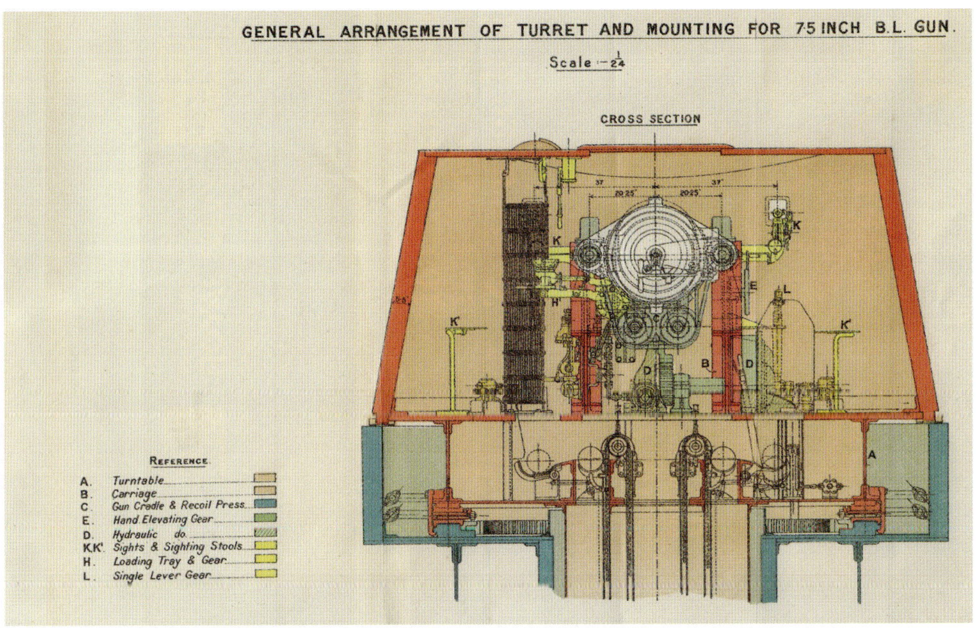


**Single 7.5in Mark I mounting**
Single 7.5in mountings were fitted in most of the armoured cruisers. The Elswick designed Mark I was hydraulically operated, with the hoist delivering shells to the left hand side of the gun, cordite to the right. *(Handbook IWM LBY 16970)*

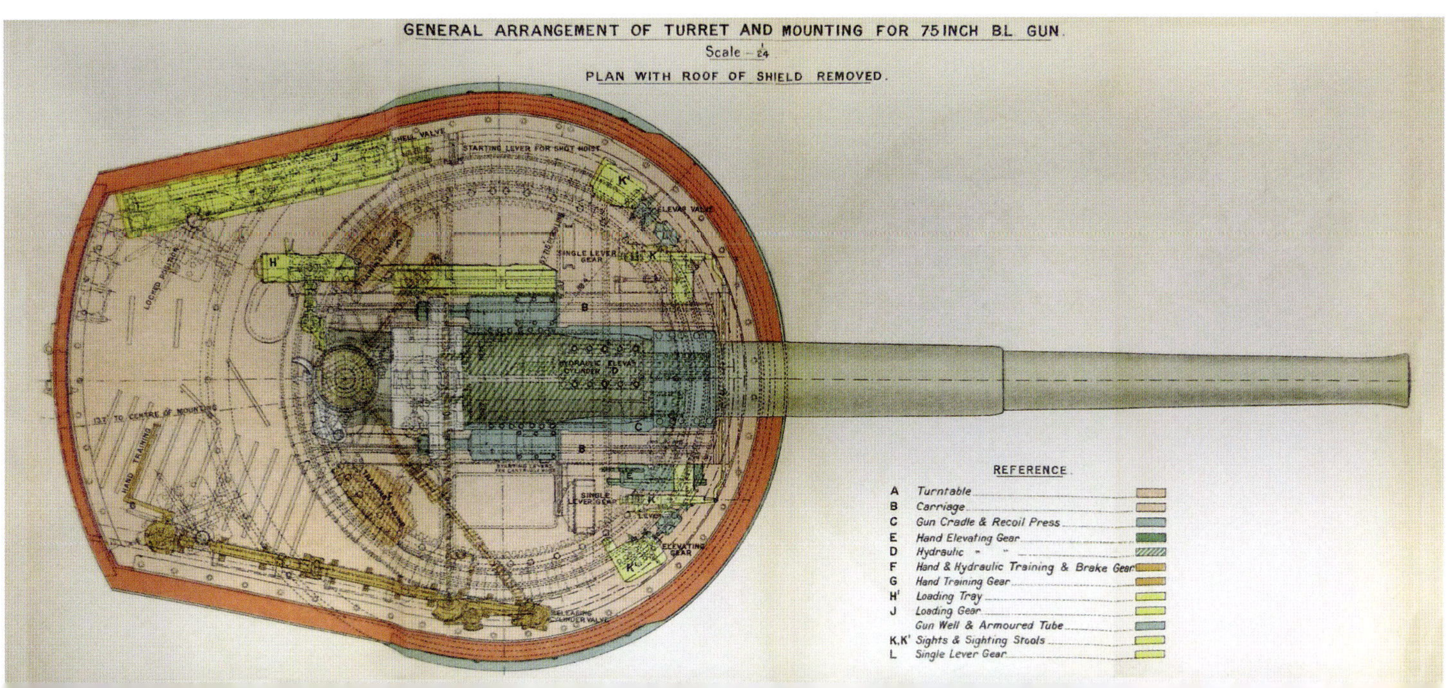

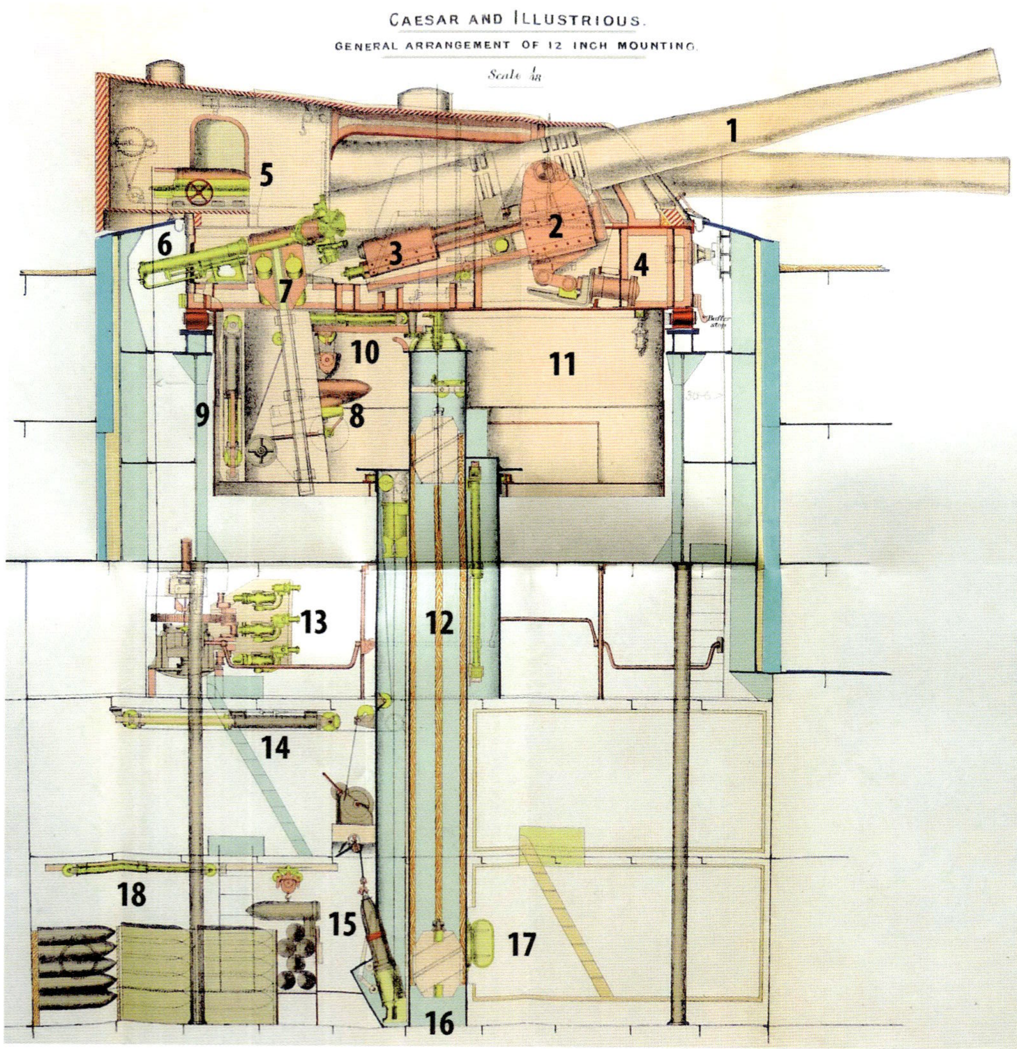


1. Gun at 13½° loading angle
2. Elevating gun slide
3. Recoil cylinder
4. Elevating cylinder
5. Shell on bogie in gunhouse
6. Rammer
7. Cordite cage tubes
8. Shell hoist cage
9. Shell hoist hydraulic press
10. Overhead crane
11. Working chamber
12. Central hoist
13. Training engine
14. Ammunition hoist hydraulic press
15. Shell lifting head
16. Cordite cage, lowered
17. Magazine
18. Shell room

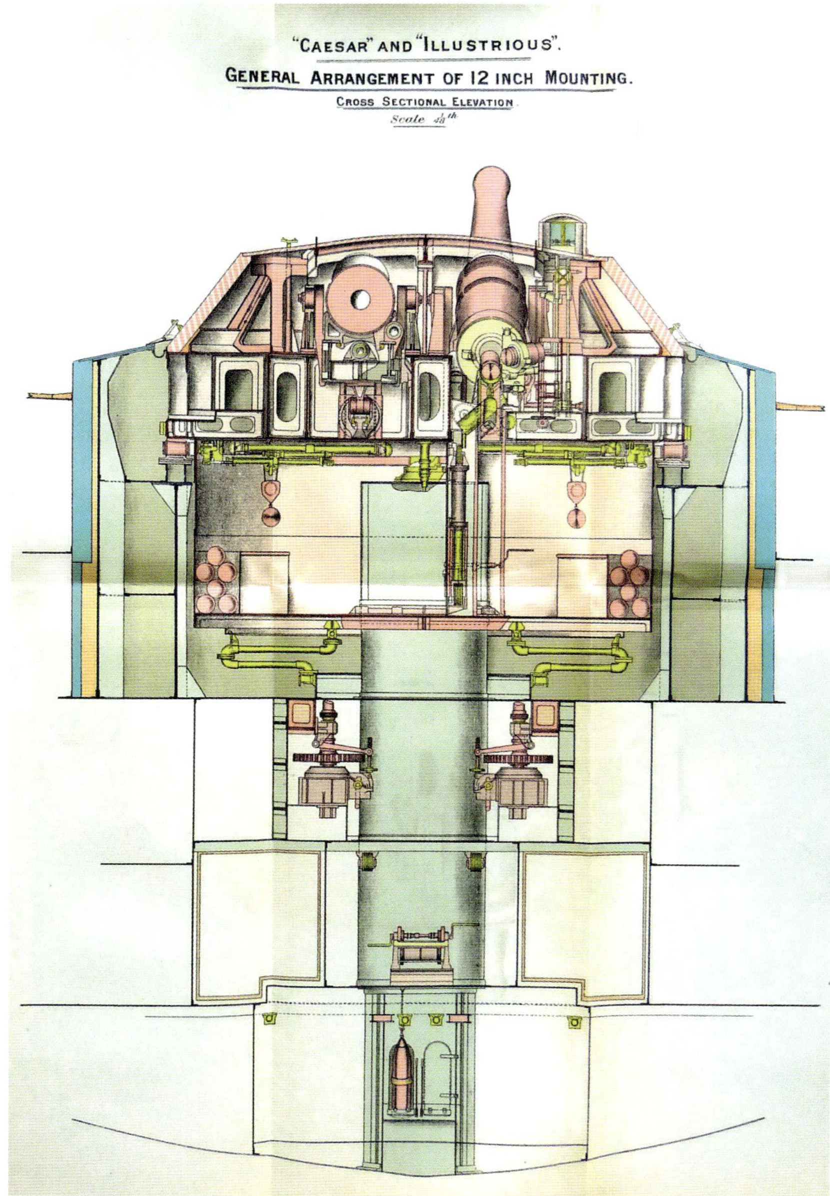


**Twin 12in B.III mounting**

*Caesar* and *Illustrious* were fitted with the Whitworth designed twin 12in B.III mounting, which featured a working chamber which permitted ammunition resupply at all angles of turret training. The two-stage process had a central hoist (12 in the figure) delivering to the working chamber (11), then cage hoists for taking shell (8) and cordite (7) up into the gunhouse for ramming (6) into the breech. *(ADM 186/172)*

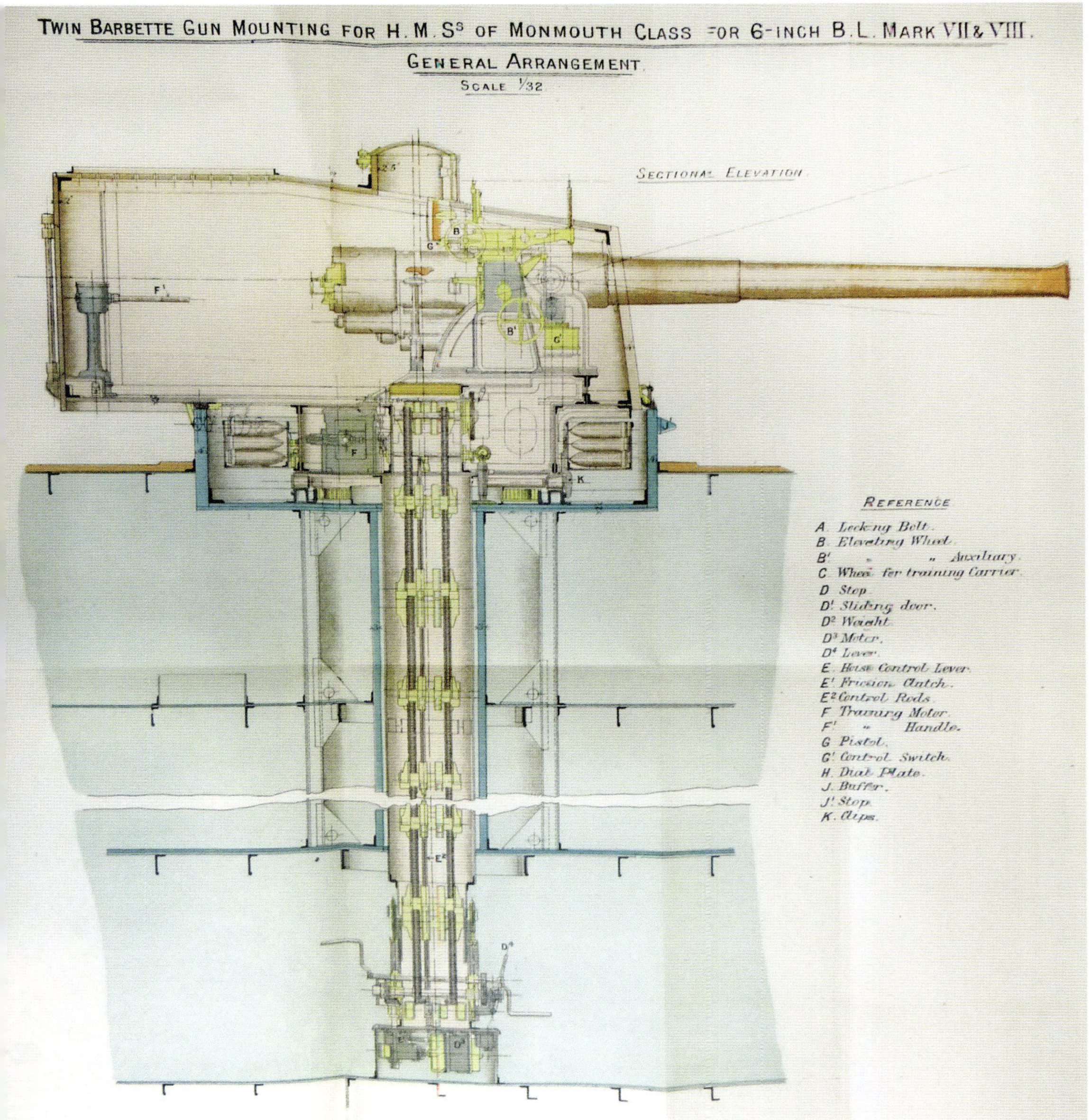


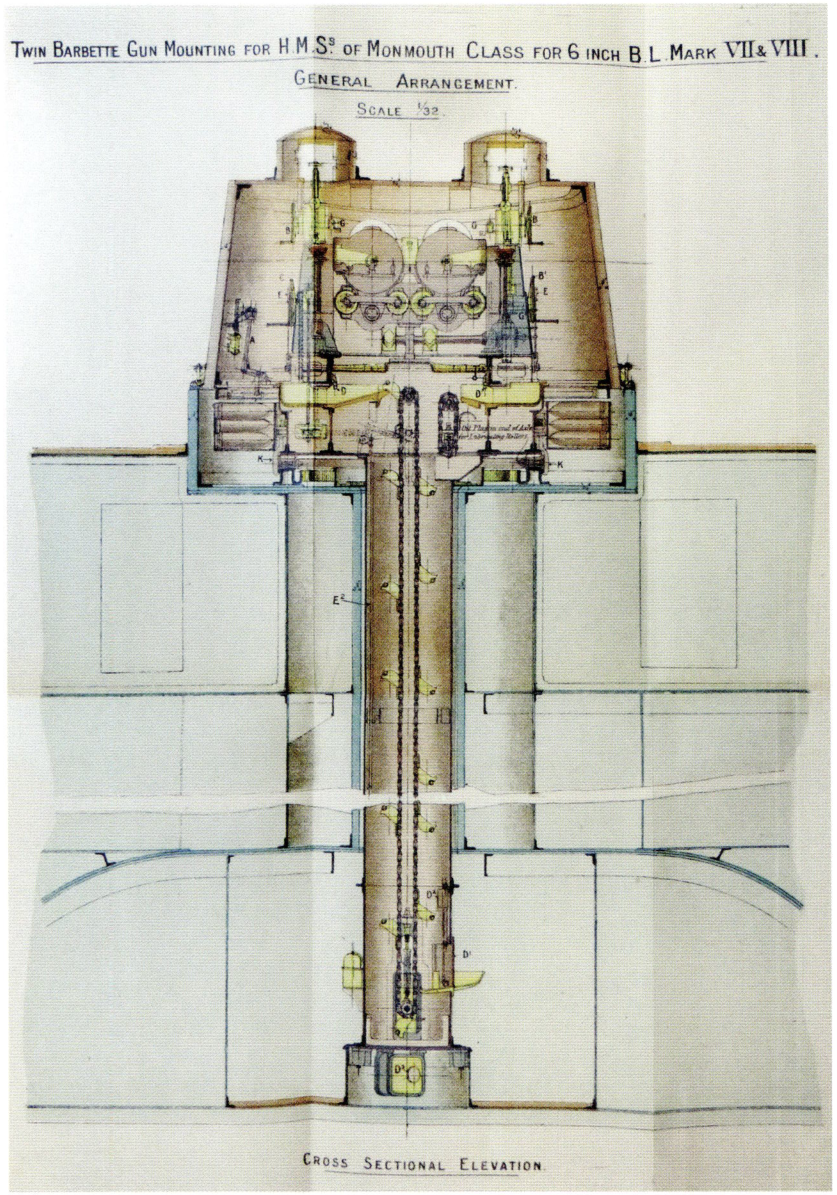


**Twin 6in Mark I mounting**

One twin 6in Mark I mounting designed by Vickers was fitted forward and one aft in the *Monmouth* class armoured cruisers (plus ten single 6in on transferable mountings); 153 shells were stowed in the circular space immediately below the turret. Charges were delivered up the central hoist to outside each gun. Training and hoisting were electrically powered but the two guns in a single cradle elevated by hand. *(Handbook IWM LBY 16695-1)*

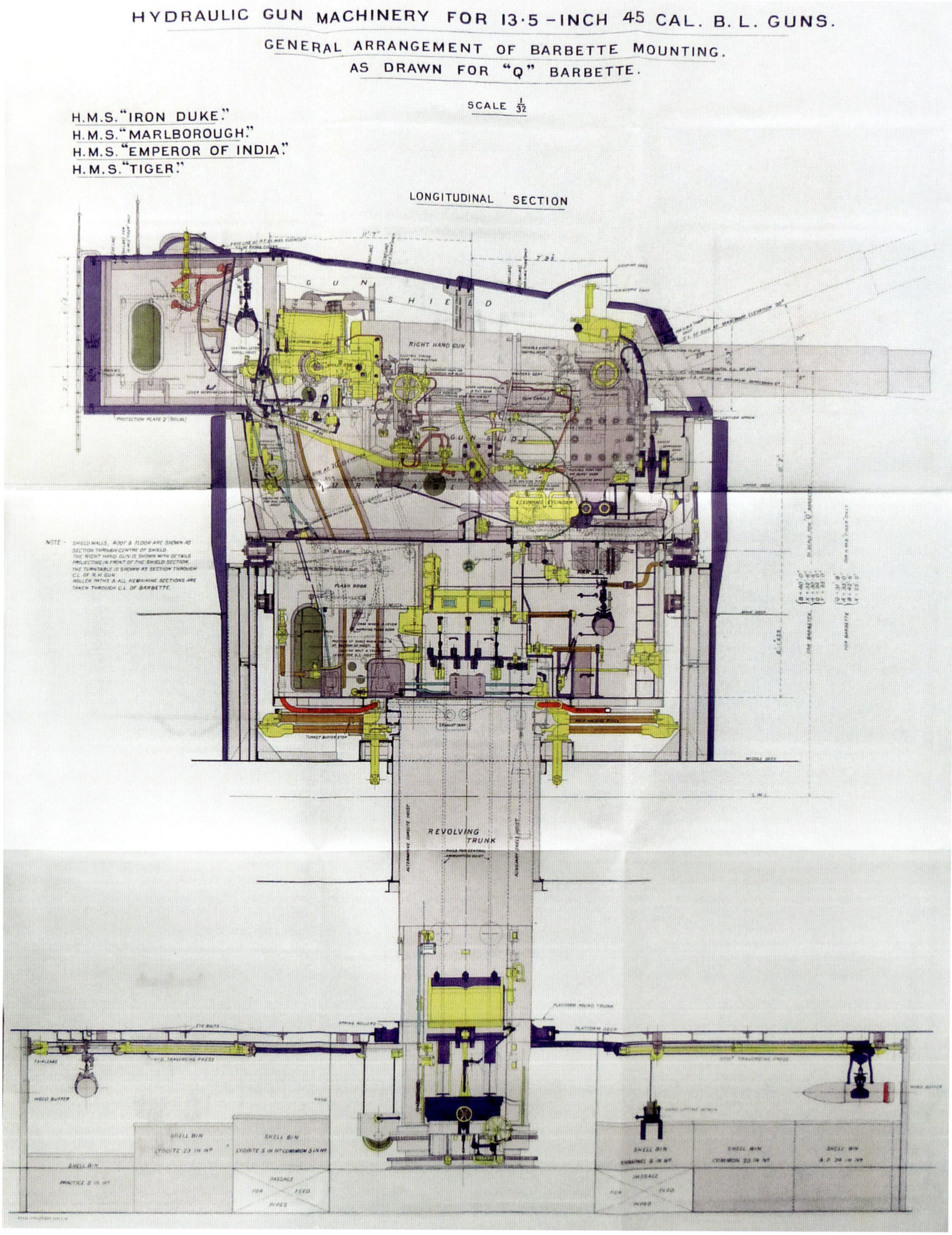
HYDRAULIC GUN MACHINERY FOR 13·5-INCH 45 CAL. B. L. GUNS.
GENERAL ARRANGEMENT OF BARBETTE MOUNTING.
AS DRAWN FOR "Q" BARBETTE.
SCALE 1/32
H.M.S. "IRON DUKE."
H.M.S. "MARLBOROUGH."
H.M.S. "EMPEROR OF INDIA."
H.M.S. "TIGER."
LONGITUDINAL SECTION
GUN SHIELD
RIGHT HAND GUN
GUN SLIDE
ELEVATING CYLINDER
FLASH DOOR
REVOLVING TRUNK
SHELL BIN
PASSAGE FOR FEED PIPES

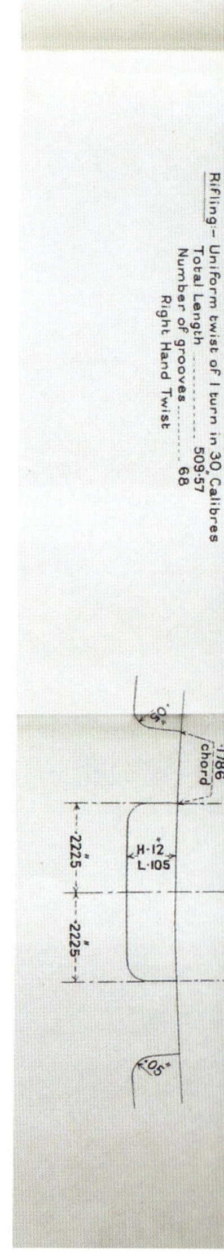
Rifling – Uniform twist of 1 turn in 30 Calibres
Total Length ........ 509·57
Number of grooves ........ 68
Right Hand Twist

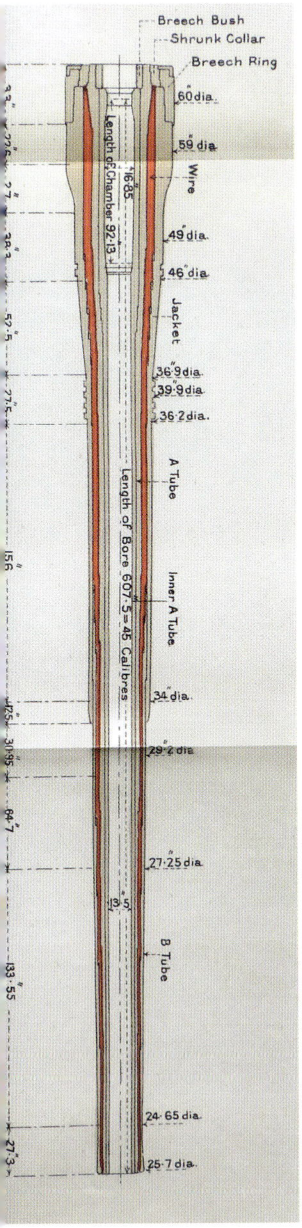


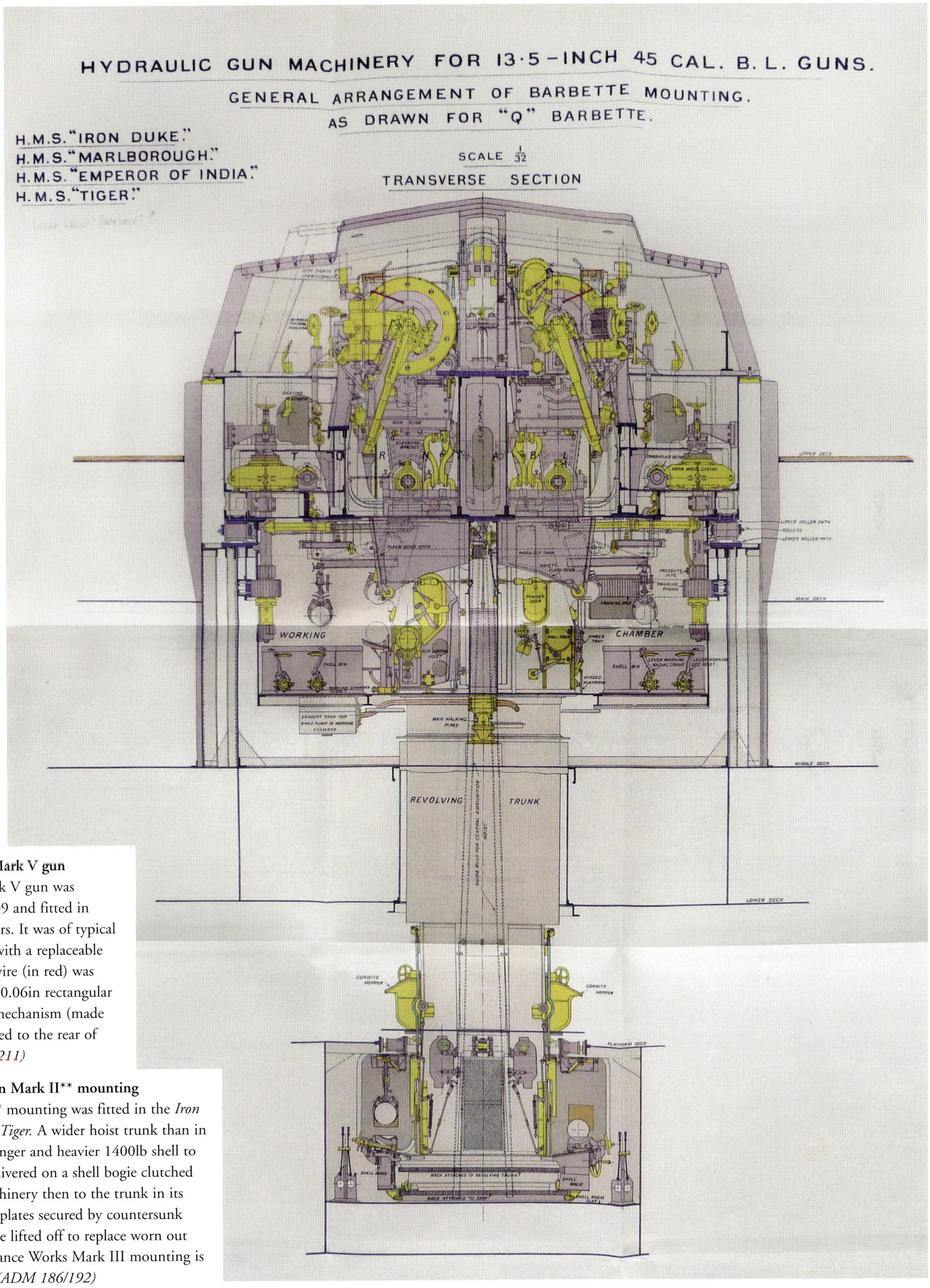


**_Above:_ 13.5in 45 calibre Mark V gun**
The 13.5in 45 calibre Mark V gun was designed by Vickers in 1909 and fitted in battleships and battlecruisers. It was of typical wire-wound construction with a replaceable inner A tube (liner). The wire (in red) was high tensile steel of 0.25 x 0.06in rectangular cross section. The breech mechanism (made separately) would be screwed to the rear of the gun body. *(ADM 186/211)*

**_Left and right:_ Twin 13.5in Mark II** mounting**
The twin 13.5in Mark II** mounting was fitted in the *Iron Duke* class battleships and *Tiger.* A wider hoist trunk than in the Mark II enabled the longer and heavier 1400lb shell to be hoisted horizontally, delivered on a shell bogie clutched first to the shell room machinery then to the trunk in its rotated position. The roof plates secured by countersunk tapped screw bolts could be lifted off to replace worn out guns. The Coventry Ordnance Works Mark III mounting is shown in the frontispiece *(ADM 186/192)*

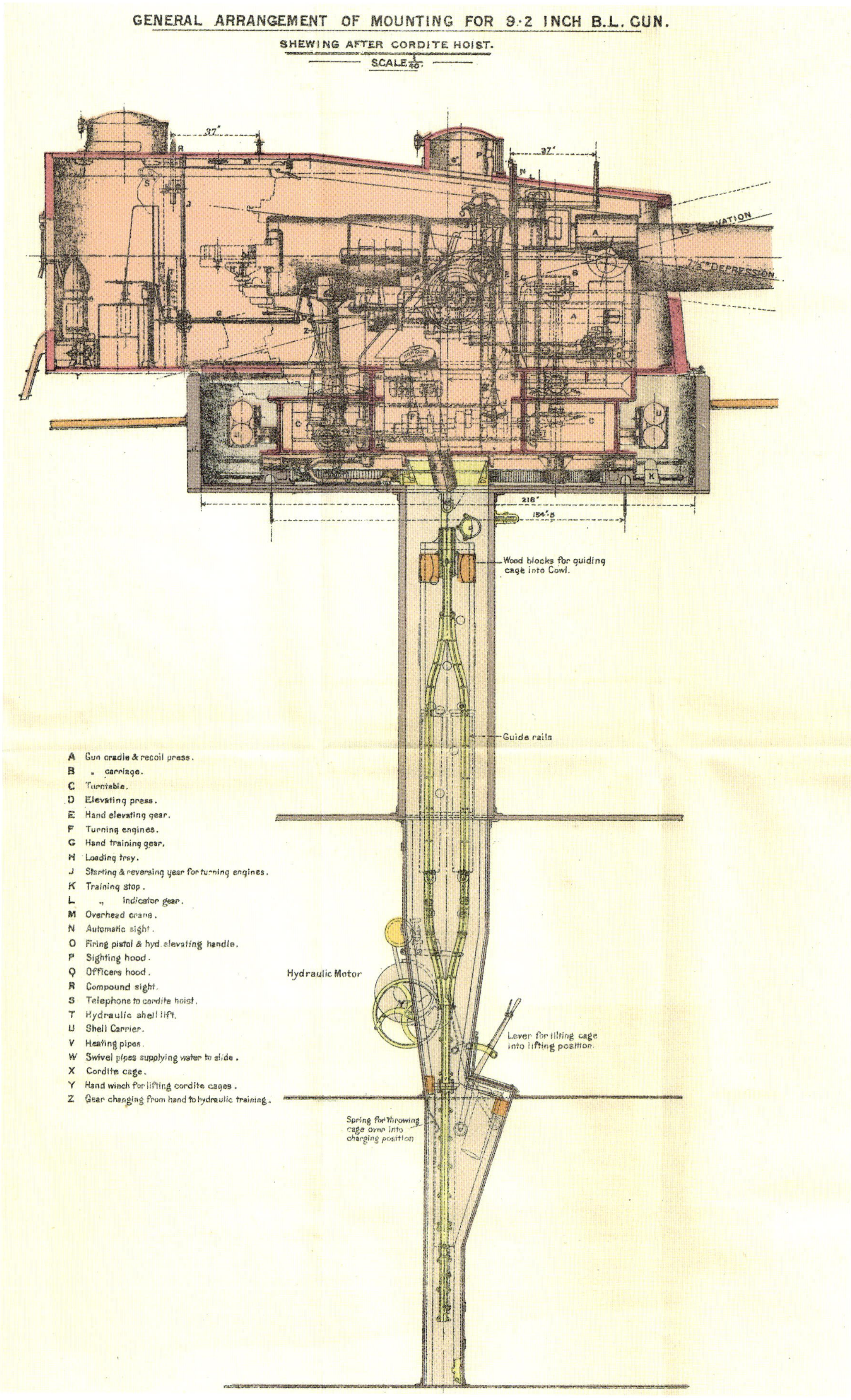


**Single 9.2in Mark V gun mounting**
9.2in guns were widely fitted in armoured cruisers and as secondary armament in later pre-dreadnoughts in both single and twin mountings. The single Mark V illustrated was an Elswick design fitted in the *Cressy* class, one forward one aft, hydraulically operated. *(Addenda to Manual of Hydraulics: 9.2-inch BL Mark X Guns on Mark V (Elswick) Mountings, 1901)*

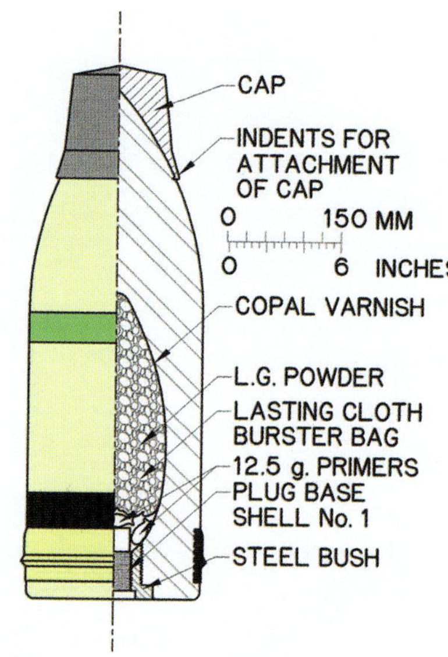


**A 9.2in common pointed capped 4crh shell without windshield.**
The soft cap distributed the impact over a larger area allowing the hard point to penetrate the armour. *(W Jurens)*

***Right and overleaf:* twin 15in Mark I* mounting**
The 15in twin mounting was the most successful British heavy gun mounting, the result of twenty years of evolution by Elswick. The drawings taken from the handbook show the Mark I* version fitted in *Royal Oak, Renown, Repulse, Courageous* and *Glorious*. Compared with the Mark I, they had a different arrangement for transferring shells from the shell room to the bottom of the revolving hoist through Kenyon doors. The plans of the Mark I in the *Queen Elizabeths* are reproduced in Robert Brown's *Battleship Warspite* (Seaforth Publishing 2017). All-angle loading was possible from 20° to -5°. *(ADM 186/315)*

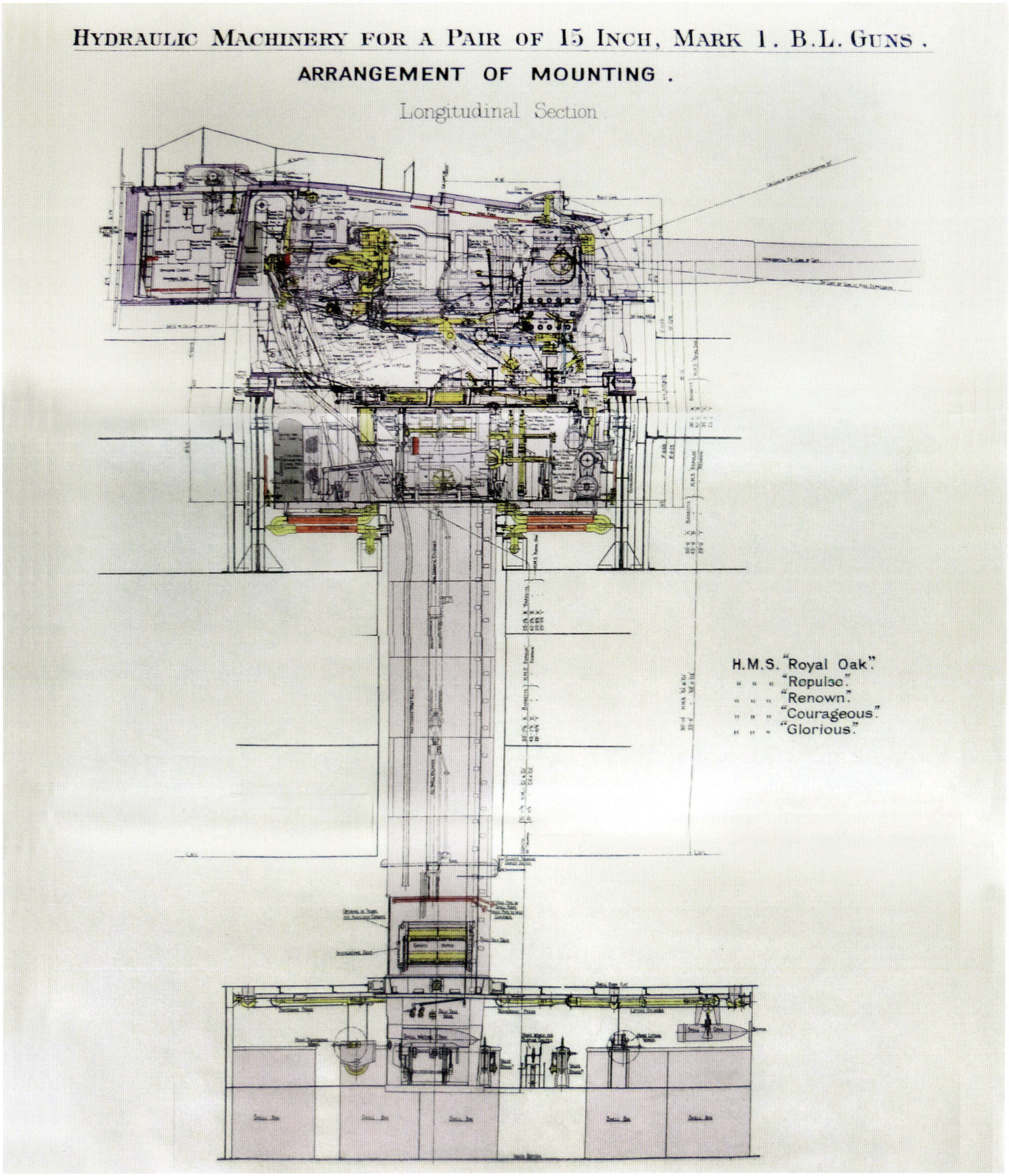
HYDRAULIC MACHINERY FOR A PAIR OF 15 INCH, MARK 1. B.L. GUNS.
ARRANGEMENT OF MOUNTING.
Longitudinal Section
H.M.S. "Royal Oak."
" " " "Repulse."
" " " "Renown."
" " " "Courageous."
" " " "Glorious."

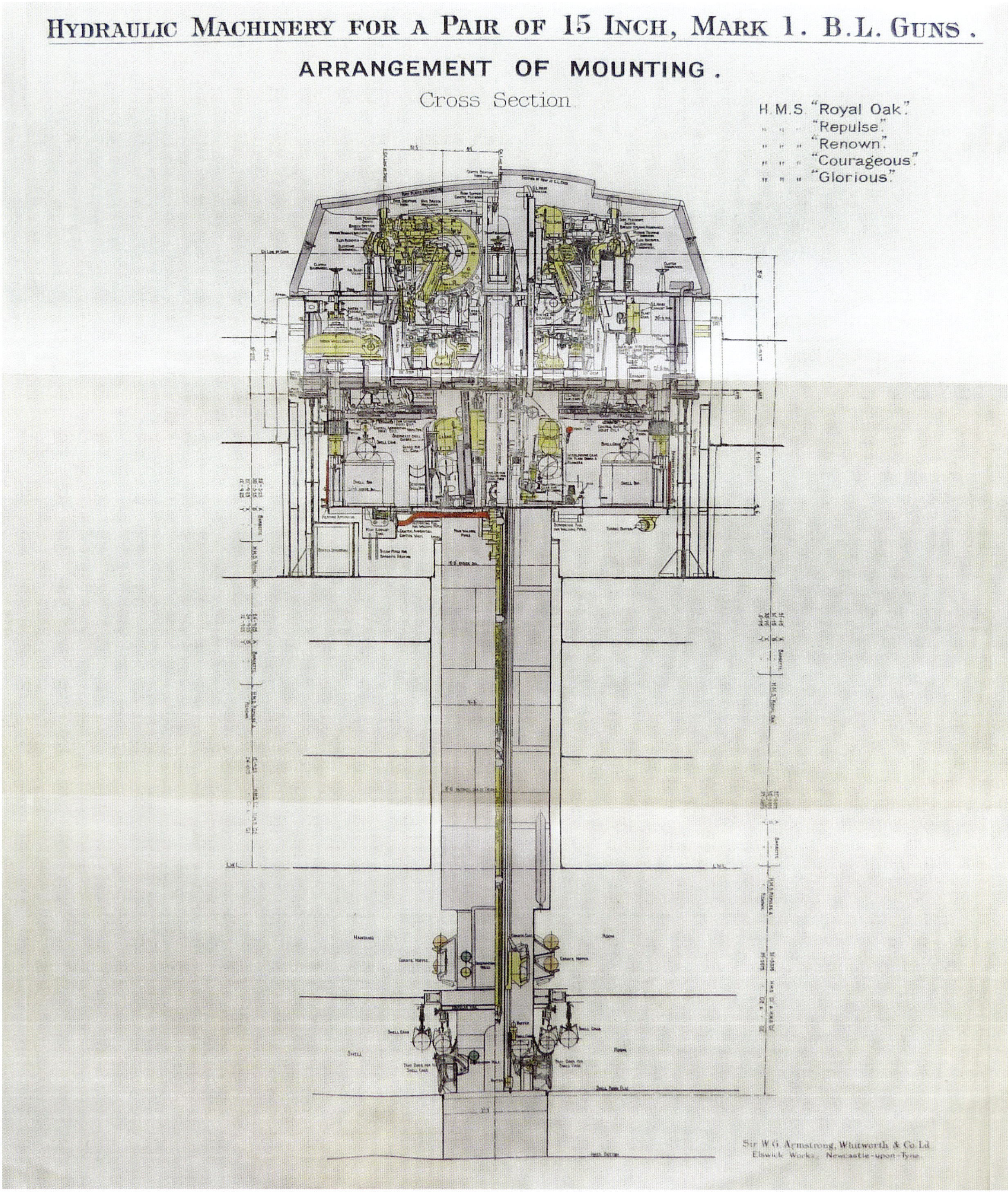
Hydraulic Machinery for a Pair of 15 Inch, Mark 1. B.L. Guns.
ARRANGEMENT OF MOUNTING.
Cross Section
H.M.S. "Royal Oak."
" " " "Repulse."
" " " "Renown."
" " " "Courageous."
" " " "Glorious."
Sir W.G. Armstrong, Whitworth & Co. Ld
Elswick Works, Newcastle-upon-Tyne

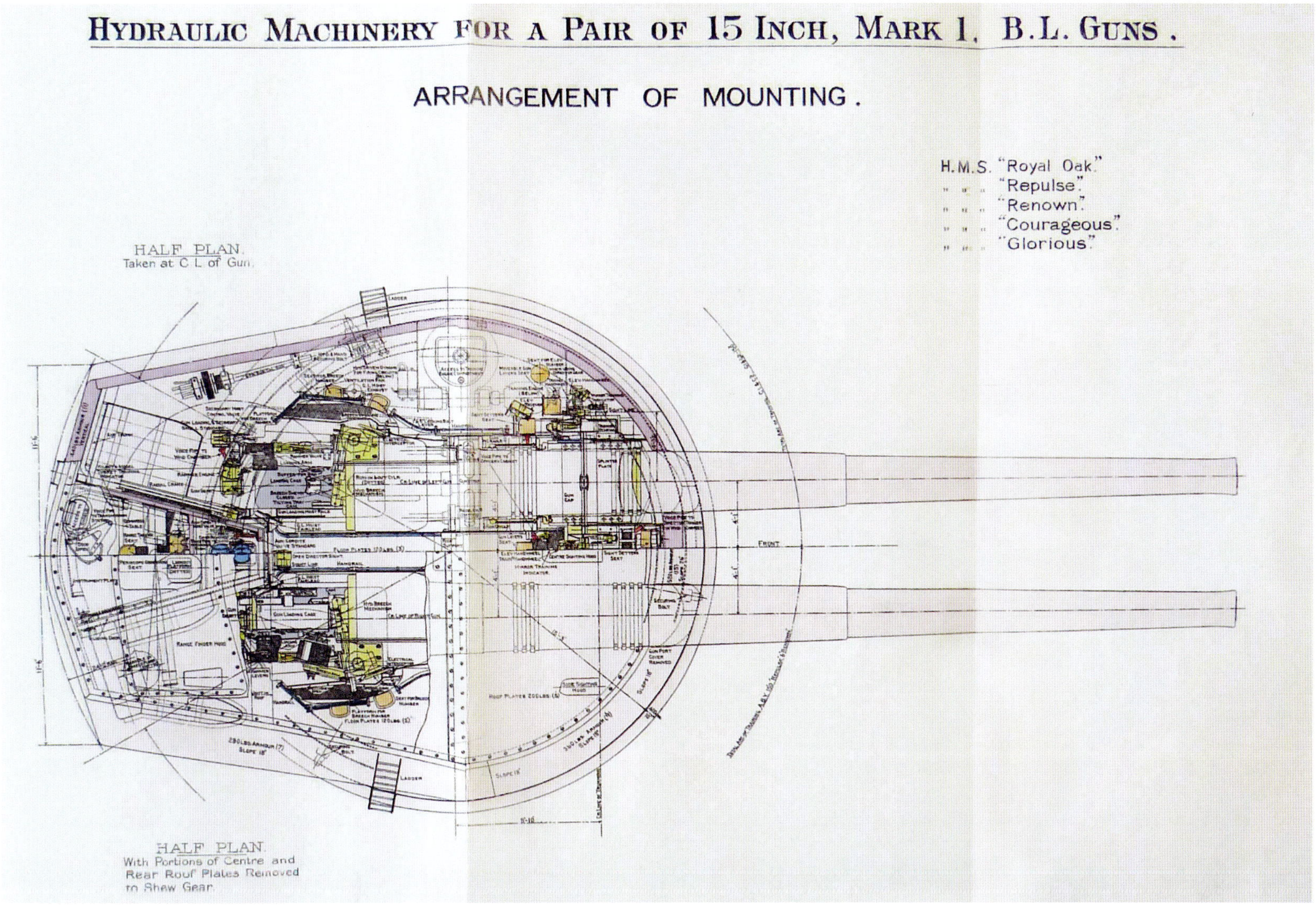
HYDRAULIC MACHINERY FOR A PAIR OF 15 INCH, MARK I, B.L. GUNS.
ARRANGEMENT OF MOUNTING.
H.M.S. "Royal Oak."
" " " "Repulse."
" " " "Renown."
" " " "Courageous."
" " " "Glorious."
HALF PLAN.
Taken at C L of Gun.
FRONT
HALF PLAN.
With Portions of Centre and
Rear Roof Plates Removed
to Shew Gear.

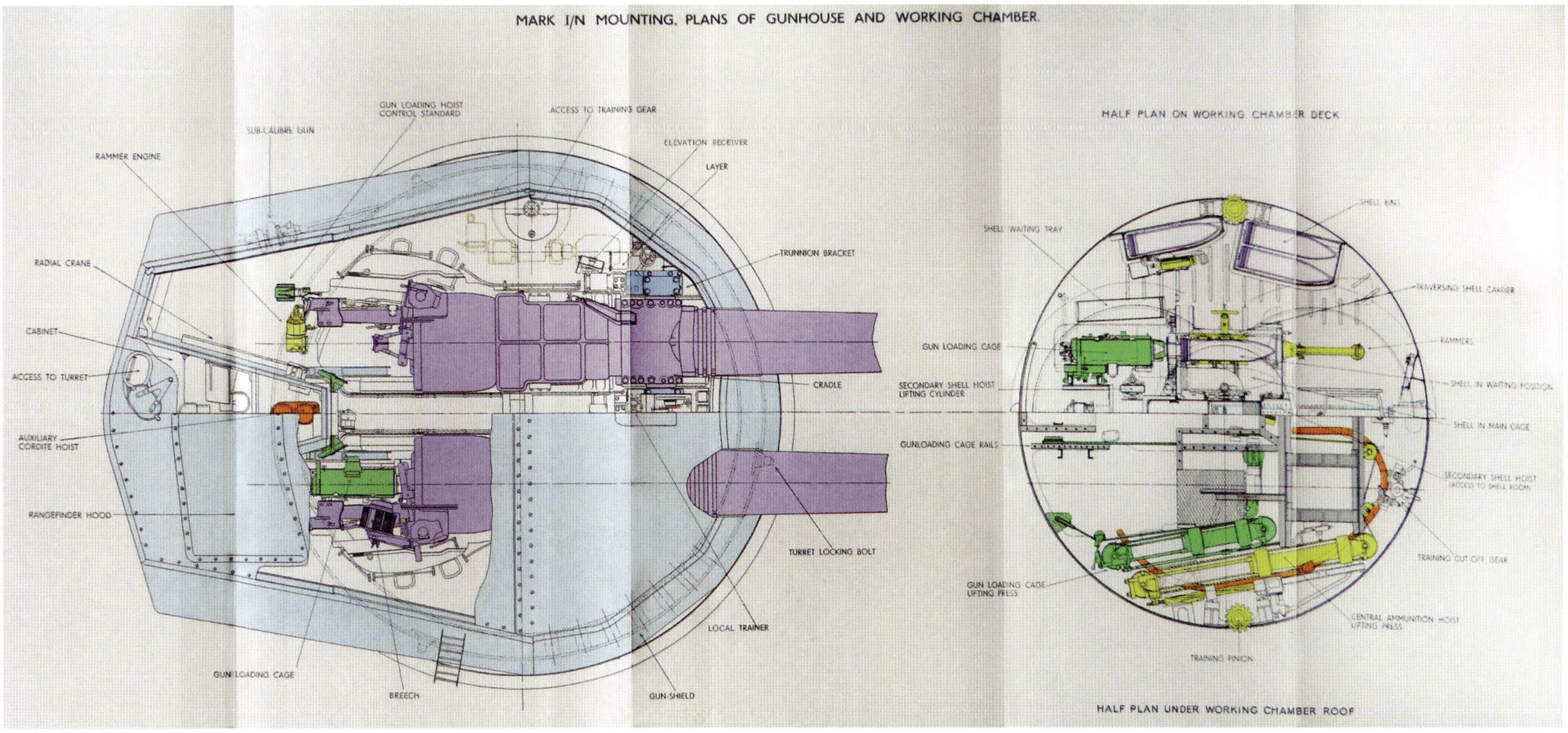
MARK I/N MOUNTING, PLANS OF GUNHOUSE AND WORKING CHAMBER.
GUN LOADING HOIST CONTROL STANDARD
ACCESS TO TRAINING GEAR
SUB-CALIBRE GUN
RAMMER ENGINE
ELEVATION RECEIVER
LAYER
RADIAL CRANE
TRUNNION BRACKET
CABINET
ACCESS TO TURRET
CRADLE
AUXILIARY CORDITE HOIST
RANGEFINDER HOOD
TURRET LOCKING BOLT
LOCAL TRAINER
GUN LOADING CAGE
BREECH
GUN SHIELD
HALF PLAN ON WORKING CHAMBER DECK
SHELL BINS
SHELL WAITING TRAY
TRAVERSING SHELL CARRIER
GUN LOADING CAGE
RAMMERS
SECONDARY SHELL HOIST LIFTING CYLINDER
SHELL IN WAITING POSITION
SHELL IN MAIN CAGE
GUNLOADING CAGE RAILS
SECONDARY SHELL HOIST (ACCESS TO SHELL ROOM)
TRAINING CUT-OFF GEAR
GUN LOADING CAGE LIFTING PRESS
CENTRAL AMMUNITION HOIST LIFTING PRESS
TRAINING PINION
HALF PLAN UNDER WORKING CHAMBER ROOF

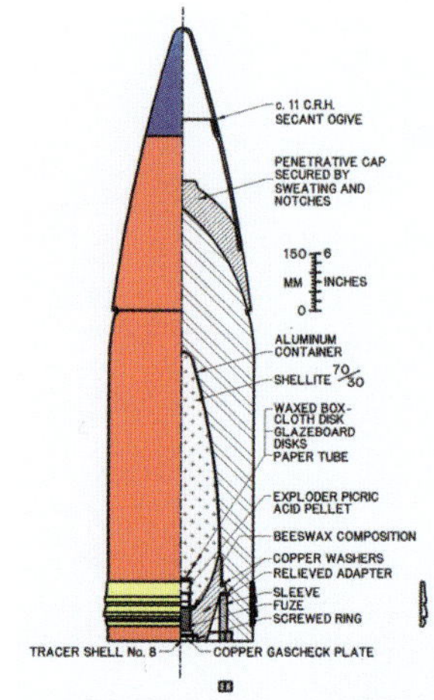


**15in shell**
A 15in armour piercing capped 6crh shell. *(W Jurens)*

***Previous page and below:* Twin 15in Mark I/N mounting**
The profile, section, plan and working chamber from the 1943 handbook BR 849 give the clearest impression of the main parts of a 15in mounting, compared with the overly detailed drawings in the WW1 handbooks. The Mark I/N mounting is illustrated with its elevation increased to 30° but retaining the shell room below the magazine. *(ADM 234/168)*

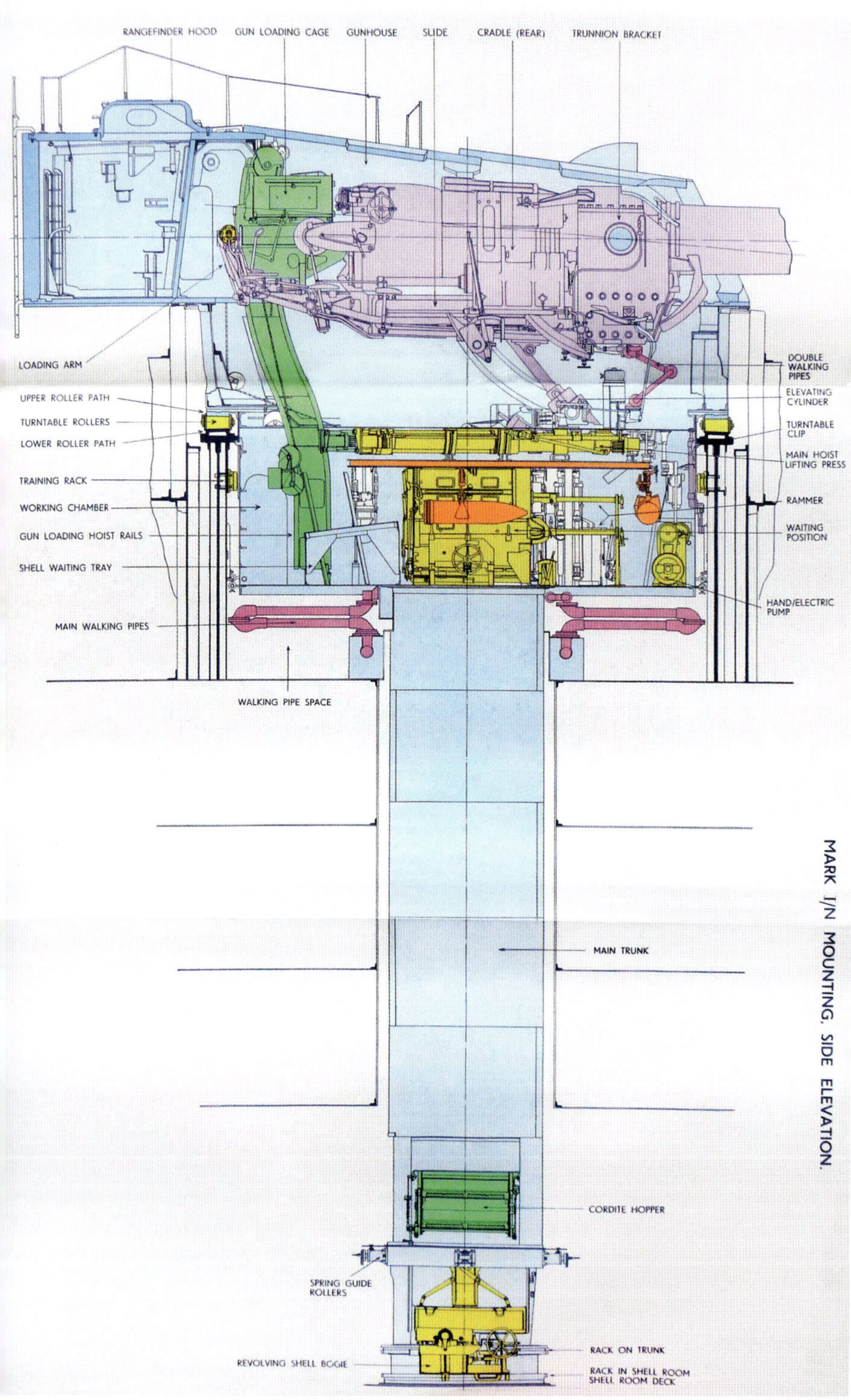


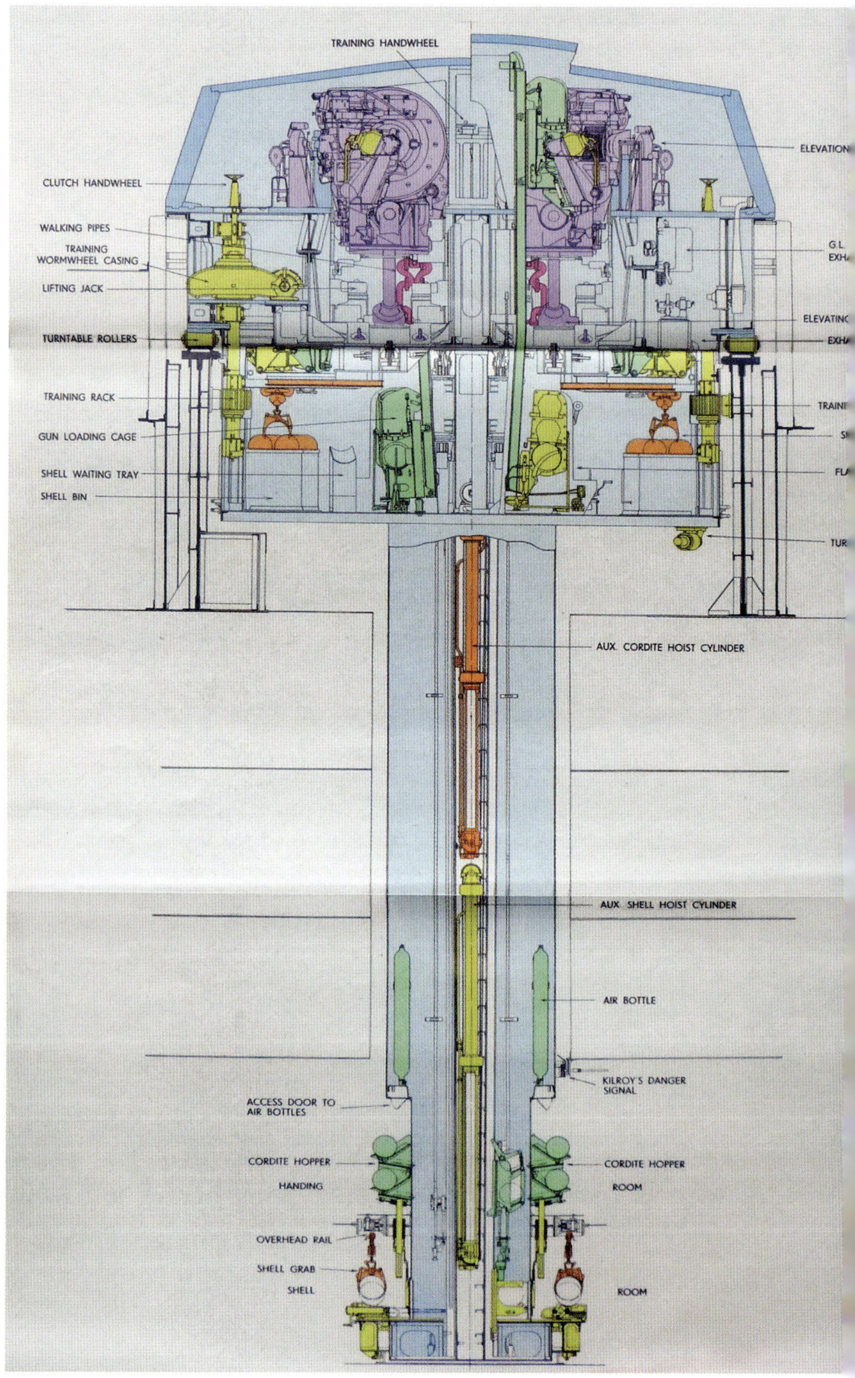

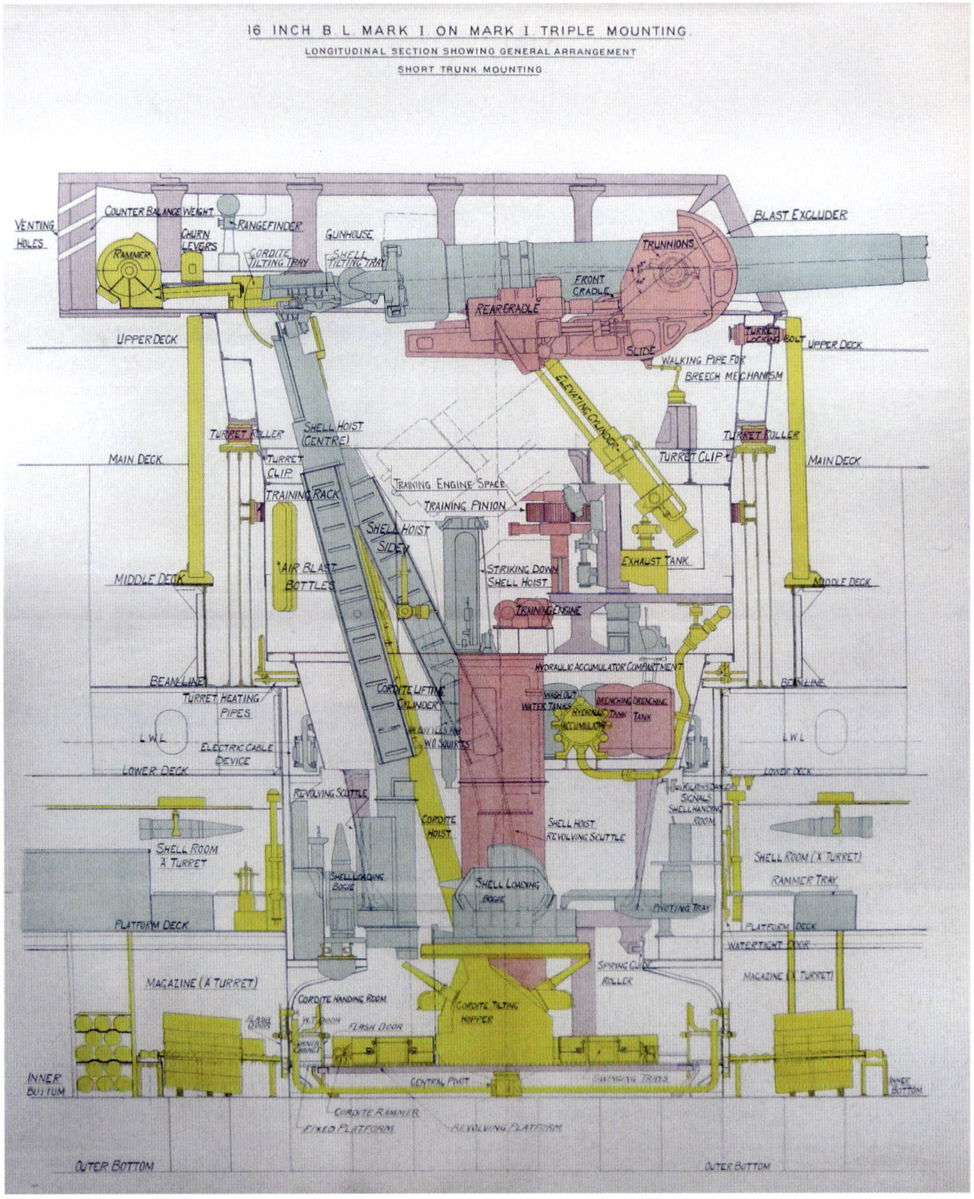


*Left and overleaf:* **Triple 16in Mark I gun mounting**
The triple 16in Mark I mountings in *Nelson* and *Rodney* designed by Elswick differed in many respects from the twin 15in. Magazines were below the shell rooms to give better protection from long range plunging fire and bombs. There was no working chamber to transfer ammunition from a central hoist to the gunloading cages. Instead, the three cordite and shell hoists were angled backwards to deliver direct to loading trays at the rear of each gun but with improved flashtightness (cordite yellow, shells blue). The more complex bogie loading system at the bottom of the hoists proved troublesome. The thick steel balanced plate fitted inside the rear armour plate is evident, both pierced by gas venting holes. *(ADM 186/285-287)*

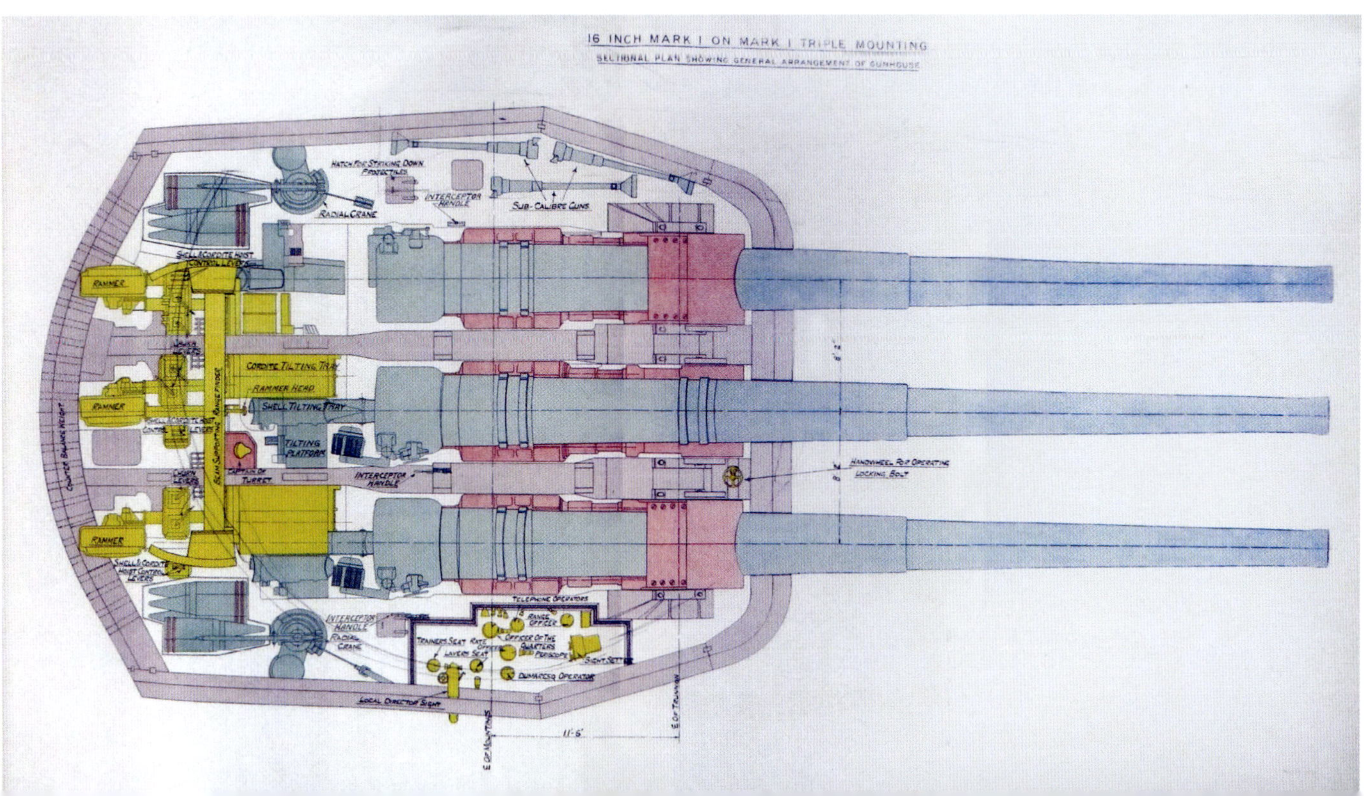

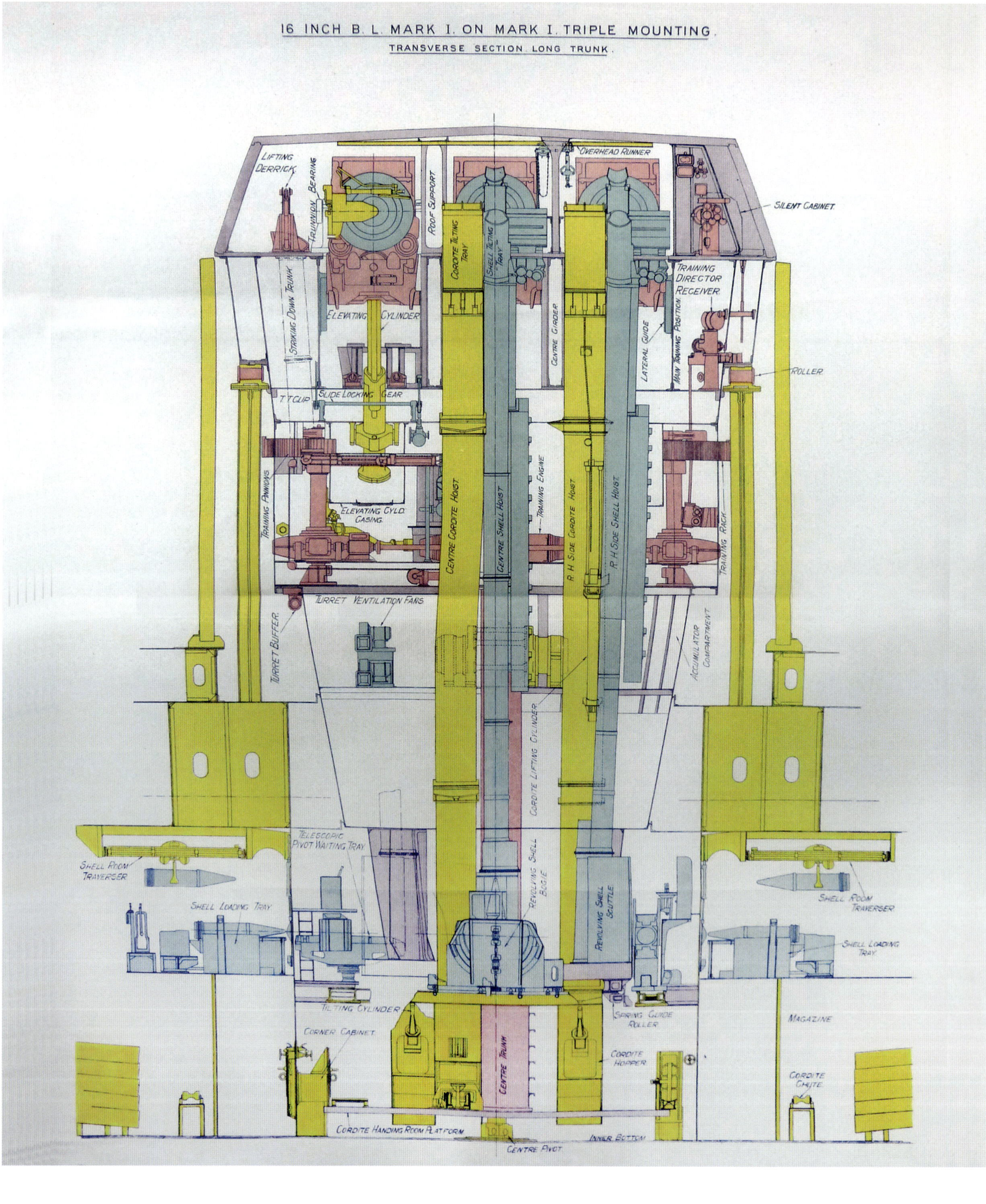

16 INCH B. L. MARK I. ON MARK I. TRIPLE MOUNTING.
TRANSVERSE SECTION LONG TRUNK.
LIFTING DERRICK
TRUNNION BEARING
ROOF SUPPORT
CORDITE TILTING TRAY
SHELL TILTING TRAY
OVERHEAD RUNNER
SILENT CABINET
STRIKING DOWN TRUNK
ELEVATING CYLINDER
CENTRE GIRDER
LATERAL GUIDE
MAIN TRAINING POSITION
TRAINING DIRECTOR RECEIVER
ROLLER
T.T. CLIP
SLIDE LOCKING GEAR
TRAINING PINIONS
ELEVATING CYLD. CASING
CENTRE CORDITE HOIST
CENTRE SHELL HOIST
TRAINING ENGINE
R. H. SIDE CORDITE HOIST
R. H. SIDE SHELL HOIST
TRAINING RACK
TURRET VENTILATION FANS
TURRET BUFFER
ACCUMULATOR COMPARTMENT
CORDITE LIFTING CYLINDER
TELESCOPIC PIVOT WAITING TRAY
SHELL ROOM TRAVERSER
REVOLVING SHELL BOGIE
REVOLVING SHELL SCUTTLE
SHELL LOADING TRAY
TILTING CYLINDER
SPRING GUIDE ROLLER
MAGAZINE
CORNER CABINET
CENTRE TRUNK
CORDITE HOPPER
CORDITE CHUTE
CORDITE HANDING ROOM PLATFORM
INNER BOTTOM
CENTRE PIVOT

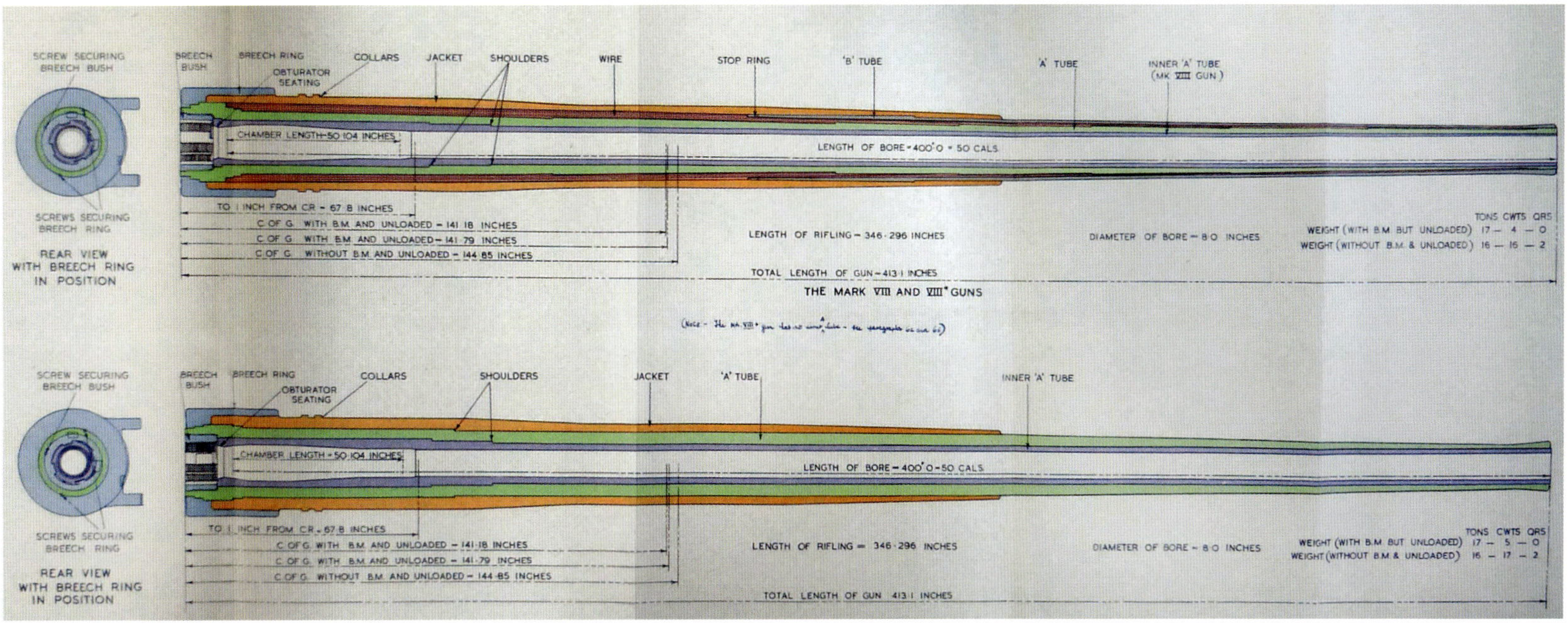


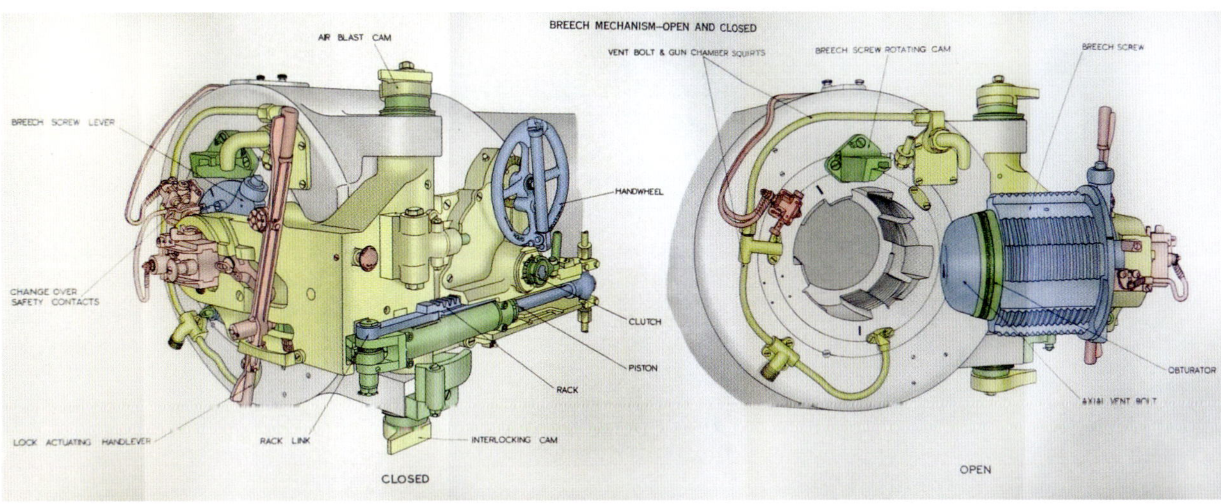


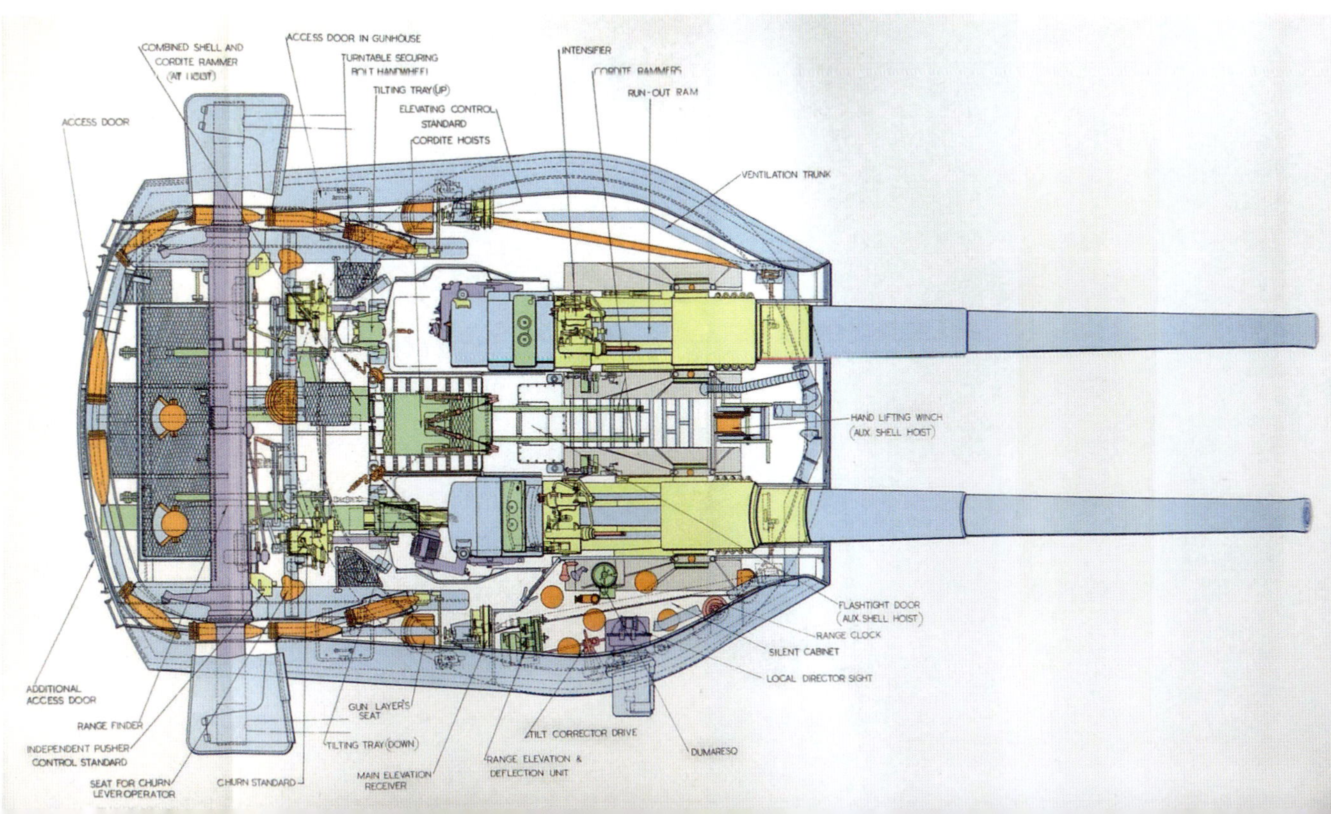


*Above:* **8in Mark VIII (top) and Mark VIII***
Early 8in guns used the well-tried wire winding construction in the Mark VIII. With improved steels and better forming techniques, all-steel guns became possible as illustrated by the Mark VIII*. The guns were interchangeable and some ships had a mix of both Marks. *(ADM 234/190)*

*Left:* **8in breech mechanism**
The 8in breech mechanism could be closed hydraulically or by hand. *(ADM 234/190)*

*Left and following two pages:* **Twin 8in Mark I mounting**
In the 8in Mark I mounting, the specified 70° elevation required a deep gun well to accommodate the recoil. The droop at the rear of the gunhouse floor was to accommodate the angled rammer. *(ADM 234/190)*

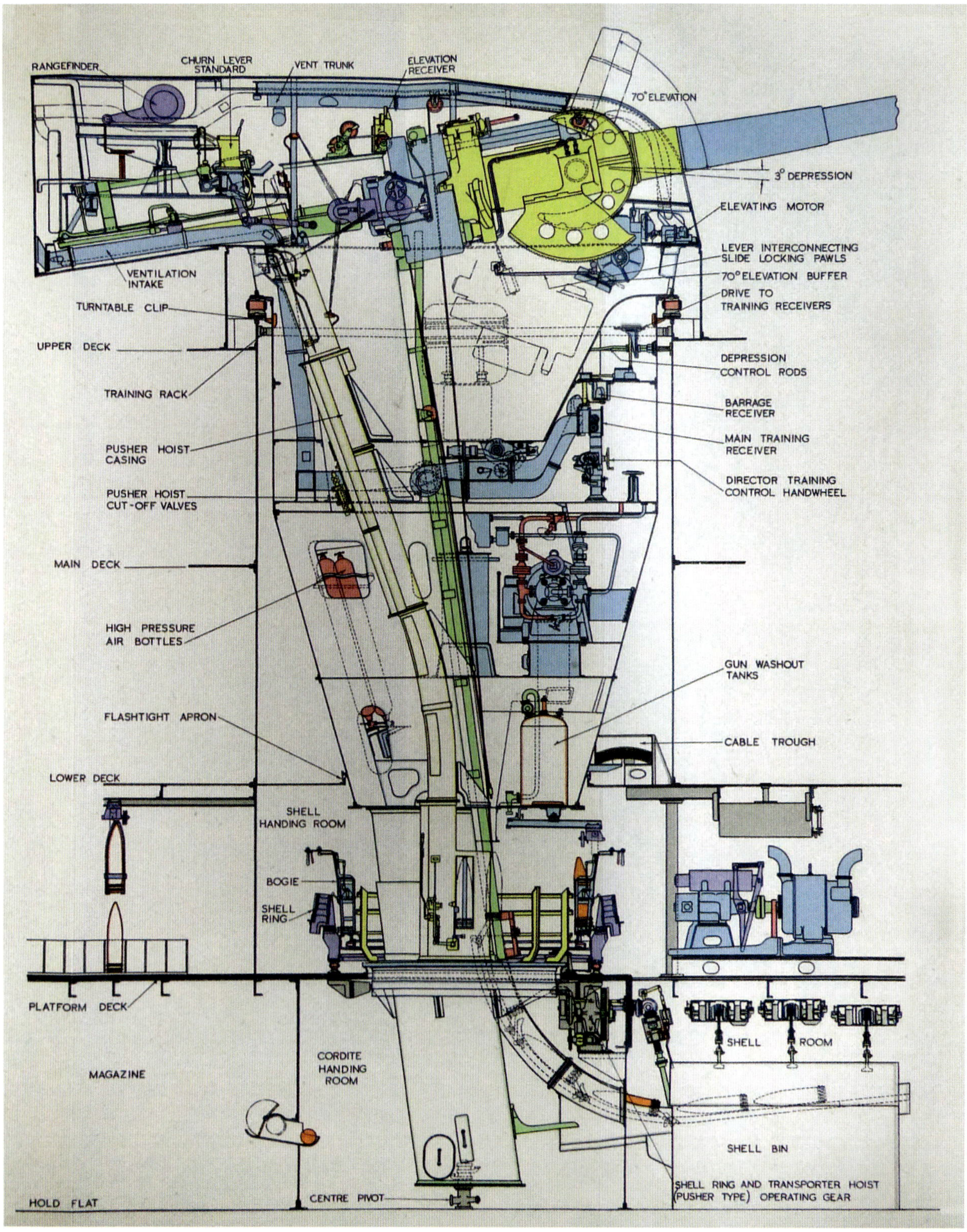


8in Mark I mounting profile

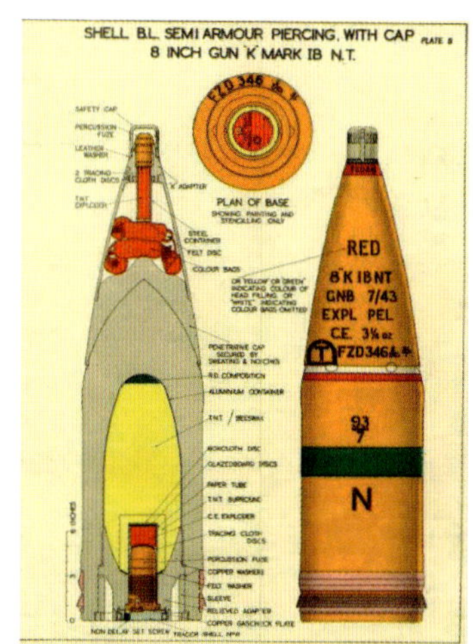


**8in shell**

8in SAP capped shell of WW2 with coloured splash dye to identify the firing ship. *(Handbook on Ammunition. BR 932, 1945. Navweaps)*

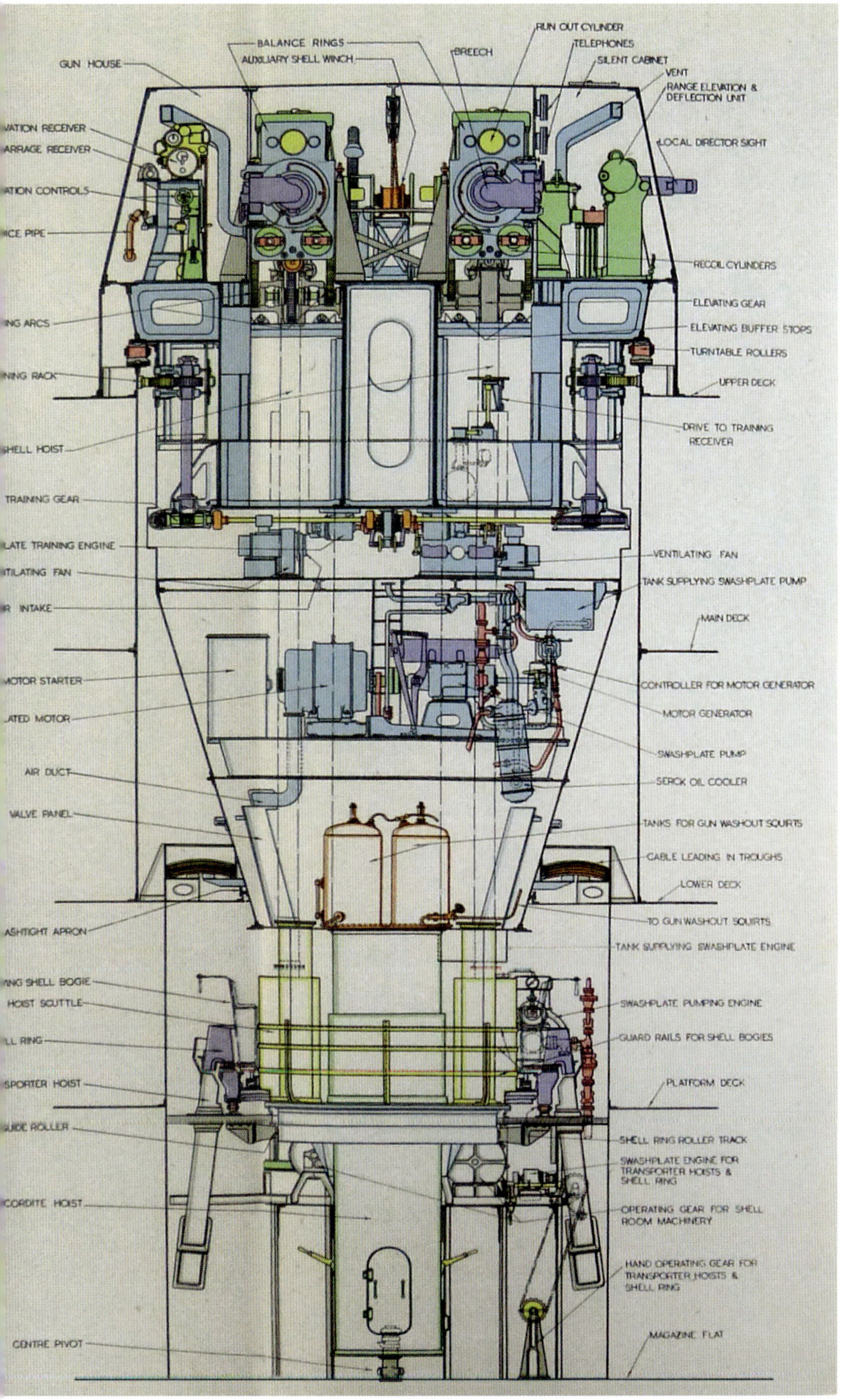


**8in Mark I mounting cross section**

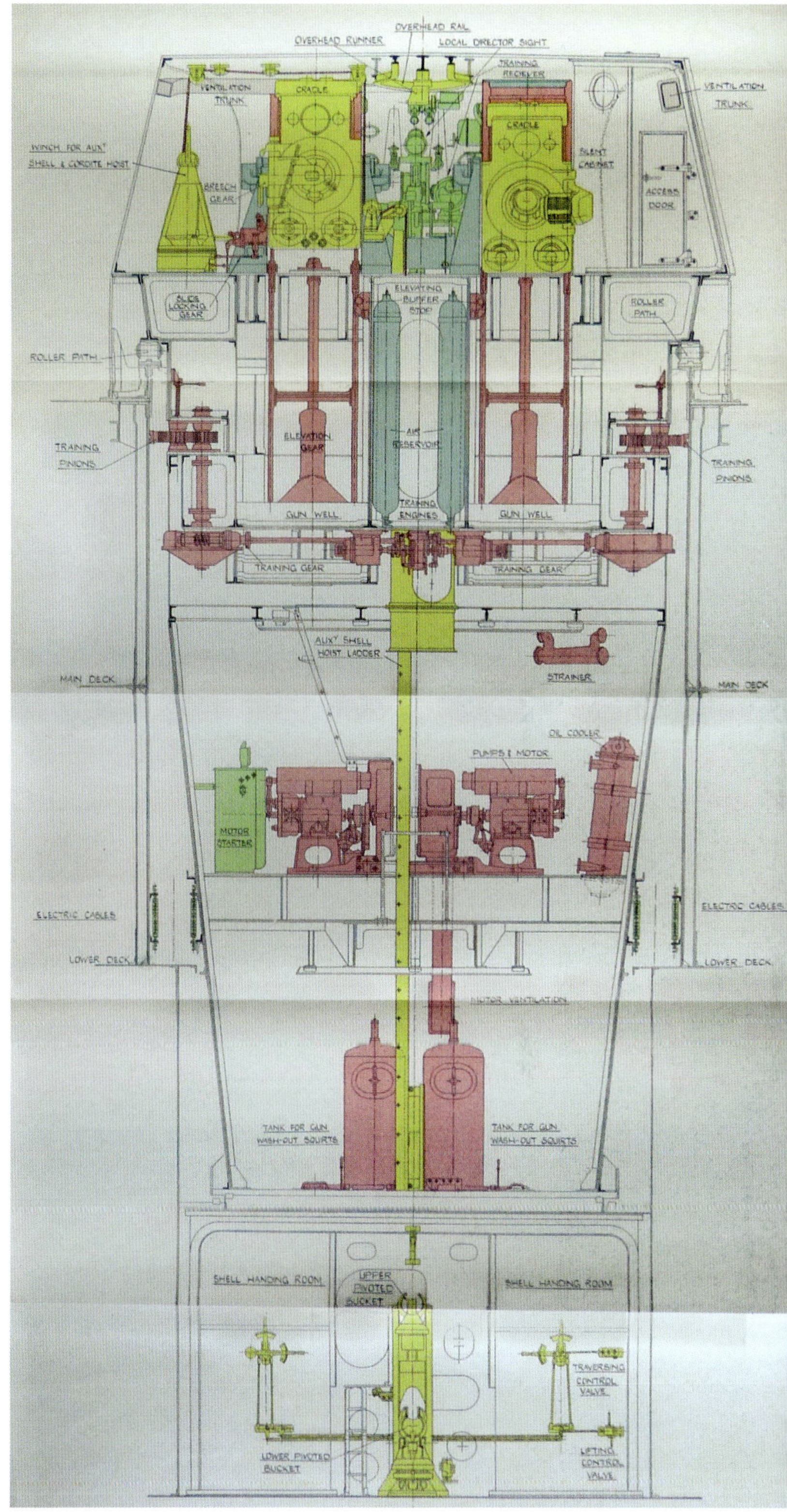


**8in Mark II mounting**

The 8in Mark II section with hoists direct from magazine to gunhouse. Angling the pusher hoists eliminated the transfer space of the Mark I. *(ADM 186/308)*

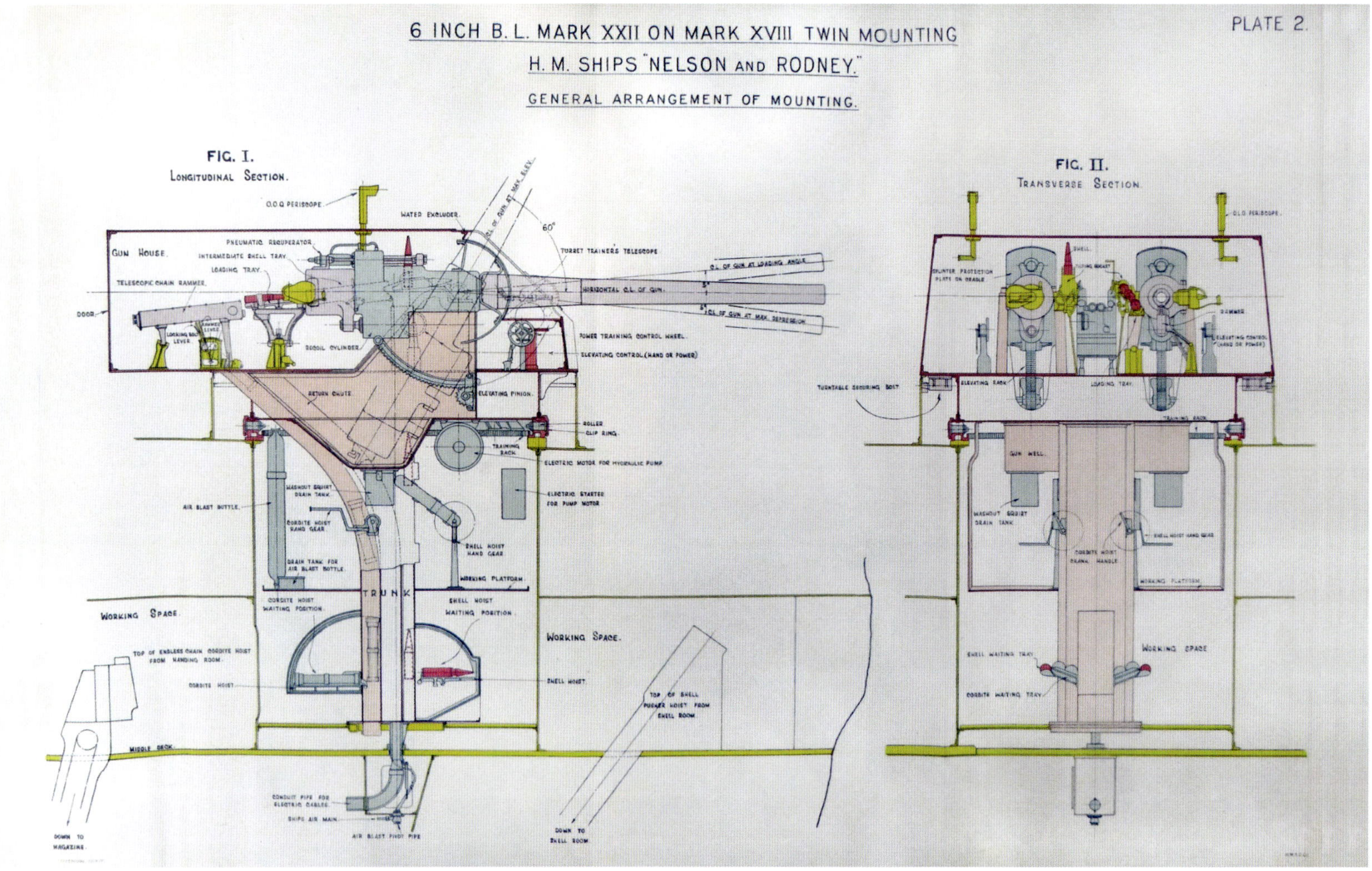


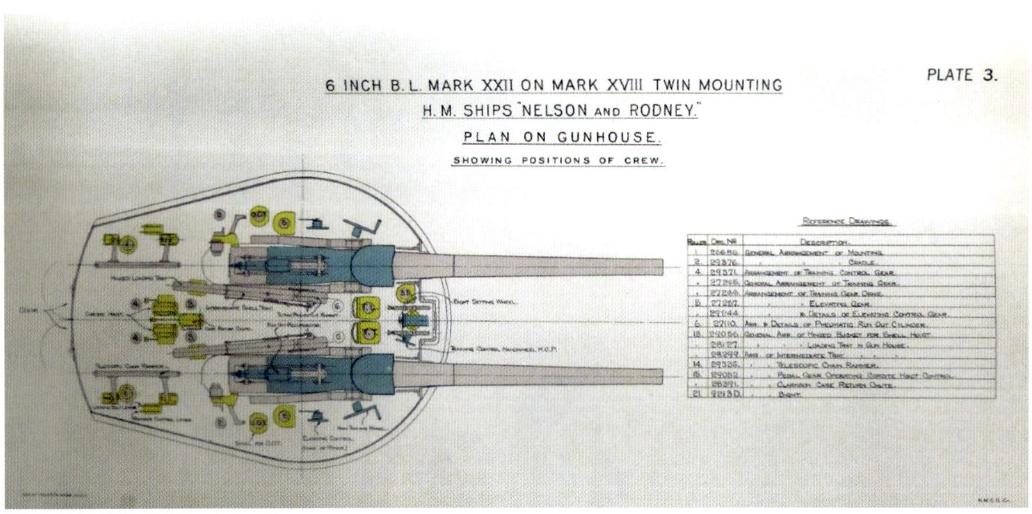


**Twin 6in Mark XVIII mounting**
*Nelson* and *Rodney* each had six Vickers designed twin 6in Mark XVIII mountings, mainly for LA fire but with 60° elevation could put up a HA barrage. They had a two-stage hoist mechanism as the magazines were not directly below the mountings, with chain driven hoists delivering to a working space before transfer to the trunk revolving with the mounting. Hydraulic motions were powered by an electrically driven pump. *(ADM 186/280)*

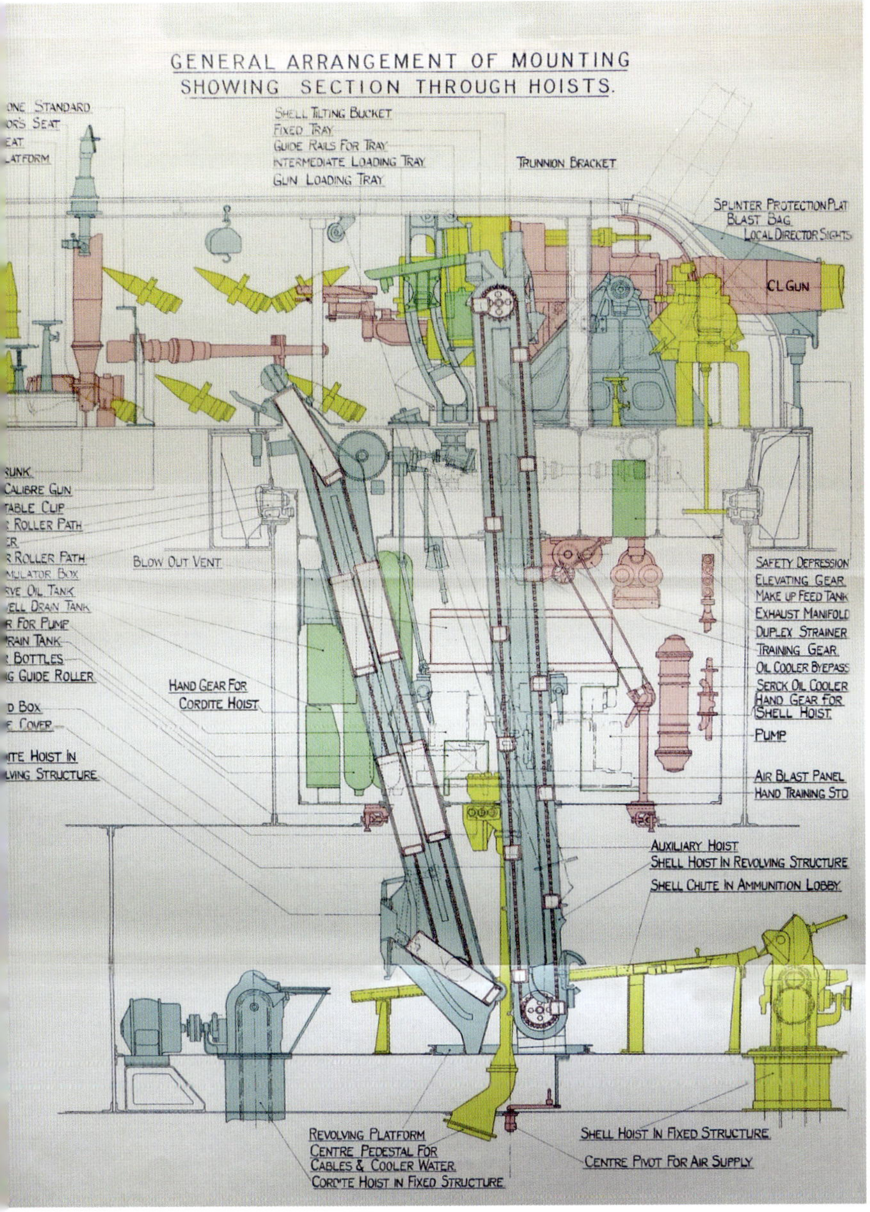


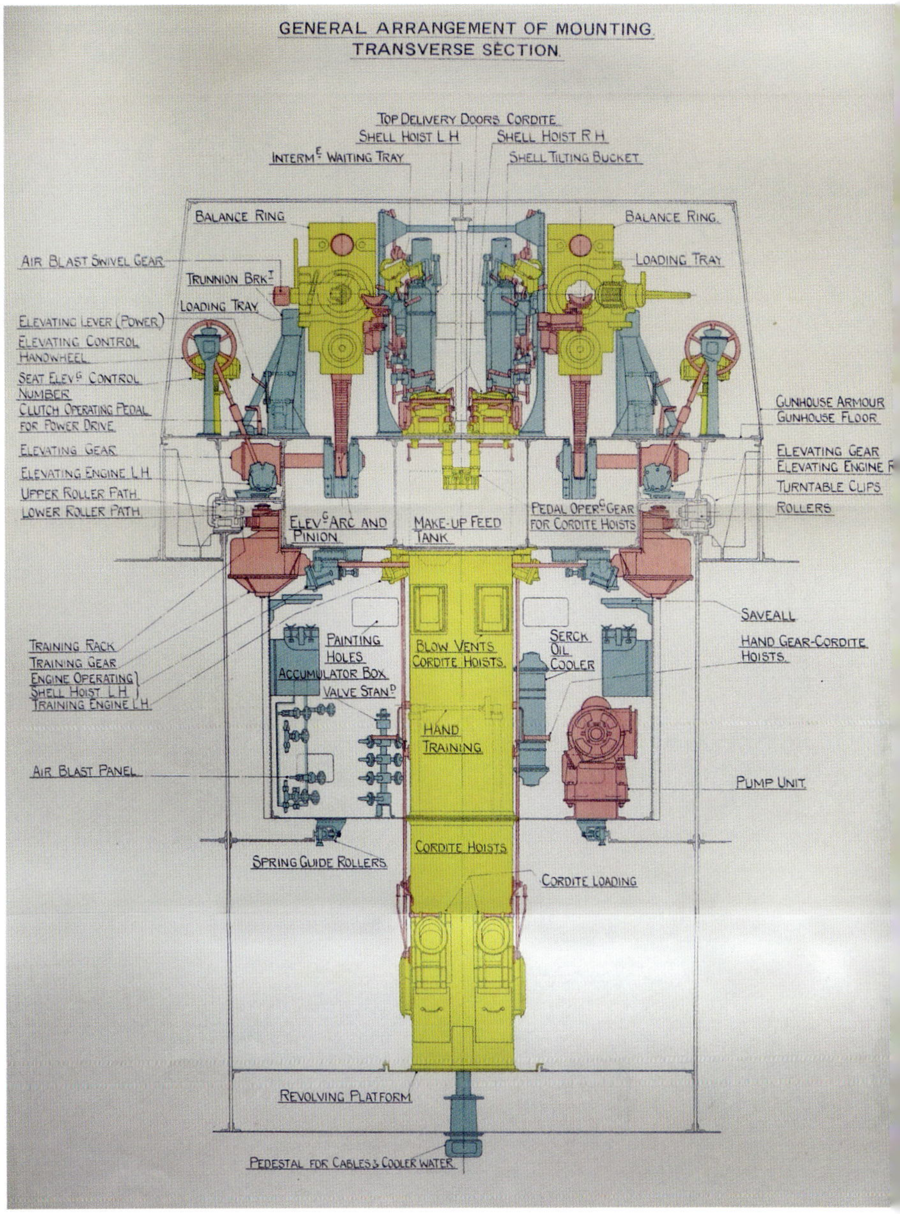


**Twin 6in Mark XXI mounting**

The *Leander* class cruisers had four twin 6in Mark XXI mountings designed by Elswick. It was of the 'short trunk' type, with the lower fixed hoists delivering from magazine and shell room to a transfer space where shell and charge were manually moved to the hoists rotating with the gunhouse. *(ADM 186/310 & 311)*

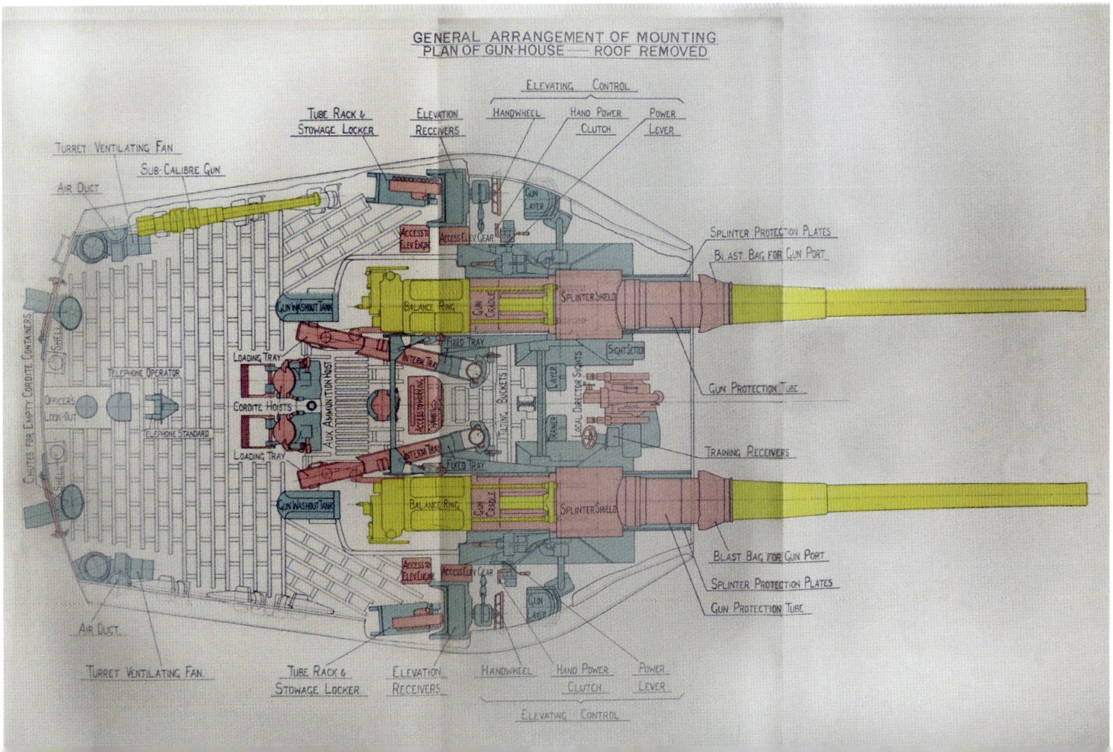

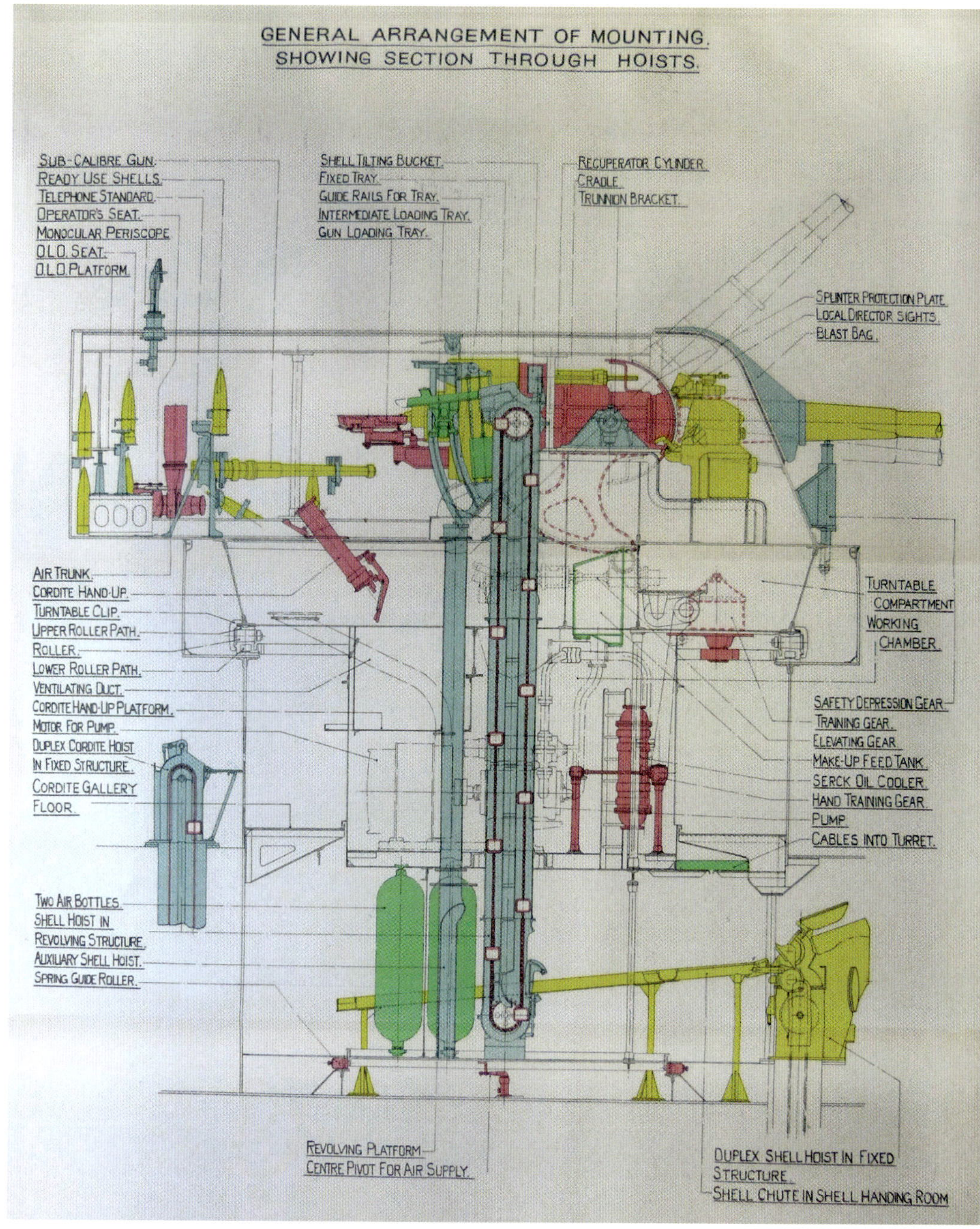


**Triple 6in Mark XXII mounting**
Four triple 6in Mark XXII mountings were fitted in the *Southampton* class cruisers, with the same 6in Mark XXIII gun as in the Mark XXI mounting. The shallow turntable compartment contained the training motors, which were powered by an electrically driven pump in the working chamber. The yellow fixed hoist from the shell room delivered to a chute, then to three vertical dredger hoists to the gunhouse. The shells were then tilted on to the fixed green trays, then on to the pink swinging trays to line up with the gun breeches. The centre gun was 2.5ft further aft than the wing guns which not only gave more space in the gunhouse for loading but reduced interference from the shells in flight. *(IWM LBY 30581-1)*

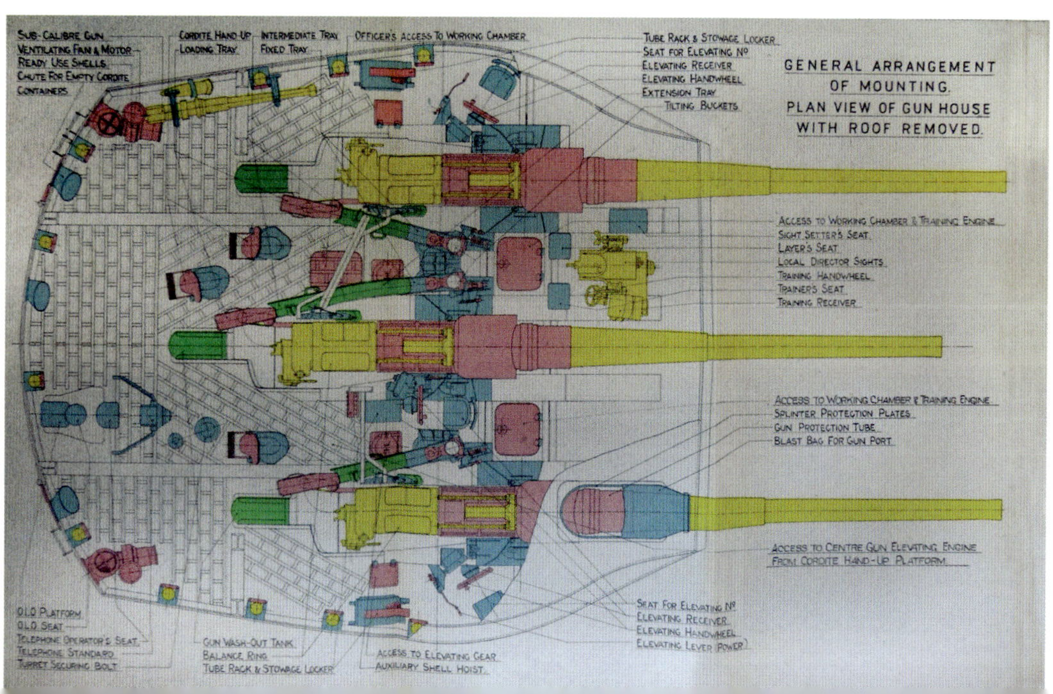

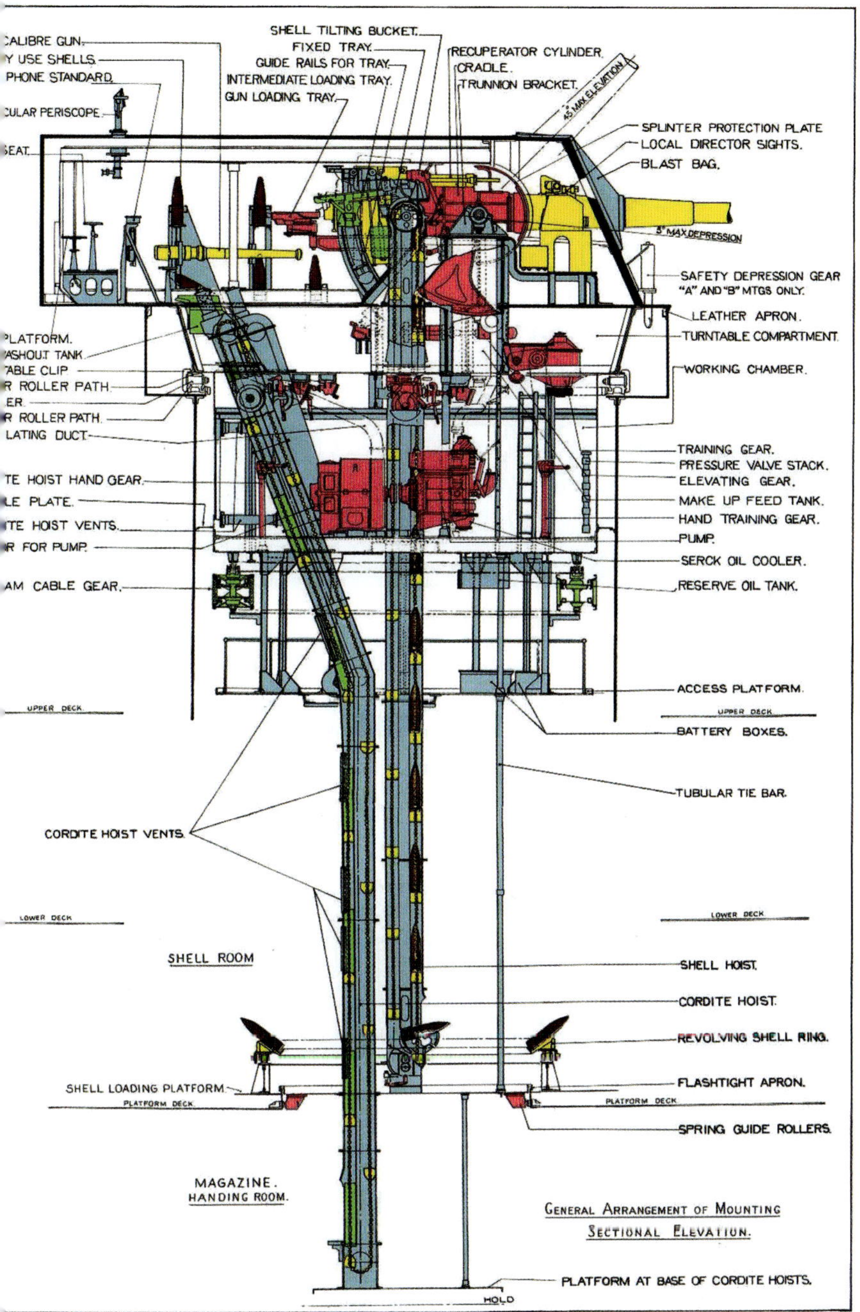


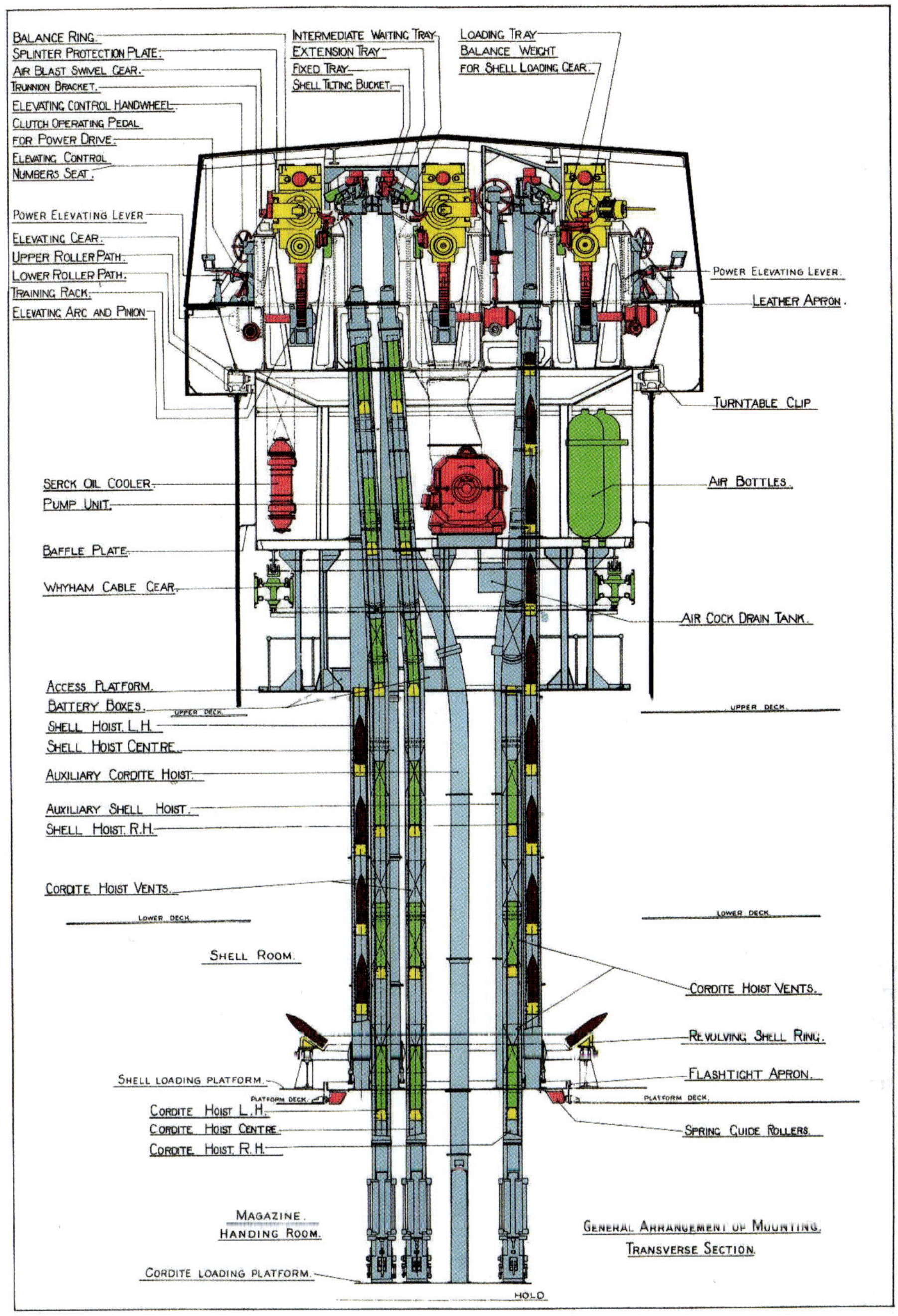


**Triple 6in Mark XXIII mounting**

Four triple 6in Mark XXIII mountings were fitted in the *Belfast* and 'Colony' class cruisers. These were of the 'long trunk' type where the hoists delivered direct to and rotated with the turret, with the magazines below the shell rooms. This made for faster loading as well as reducing the number of ammunition handling crew. The shells arrived between the guns and were tilted on to transfer trays to be swung in line with the breech. The cordite charges arrived behind the guns and were lifted manually on to the loading trays before hand ramming. *(ADM 234/191)*

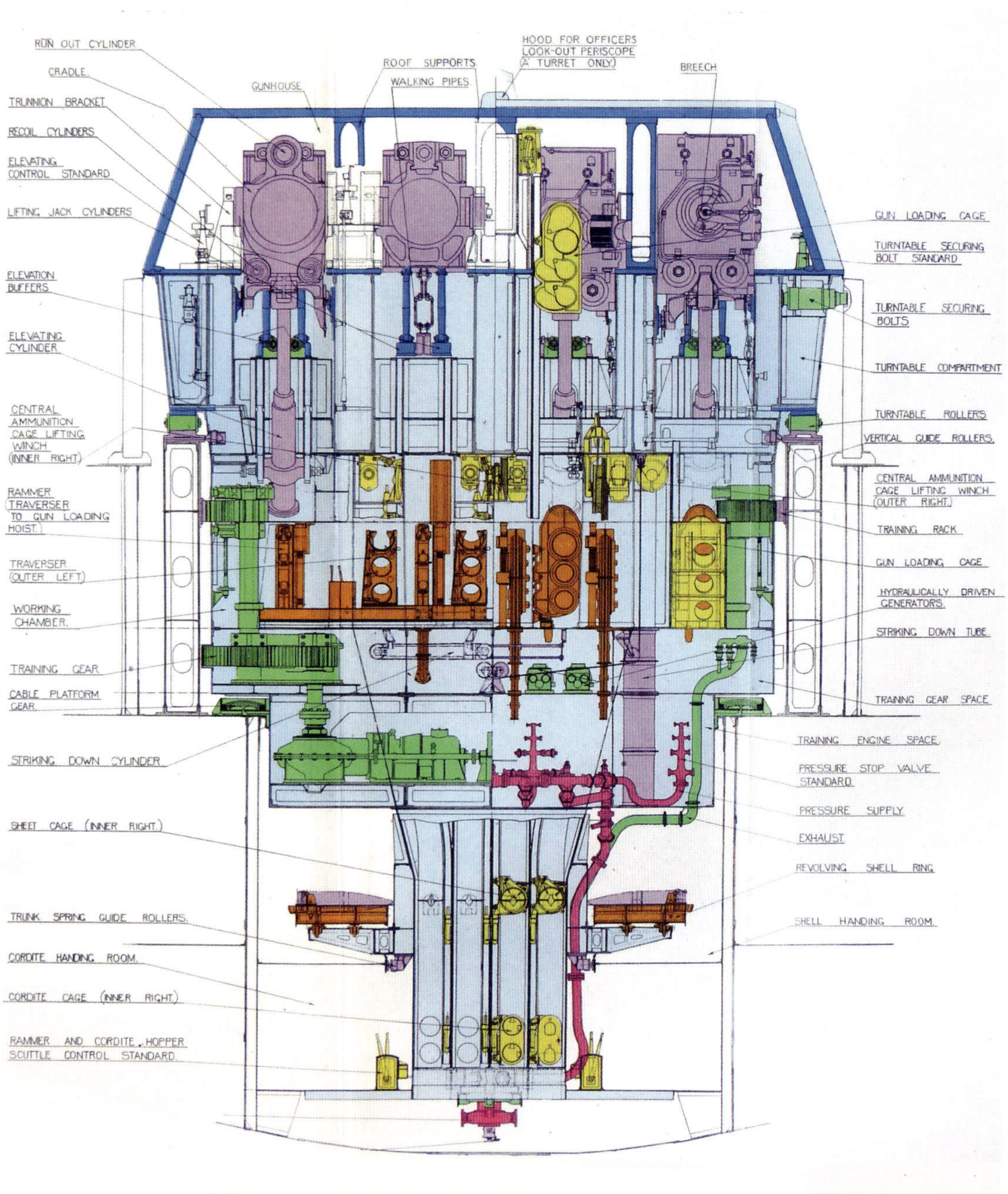


***Page 171 top:* 14in Mark III quadruple mounting**

The side elevation of the 14in Mark III quadruple mounting shows all the rotating parts coloured. From top to bottom they are: gunhouse; turntable; working chamber; training gear space; ammunition hoists. The entire 1550-ton revolving weight rests on 64 tapered turntable rollers (green), with the lower roller path (purple) supported by the ring bulkhead (white). Close inside the roller path were the vertical guide rollers (purple, to centre the mounting) and the turntable holding down clips (blue) which prevented the mounting lifting if the guns were fired or the ship was pitching or rolling. Outside was the fixed barbette armour (white). The mounting was operated hydraulically, with water under 1100lb/sq in pressure fed in at the very bottom (mauve centre pivot connection – see page 146) from the hydraulic pumping engines in the action machinery rooms. The entire gun slide, cradle and gun (purple) were elevated by the angled hydraulic cylinder beneath (purple), allowing the gun to fire from 3° depression to 40° elevation. The thick gunhouse armour and massive roof supports are shown in blue. The gun loading cage (yellow) is shown in line with the breech at the loading angle of 5° with the chain rammer (orange) behind. Its upper level held one 1590lb shell, its two lower levels two 84½lb cordite quarter charges each (338lb total). It was loaded at its lower level from the traverser (orange angled) in the working chamber before being lifted up rails by the vertical cage lifting cylinder (yellow at left). In the central ammunition hoist, the shell and cordite cages (yellow) are shown at the level they are fed from the magazines, similar for each gun. At the bottom in the cordite handing room are the paired hoppers (brown) fed via flashtight scuttles from the magazine, which feed into the cordite cages. Above in the shell handing room is the shell revolving ring with four waiting trays on the four quadrants. The ring enabled the shell waiting trays to be lined up with scuttles in the shell room, before being rotated to line up with the central hoists. In the gunhouse is the 42 ft rangefinder (green). The apparent double thickness of rear armour (blue) is due to fitting a 10in thick balance plate to bring the centre of gravity of the whole (A) mounting in line with its axis.

Similar drawings of the Mark II twin mountings are in *Battleship Duke of York. (ADM 234/271)*

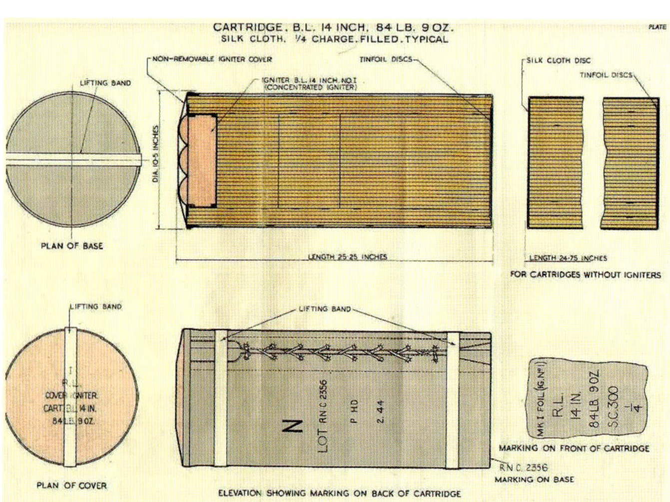


**14in cartridge for the Mark VII gun**

14in cordite for the Mark VII gun came in four 84½lb quarter charges, with the 0.3in diameter SC300 sticks in silk bags, which were completely burnt up when the gun fired. The final charge loaded had an igniter which was set off by an electrical or percussion tube placed in the lock in the breech. *(Handbook on Ammunition. BR 932, 1945. Navweaps)*

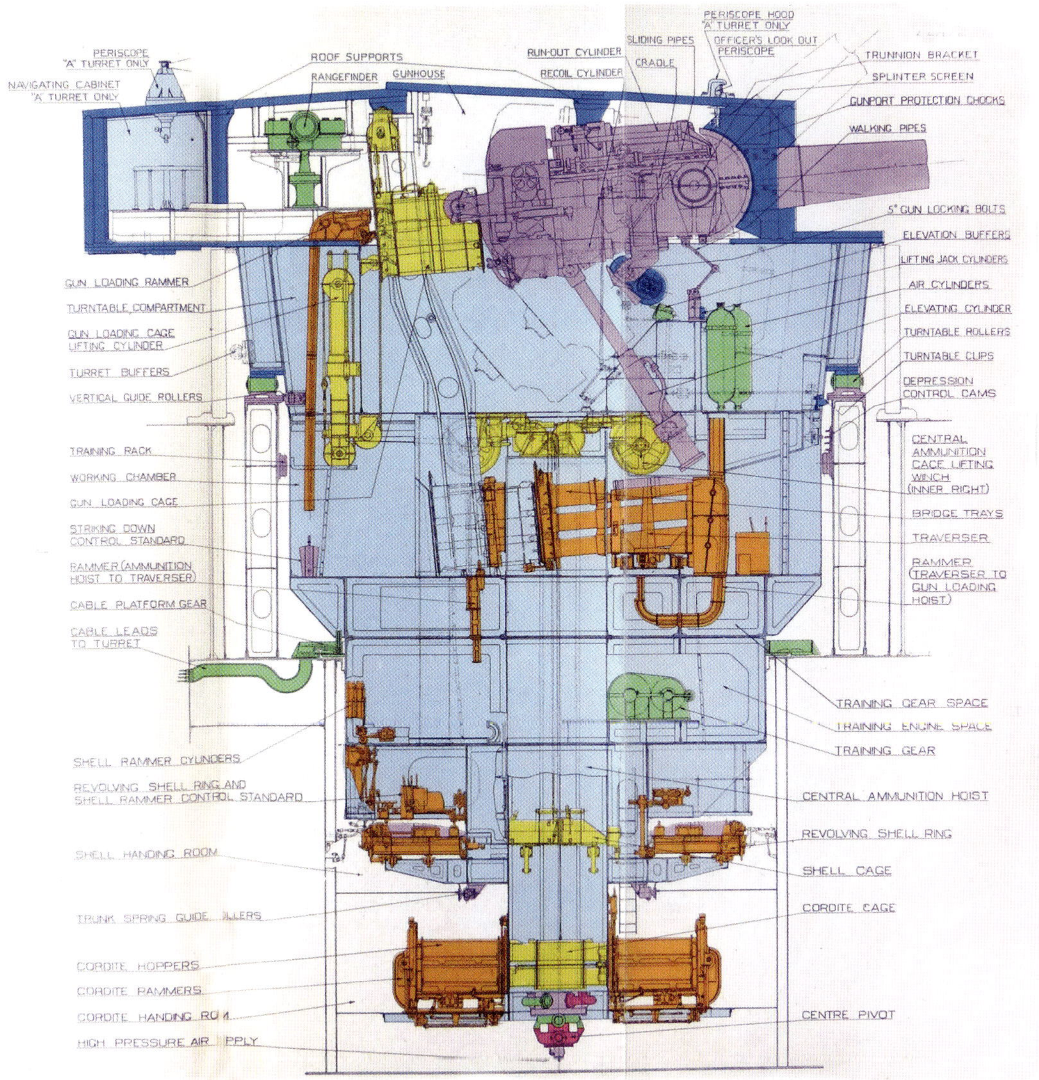


**Page 170 top:** **14in Mark III quadruple mounting**
The cross section of the Mark III mounting shows the hydraulic training gear (green, two for redundancy) engaging with the training rack (purple). The gun loading cage (yellow) is shown both at its upper position (right inner gun) and lower position (right outer gun). The twin recoil cylinders are shown beneath the left outer gun (purple). The run-out cylinder on top of the gun (purple) was used to push the recoiled gun back into the loading position in 2 seconds. The traverser (brown) moved the shell and charges outwards from the central hoists to line up with the gun loading cages. The securing bolts (green) just below the gunhouse floor locked the mounting to the barbette armour, preventing it from rotating when not in operation, without stressing the training gear. *(ADM 234/271)*

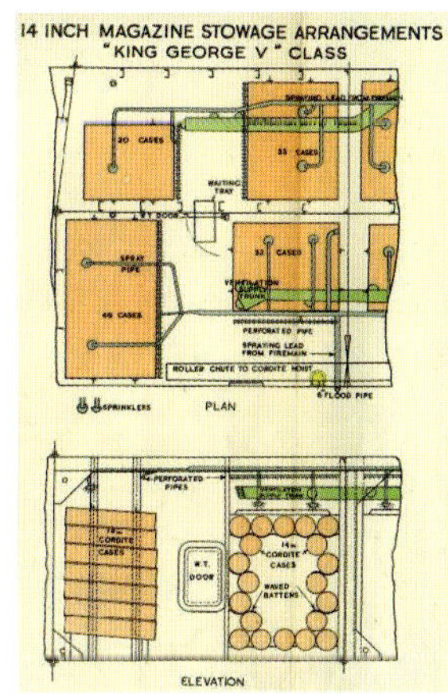


**14in magazine stowage arrangements**
14in magazine stowage with two half charges per case. Ventilation trunks cooled the space while overhead pipes could spray salt water on the cases. *(Handbook on Ammunition. BR 932, 1945, Navweaps)*

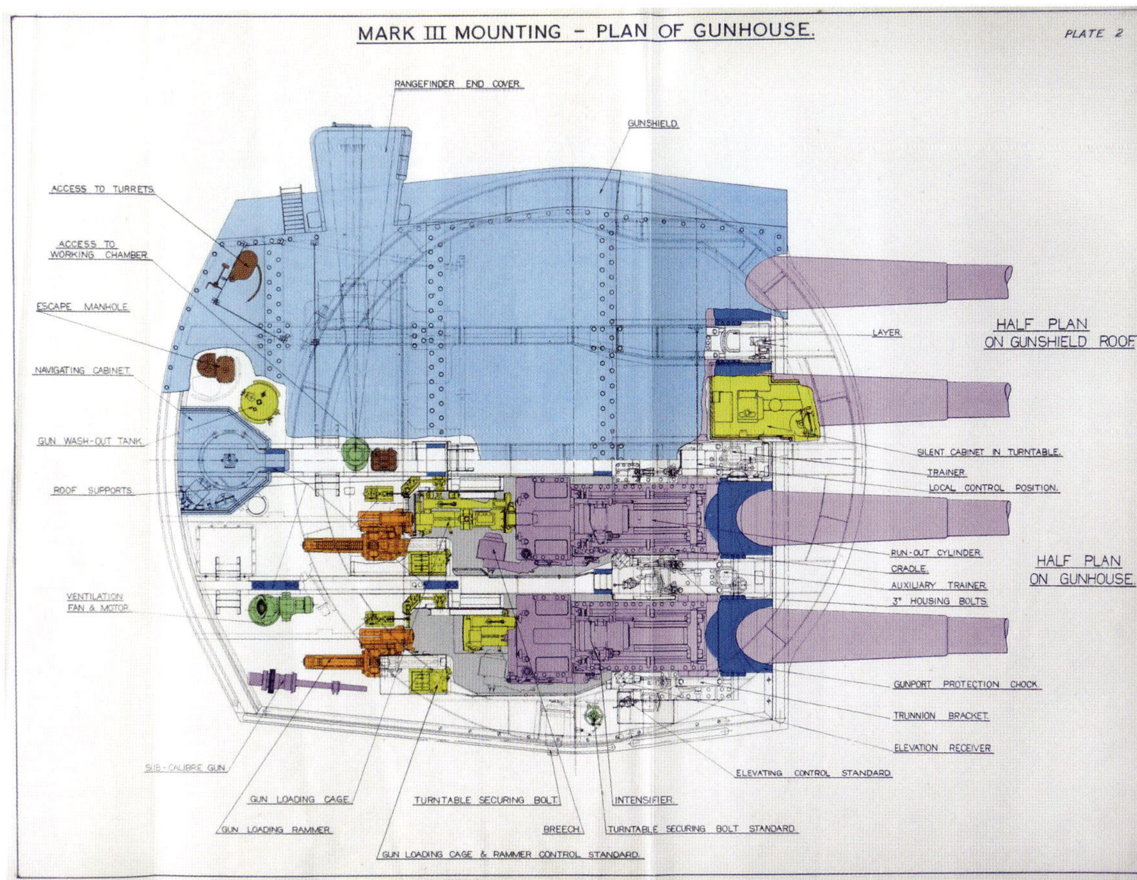


***Right:*** **14in Mark III quadruple mounting**
This composite plan view of the Mark III mounting shows at its left side the gunhouse roof made up of four 240lb NC plates. The securing bolts shown enable the gunhouse to be opened up to replace worn out guns or other equipment. The right side inside the gunhouse shows the gun loading cages (yellow) and rammers (orange). The breech (purple) is shown open in the right inner gun. The gunlayers (each gun could elevate independent of the others) normally followed their indicators from the director control tower and transmitting station, but in emergency could use the sight through a port in the face of the gunhouse. The trainer in the centre also followed the indicator from the TS. At the side is a 6pdr sub-calibre gun (purple) which fitted into the breech for drill and training purposes. *(ADM 234/271)*

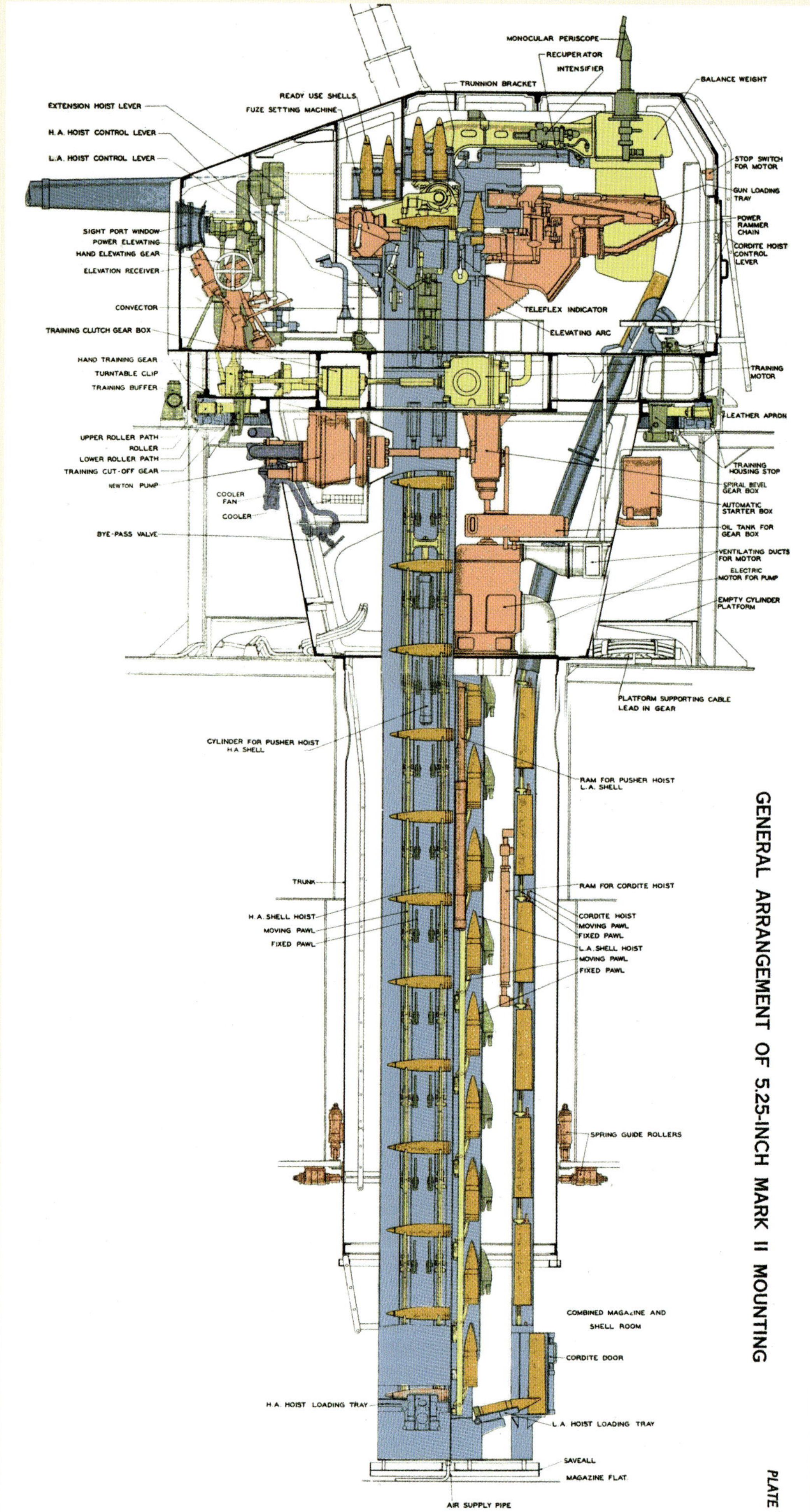


**Twin 5.25in Mark II mounting**

Although the gunhouses in the twin 5.25in Mark I in the *King George V* class battleships and in the Mark II in *Dido* class cruisers were similar, the hoists were completely different. In the Mark II shown here, the magazines were beneath the turrets so the endless chain hoists served them directly. The low angle hoist carried the shells vertically, the high angle one horizontally so the shells could be fed to the fuze setting machine before transfer to the loading tray; cordite cartridges vertically. In the battleships, the hoists had to bring ammunition up from the magazines in separate hoist tubes then transferred horizontally to a gunbay directly below each turret, as illustrated in the Mark I mounting drawings in *Battleship Duke of York*. *(ADM 234/183)*

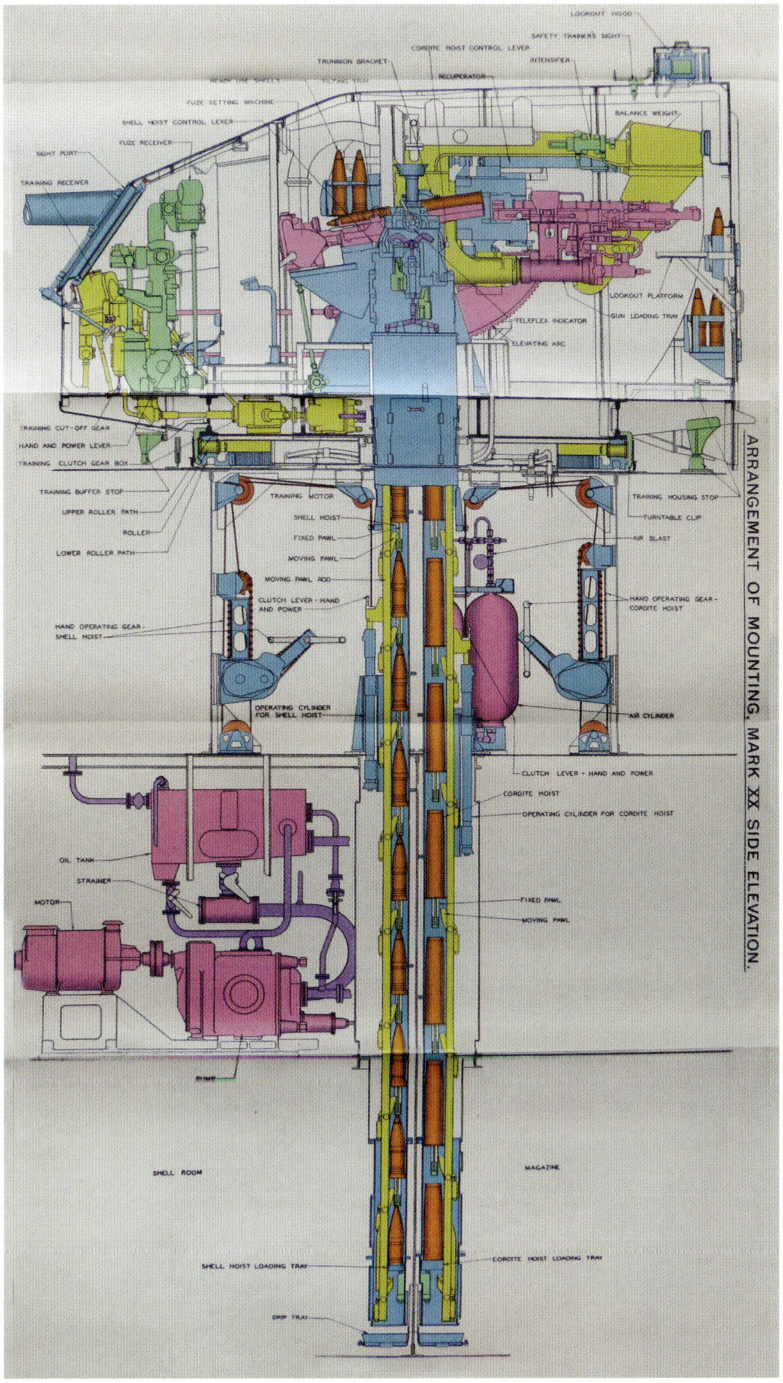


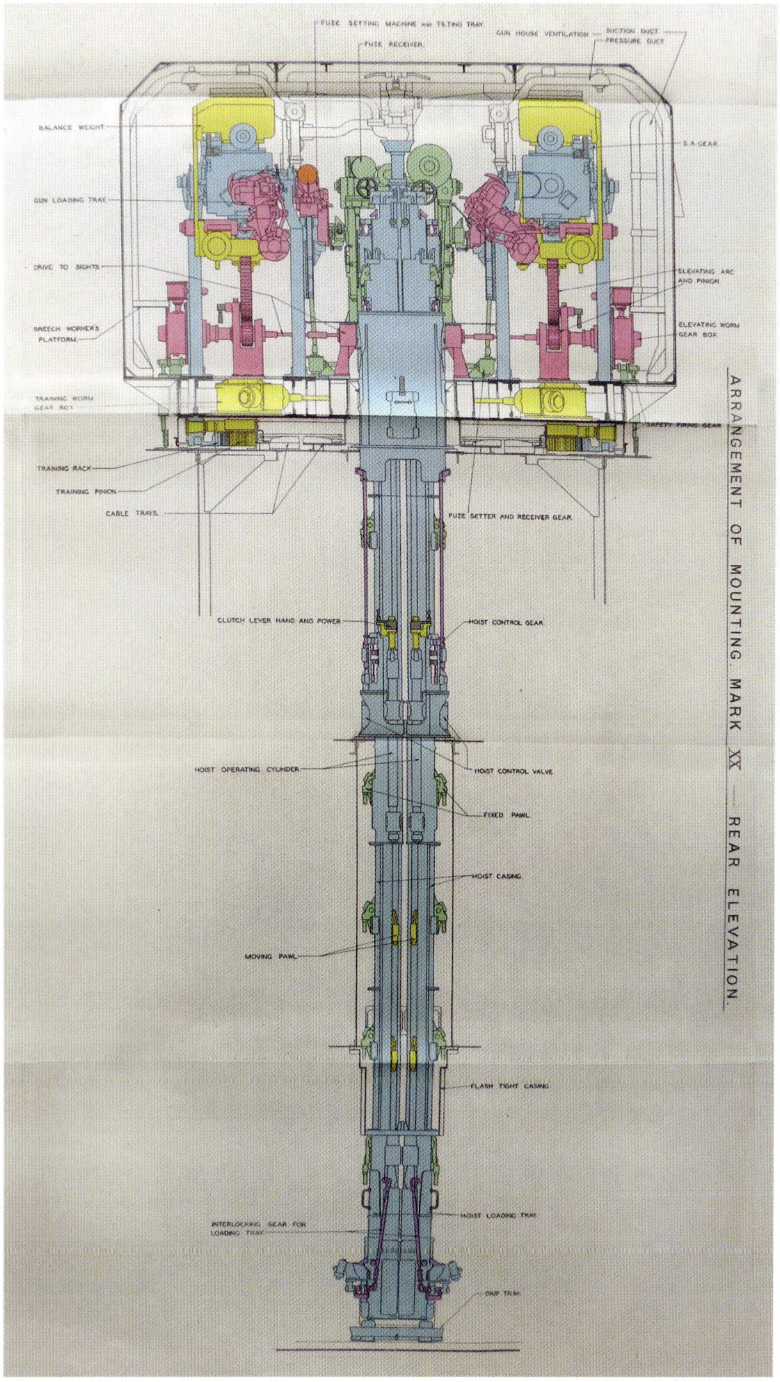


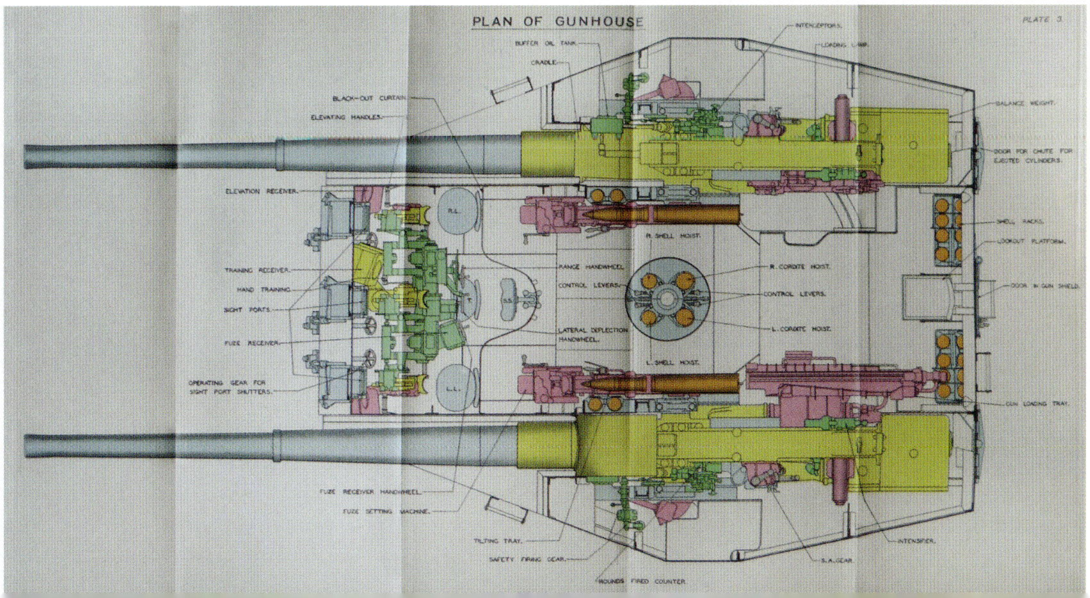


**4.7in Twin Mark XX mounting**

Unlike the transferable twin 4.7in mountings, the twin Mark XX had a centreline hoist, each half serving one gun, but it did not rotate with the mounting. Shells and cartridges were then transferred manually to the loading trays, with the HA shells first placed in the fuze setting machine via a tilting tray. Hand operating gear was also fitted for shells and cartridges. A heavy balance weight was needed to keep the CG of the elevating parts in line with the trunnions. *(ADM 234/185)*

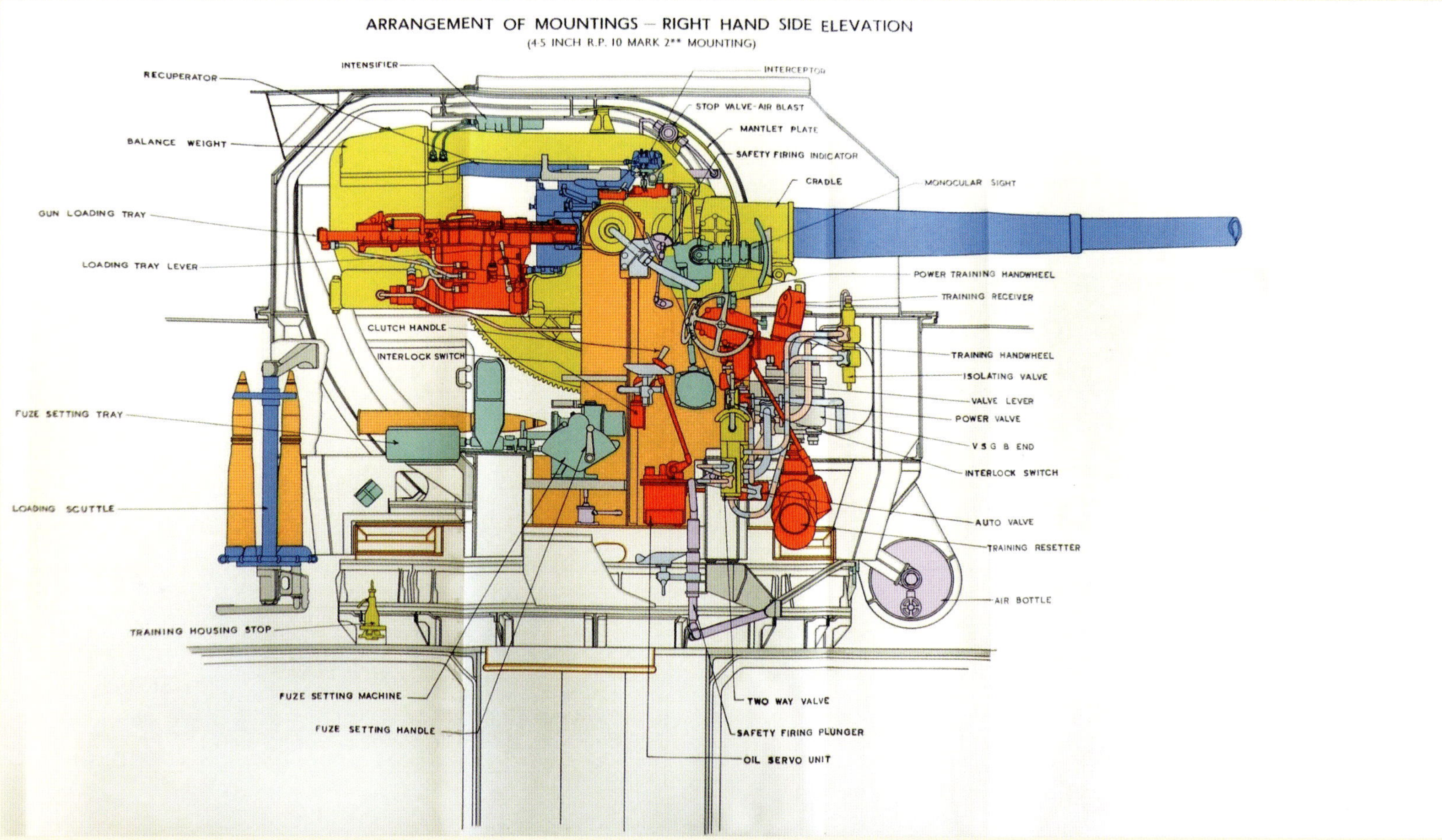


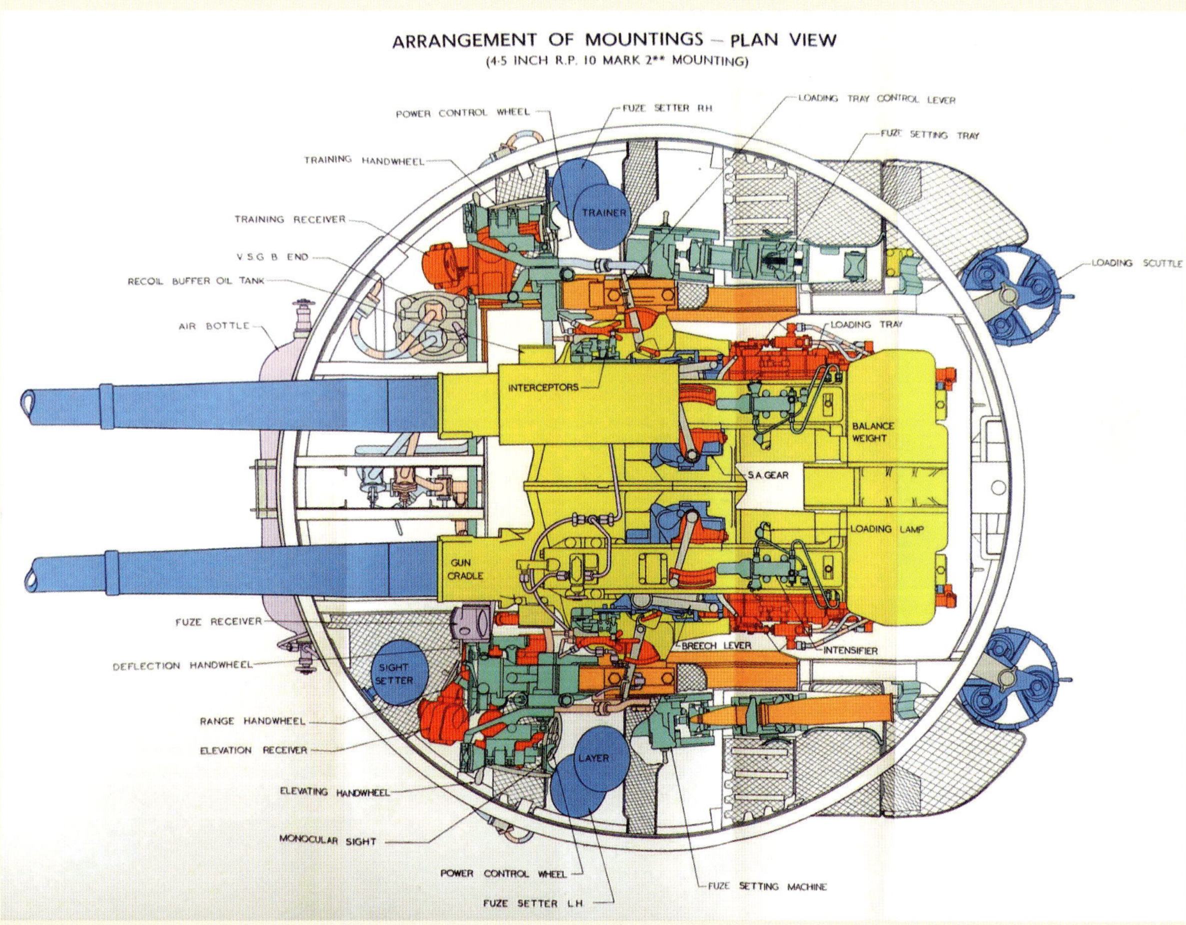


**Twin 4.5in Mark II BD mounting**
The twin 4.5in Mark II BD mounting was fitted in modernised battleships and new fleet carriers with 80° elevation. Most of the mounting was 'between decks' inside a gunbay to which angled hoists delivered from the deep magazines for manual transfer via the loading scuttles to the rotating portion. These two drawings show the Mark II** RP.10 fitted in *Indefatigable* and *Implacable.* The Mark III UD (upper deck) version was 'transferable' with an open rear relying on the traditional manual supply from nearby external hoists. *(Handbook for Q.F. 4.5 Inch Mark I(N), III(N) Guns on the Twin Mark II BD and Mark III UD Mountings. BR.917)*

**Twin 4.5in Mark IV mounting**
The twin 4.5in Mark IV mounting in the 'Battle 'class destroyers was adapted from the Mark II BD mounting. For easier handling at sea, the shells and cartridges were separate with their own hoists.
*(Handbook for 4.5-inch Mark 3 Gun on Twin R.P.10 Marks 4 and 4* Mountings. BR 1635(51))*

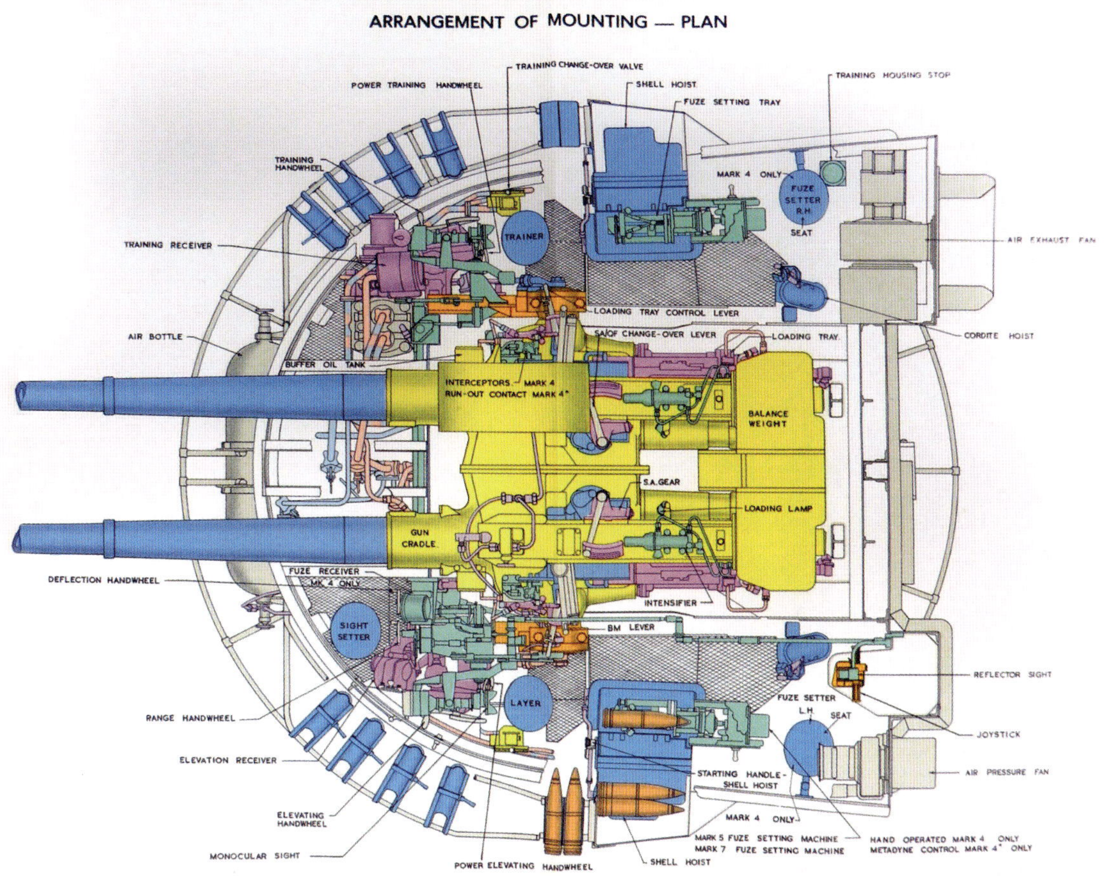


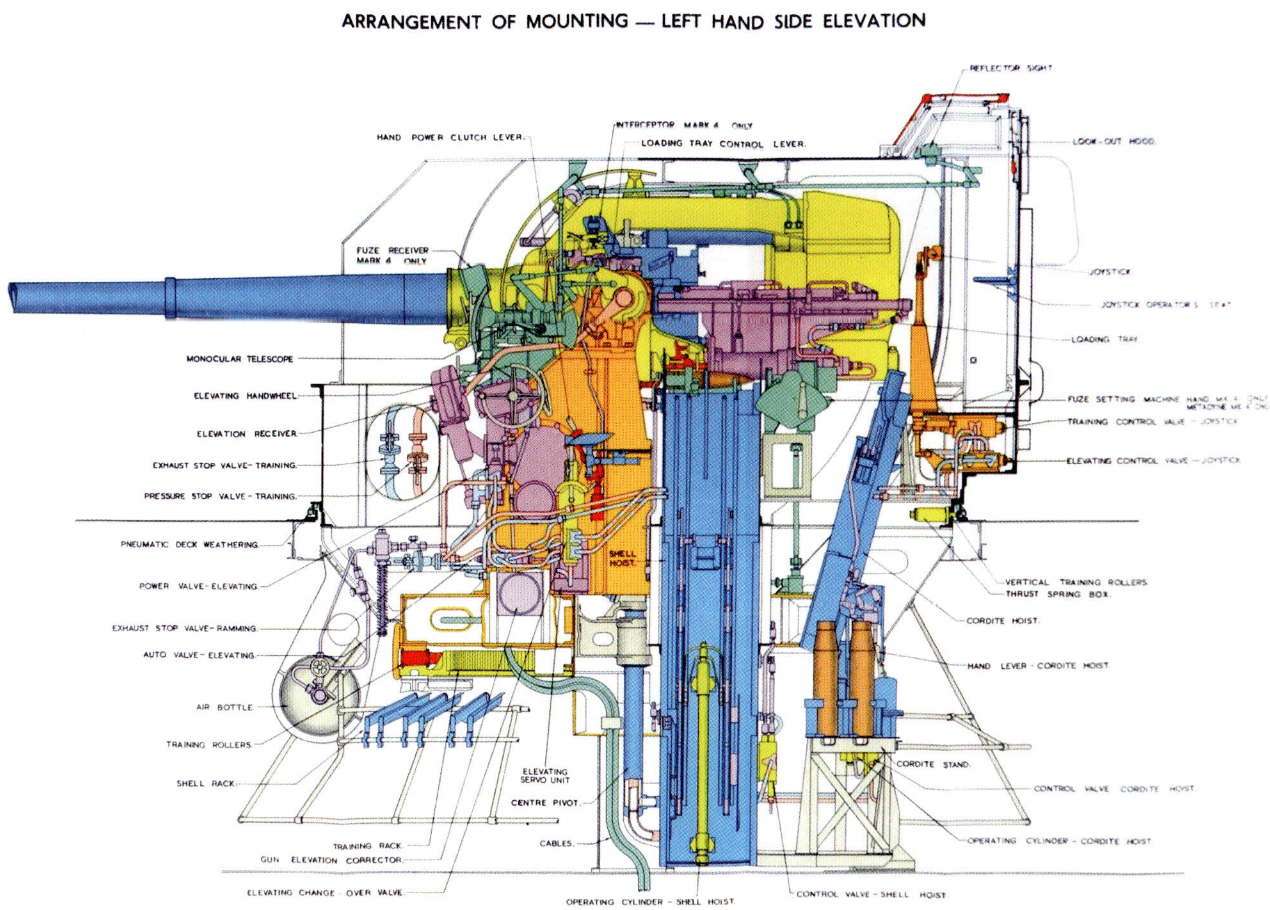

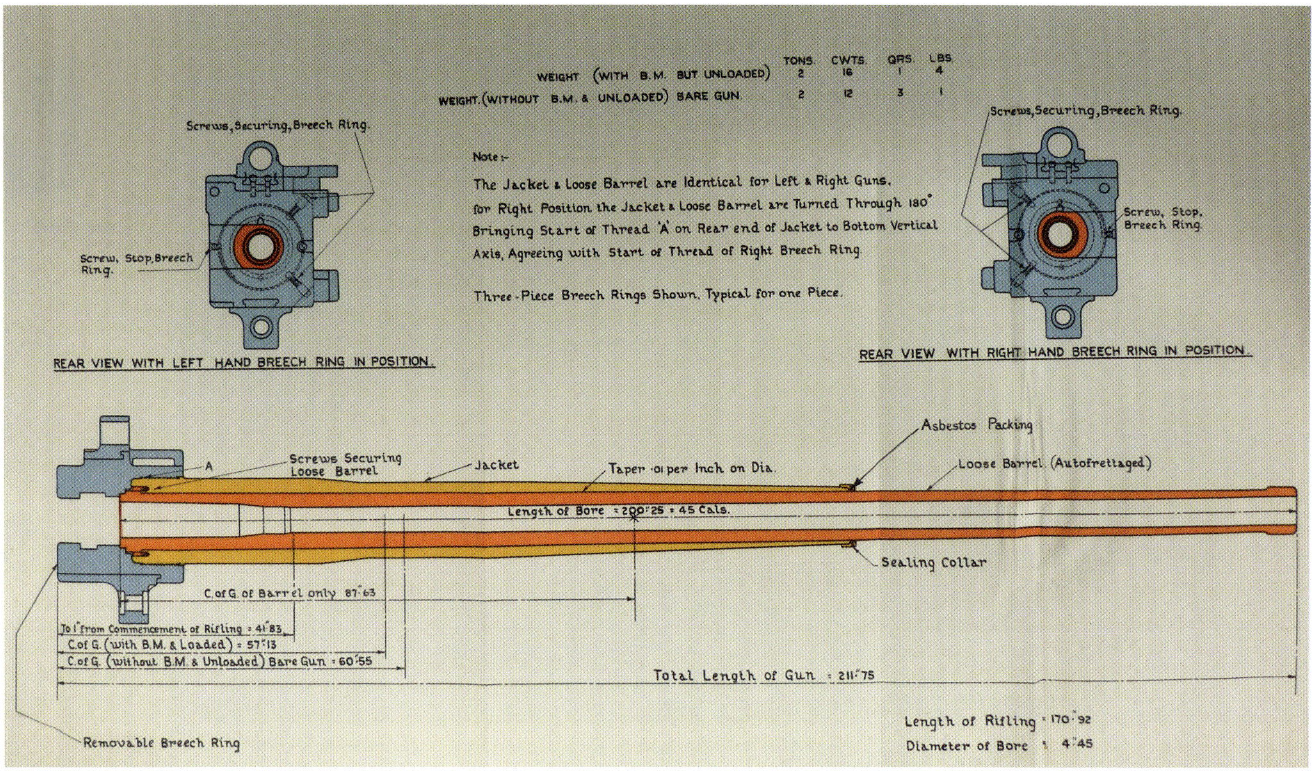


**4.5in Mark I and III gun**
The new 45-calibre 4.5in Mark I and III gun was widely fitted as a HA gun and had a replaceable barrel. *(BR 1635(51)).*

**Twin 4.5in Mark 6 mounting**
The twin 4.5in Mark 6 incorporated all the wartime experience including 80° elevation. With later modifications it remained in service in overseas navies into the 2010s. A more detailed drawing is shown on page 211. *(Handbook for Q.F. 4.5 inch Mark 5 gun on RP.41 Twin Mark 6 and 6* Mod 1 Mountings. BR 2014(B), BR 2014A 1957)*

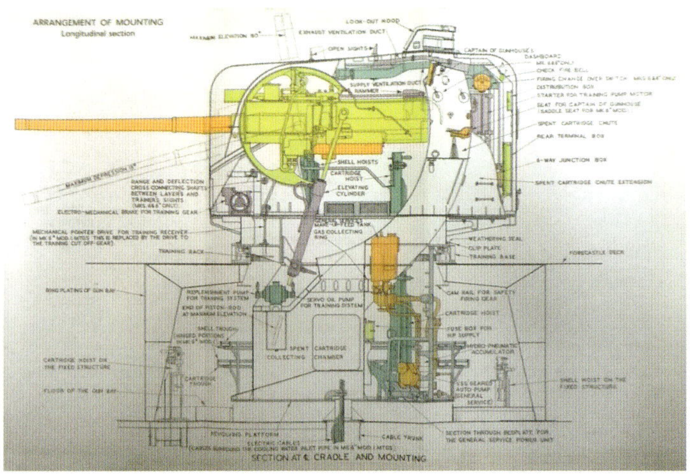

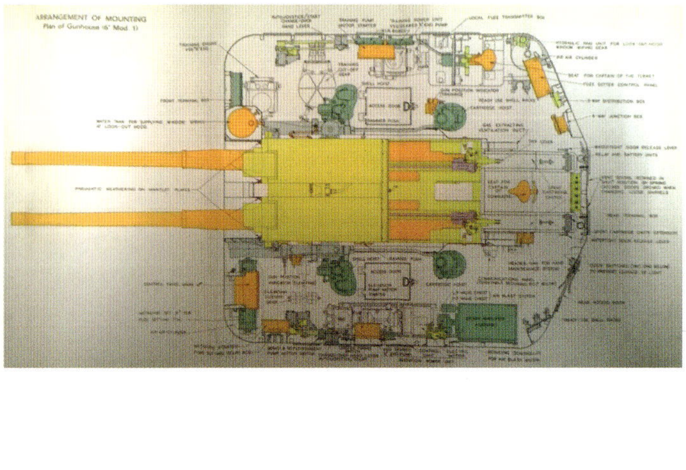

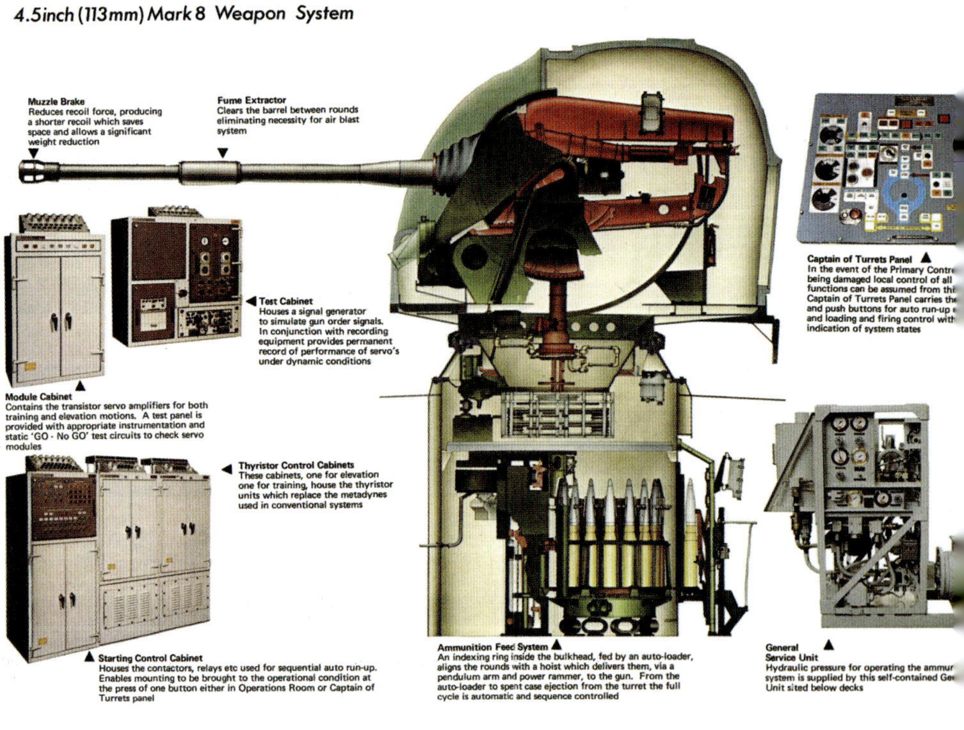


**Single 4.5in Mark 8 mounting**
A sales brochure for the single 4.5in Mark 8 shows the pre-loaded ammunition indexing ring which could be automatically rotated to bring the required round (HE, VT, star etc) into position for moving it direct up into the gunhouse. It also shows the range of control equipments, which in turn were fed with the fire control solution from the GSA system and radars.

## CHAPTER SIXTEEN

# The 14-inch Twin and Quadruple Mountings

### DESIGN

With the Admiralty keen to resume battleship construction as soon as permitted by Treaty (earliest laying down 1 January 1937), they had settled for guns of 14in calibre as best suited to a balanced design meeting the current 35,000-ton displacement limit. Although the original plan had been to fit twelve guns in three quadruple 14in turrets, that would have meant sacrificing some armour protection, so they decided on ten guns. There was now only one company with the experience and facilities to make such heavy mountings although there had been three companies prior to WW1. Vickers-Armstrongs selected their Elswick works as having the most recent experience of battleship mountings to undertake the design from October 1935 as well as being the lead works for the first ship. Their Barrow works would be the follow-on works for the second ship. Barrow would need a whole new erection shop with three pits to handle the large quadruple mountings, while Elswick would need an extension to its Nº 24 Shop with a fourth large pit, which would allow them to make a start on a third ship while the first was still under construction.

The original timescale envisaged the first two ship sets being ordered six months before the hulls as they would be on the critical path to

Erecting to turntable level *King George V*'s quadruple 14in A mounting in Elswick's Nº 24 Shop early in 1940. The jumble of components in front includes one of the guns. Beyond is B mounting looking nearly complete. *(Tyne & Wear Archives)*

completion by September 1940, *ie* in April 1936, although the formal order was placed on 15 May. The plan envisaged the first mounting shipped in March 1939, the last in December.[1] That four-year time scale was nearly double the comparable time for battleships pre-WW1, but with the rundown of staff and facilities in all three key industries – shipbuilding, gun mounting and armour manufacture – that was the best that could be achieved. But substituting one twin mounting for the third quadruple required additional design effort, since little more than the gun slides, loading and elevating gear would be common to both.

A new design of 14in gun was also required, 45 calibres long, the Mark VII. Unlike the preceding 16in, it was all steel, comprising an A tube, an inner A and a jacket. Compared with a wire-wound gun, it was lighter, less liable to droop and had a longer life. The bare gun weighed in at 79.55 tons including breech mechanism (1.85 tons), which could be operated by hand or by power. An 11.5-ton balance weight was fitted above and below the breech to allow the gun to be elevated to 40° and with less space required in the gun well to accommodate the 45in recoil, bringing its centre of gravity in line with the trunnions. The lower weight housed the two recoil cylinders while the run-out cylinder ram (recuperator) was fixed to the upper. The elevating mass of gun, balance weight and cradle was 125 tons, of which 97 tons recoiled, with a force of 352 tons at 40° to be absorbed. The guns were to be spaced 8ft apart and bodies be interchangeable, although breech mechanisms were fitted for right or left guns. [2]

There were still three big gun manufacturers, Elswick, Beardmore and Woolwich. *The Battleship Builders* [p178] notes 78 made from 1938, with respectively 38, 15 and 23 made plus two more from Elswick but completed at Woolwich in 1947, with one of the last, S.143, preserved at Fort Nelson near Portsmouth. Elswick also made the experimental SE.60, followed by a production contract for fourteen (S.61–74), which were all completed by June 1939 at a price of £19,050 for each gun body plus fourteen breech mechanisms at £3200 each, a total of £22,250 for a finished gun. Eighteen later guns (S.113–130) came out slightly cheaper at £21,077 each.[3]

The 15in Reg Nº 59 (ex-*Malaya*) was relined to 14in in 1937 for trials. The first new gun proof trial was of S.61 (a left gun) carried out at Ridsdale with five rounds on 12 April 1938; one of the last S.130 on 27 March 1942. In July 1938 S.82 (probably one from Woolwich) was fired from Yantlet battery, Isle of Grain, shooting across the Thames Estuary to the Shoeburyness mudflats where shells could be recovered at low tide. Beardmore's 14in were also proofed at Ridsdale, the first (S.75 a right gun) on 13 July 1939 and last (S.112) on 7 August 1942. Two Woolwich guns were also tested there. Generally, a 1590lb projectile was fired at 2495ft/sec using 340lb SC300 propellant.[4]

The first three battleships required 9 mountings, 40 gun bodies and 33 breech mechanisms:[5]

| | *Gun bodies* | *BM* |
|---|---|---|
| Elswick | 18 | 15 |
| Beardmore | 9 | 7 |
| Woolwich | 13 | 11 |
| Total | 40 | 33 |

The 14in mounting was designed on the well-tried principles of the 15in and 13.5in, but with the magazine now sited below the shell room. This not only gave better protection to the more vulnerable charges from long range plunging fire or bombs but allowed quicker flooding as they were further below the waterline. The magazines could be flooded directly from the sea via seacock intakes in the bottom with 10in and 12in diameter pipes with 6in branches to the magazine spaces and 4in to the shell rooms. Drenching of the cordite cases in case of overheating or spontaneous ignition could also be made from the ship's pressurised salt water main with 4.5in pipes leading to multiple spray heads above each case. All the flooding control valves could be actuated by rods and gearing from the middle deck outside the ring bulkhead. *(See colour plate on page 171)*

A fixed loading angle of 5° was selected as being less complex than all-angle loading, albeit slightly slower. The arrangement of the gun loading cage (with the shell above two levels each holding two quarter charges) and rammer was thus simplified. Great attention was paid to flashtightness with gunhouse and working chamber separated from the traverser space (in which shells from the central ammunition hoist were moved sideways to line up with the gun loading cages), from the central hoist, from the shell room and from the magazine with its handing room. Numerous safety interlocks were required to prevent openings between more than two zones at a time, operated by levers and rods rather than the less secure pressure piping. The result was a more complex mounting than the 15in, especially the quadruple which required four transfer points between the magazine, the central hoist and then distribution to the four gun loading cages in the working chamber.

Gunhouse protection was formed mostly of flat plates and was similar for both quad and twin, namely front 520lb (12.74in actual) cemented armour, side front 360lb (8.82in) C, side rear 280lb (6.86in) non cemented, rear 280lb NC and roof 240lb (5.88in) NC, the latter in four bolted plates, two of which could be removed to allow gun changes. Those for *King George V* were ordered from English Steel Corporation in May 1936; ESC also made the plates for *Duke of York*'s A and B and all of *Anson* and *Howe*'s (both ordered in December 1937).[6] The Admiralty contract had agreed a standard price of £195 per ton for C and £170 NC, although these prices were later renegotiated down to £97.25 and £75.75. In addition, thick balance plates were fitted inside the rear armour to bring the mounting centre of gravity in line with its centre of rotation, 10in for A, 3in for B and 8in for Y which had a 16-ton octuple 2pdr mounting on its rear roof. The rear plates were perforated with vent holes to allow escape of gas build up from any burning charge. The rear gunhouse floor outside the barbette had an additional 3in plate to the 3in floor, with holes for ventilation

[1] *King George V* class Ships Cover, NMM

[2] *Handbook for the 14-inch B.L. Mark VII Gun on Twin Mark II and Quadruple Mark III Mountings, 1942.* ADM 234/271

[3] Tyne & Wear Archives DS.VA/3/5/15

[4] Tyne & Wear Archives DS.VA/6/35/10

[5] National Archives T 161/931/8

[6] Prices for Armour Plate. 1939–43. ADM 1/12584

*King George V*'s A mounting in Nº 24 Shop on 26 April 1940 being tested prior to erection in the ship. The hole in the rear top plate is for access; there was a similar one in the overhanging gunhouse floor plate. A ventilation supply duct snakes around the side. Its intake was through holes at the rear of the gunhouse floor; exhaust through vent holes in the rear armour plate. *(Tyne & Wear Archives)*

intake and an access hatch. Two fans gave an air change every 5 minutes. At the forward end of each cradle was fixed a semi-circular splinter screen (mantlet) which sealed the gun port openings as the gun elevated.[7] *(See colour plates on pages 170-171)*

The quad had not only a large mass but also inertia, which affected the power required to accelerate and train the mounting; a speed of 2° per second was required. Revolving mass including guns was 1550 tons for the quad and 898 tons for the twin, but 'mounting' weight was a key factor in ship design calculations so included the lower roller path fixed to the ring bulkhead supporting the whole rotating mass as well as shell room loading gear. The corresponding 'armament' weights then became 1582 tons for the quad and 915 tons for the twin.

Two training engines were fitted below the working chamber, hydraulically driven VSG swashplate pumps (two 160hp for quad, two 70hp for twin) for smoothness, with worm gears, shafts and pinion drive to engage with the toothed training rack attached to the ring bulkhead. Only one engine could be clutched in at a time, providing some redundancy. As well as the two main gunnery directors, rangefinders were fitted at the rear of each gunhouse, 42ft for A and Y, 30ft for B. All ten guns could be fired from the directors, or from B turret or individually from each turret in local control. Voice pipes were used for reliable communication, particularly between gunhouse, working chamber and magazines: 19 in the quad, 13 in the twin. The quad required 107 men in total to operate.

To achieve two rounds per gun per minute, the loading cycle (in seconds) had to be: [8]

| | |
|---|---|
| Depress fired gun from 20° and open breech | 4 |
| Raise gun loading cage | 4 |
| Ram shell | 2.8 |
| Withdraw rammer | 2.2 |
| First cage lift | 1 |
| Ram first half charge | 1.5 |
| Withdraw rammer | 1.5 |
| Second cage lift | 1 |
| Ram second half charge | 1.5 |
| Withdraw rammer | 1.5 |
| Lower cage | 4 |
| Close breech, unlock slide, lay gun to 20° | 3 |
| Recoil and run-out | 2 |
| Total seconds | 30 |

The cycle in the central ammunition hoist was slightly shorter at 26 sec, with the traverser cycle at 17 sec. The fully loaded cage weighed 7870lb (3.5 tons).

Manufacturing tolerances were demanding: clearance between gun and cast steel cradle only 0.04in, roller flanges and roller path 0.02in, holding down clips and roller path 0.025in. High pressure air at 1250lb/sq in was used to charge the run-out cylinder and low pressure air to operate the gun air blast system. A steam turbine driven hydraulic pump was fitted in each of four Action Machinery Rooms, with one pump normally connected to each turret, with one in reserve, but all with cross connections. The hydraulic fluid was one part Argolene soluble oil to 60 parts distilled water.

Ammunition stowage was nominally 100 rounds per gun, but actual shell room capacities were A 432, B 216, Y 452, nearly all APC. By 1948 the last *KGV* in commission *Duke of York* carried 46 APC and 69 HE per gun, recognising that shore bombardment was as important as anti-ship fire. The magazine and shell room layout and ammunition embarking plans are included in *Battleship Duke of York*. *(See colour plate on page 170)*

Gun mounting trials were carried out both before and after ship handover, the latter firing typically four rounds per gun at various angles of elevation and training. The trials would be made with a team from gunnery school *Excellent* whose detailed report usually listed dozens of minor adjustments to be made.

[7] ADM 234/271, *op cit*

[8] ADM 234/271, *op cit*

*King George V*'s A mounting nearly ready for shipment on 16 April 1940, having completed pit trials. *(Tyne & Wear Archives)*

**PRODUCTION**

The mountings for the two battleships of the 1936 Programme were ordered from Vickers-Armstrongs on 15 May 1936, the official letter being sent to their London office by the Admiralty Director of Contracts.[9] Elswick got the order for two quad and one twin for *King George V* to be built at V-A's Tyne yard plus one spare mounting, which was little more than the elevating portion. Barrow got two quad and one twin for *Prince of Wales* to be built by Cammell Laird at Birkenhead. The hulls were ordered only two months later, by foregoing the normal tendering process. No prices for the mountings were specified but later a provisional price was agreed on which to base progress payments. Elswick made quarterly progress reports to the V-A board summarised below.[10] These show the programme slipping by about a year, with the first mounting (Y for *King George V*) only being installed between 3 and 7 April 1940. A succession of barges were towed the five miles downstream from Elswick with the various components, unloaded by V-A's 250-ton cantilever crane.

Quarterly reports by Elswick to the V-A board on 14in mountings summarised:

1Q 1936 Drawing office actively engaged on design of heavy gun mountings for capital ships.

2Q 1936 Order for two quadruple and one twin 14in mounting for *King George V* received. Material being ordered.

1Q 1937 Comparatively slow progress, due to lack of floor space and interruption for erection of extension to 24 Shop. Large volume of work in hand for machine shops. Order received for one quad and one twin for *Duke of York* and two quad and one twin for *Anson*. Order for 18 14in gun bodies and 14 breech mechanisms.

2Q 1937 24 Shop extension and fourth pit ready.

3Q 1937 Progress on structural work for quads is reasonably satisfactory. Many of the larger components such as cradles, balance weights, trunnion brackets etc are making good progress. It is difficult to forecast whether the progress on these mountings is sufficient to enable us to meet delivery dates. The percentage of drawings approved and issued to the shops is still small compared to the total required.

4Q 1937 The first gun, breech mechanism and cradle were delivered to Shoeburyness and carried out preliminary firing trials.

2Q 1938 We have been successful in sub-contracting a considerable number of components of these mountings, which will ease the pressure on Elswick.

4Q 1938 First quad (Y) started erecting in pit. Twelve 14in guns made to date.

1Q 1939 Second quad (A) started in pit.

2Q 1939 Twin (B) started erection in pit.

3Q 1939 Y mounting has three guns and cradles assembled. There remains a large mass of control standards, interlocks, control valves, etc to be fitted. B mounting gunshield not yet arrived from Beardmore. Revised dates for delivery to *King George V:*

Y mounting February 1940

A mounting April 1940

B mounting May 1940

4Q 1939 A mounting on pit has two guns installed. First quad (A) for *Duke of York* on pit [probably the new fourth pit]. Progress is not satisfactory owing to shortage of men, concentrated on *King George V.* Eleven more 14in guns completed. [Enough now for first two ships]

1Q 1940 Y mounting pit trials satisfactory. Whole energies of gun mounting department devoted to getting *King George V* completed. Mountings [for *Anson*] making slow progress owing to shortage of men.

[9] *King George V* class Ships Cover, NMM

[10] Cambridge University Library, Manuscripts. Vickers Archive 180–207

The 39ft diameter turntable for *King George V*'s A mounting being lifted on board on 7 June 1940 by Vickers-Armstrongs' 250-ton crane to rest on the roller path. In the background the barge on which it arrived is being towed back to Elswick. *(Vickers)*

2Q 1940 Y mounting delivered to ship. A mounting pit trial completed and delivered to ship.
3Q 1940 All three mountings delivered. B mounting for *Duke of York* on pit vacated by *KGV* twin. Work proceeding on three mountings for *Anson.*
4Q 1940 Gun trials on *KGV* satisfactory. A few men aboard clearing up odd jobs such as stowage of spares.
1Q 1941 *Duke of York* A mounting satisfactory progress. Pit trial of B.
2Q 1941 A and B mountings delivered to ship [by coaster to John Brown].
4Q 1941 *Duke of York.* Satisfactory trials. *Anson* mountings manned day and night. Y delivered.
1Q 1942 Equipment trial on Y satisfactory. A and B delivered.
2Q 1942 The gun trials of *Anson* were carried out on June 19th with very good results. Letter of appreciation received from Secretary of Admiralty:

'I am commanded by My Lords Commissioners to express their appreciation of the work done by your firm to bring the A turret mounting of HMS *Anson* to such a satisfactory state of completion, cleanliness and efficiency in the short time available and without the assistance of a pit trial.'

The mountings for *Prince of Wales* were shipped to Birkenhead in 2950 grt coaster *Sea Fisher,* which had been adapted with wider hatches to carry these heavy gun mountings. Loading was by V-A's new 250-ton crane at Devonshire Dock at Barrow and discharging by Cammell Laird's 200-ton floating crane. The first (Y) arrived at the end of June 1940.

Three more battleships followed in the 1937 programme with mountings ordered on 7 January 1937, three months before the hulls. Elswick got *Duke of York*'s A and B and all of *Anson*'s, while Barrow got

The turntable of *Duke of York*'s A mounting being lowered into the widened Nº 2 hatch of coaster *Sea Fisher* at Elswick on 5 June 1941. It has been stripped of non-structural components to keep the lift weight below 200 tons. *(Vickers)*

*Duke of York*'s B mounting receives its first 14in gun and cradle at Clydebank in July 1941. A mounting is less advanced but trunnion supports are clearly visible. Vent holes are visible in both rear plates. *(UCS 118-554-401)*

*Duke of York*'s Y and all of *Howe*'s plus a spare, later cancelled. When completion of *Duke of York* was given priority over *Anson* and *Howe*, *Anson*'s A and B were switched to *Duke of York* as was *Howe*'s Y. The final shipment was *Howe*'s B in new 2931 grt coaster *Sound Fisher* arriving from Barrow at Fairfield's Glasgow yard on 27 April 1942, her first (Y) having arrived in *Sea Fisher* on 10 February. John Brown and Fairfield needed their existing large cantilever fitting out cranes upgraded in 1937 to lift the largest component which was the 187-ton turntable for A and Y (B's was 114 tons) weighing with lifting gear 200 tons; the working chamber and training gear about 170 tons. *Anson* used V-A's 250-ton crane to lift her mountings as floating crane *Titan II*'s capacity was only 150 tons; the last (A) was shipped in April 1942, with gun trials on 19 June.

With no fixed price contract and under wartime pressure, the Admiralty was no longer concerned about extra costs if delivery could be expedited, *eg* from working more overtime or more costly sub-contracting. Provisional prices under the normal headings were agreed with the Admiralty, final figures being subject to adjustment as actual costs were incurred (checked by Admiralty accountants), and modifications agreed. There is no clear picture of the costs of the 14in mountings with payments being spread over a decade, although Barrow records do contain more detail of contract payments than Elswick, but

The end result: *Duke of York*'s B mounting is complete in early September 1941, with an 8-barrelled 2pdr mounting on top. The trapezoidal shape of both rangefinder casings is to allow the Barr & Stroud instruments to swivel ±5° from the turret bearing. The roof plates on A are still being bolted down. The monitor *Roberts* lies on the other side of the basin. *(UCS 118-554-509)*

the latter did sometimes record total costs and profit. Indeed, Barrow even recorded the actual weight and price of each individual plate of gunhouse armour, *eg Prince of Wales* twin from Beardmore: 7 cemented plates 53.16 tons costing £10,366, 17 non cemented plates 182.75 tons £31,067. Interestingly, across 470 tons, the actual weights averaged 3.1% less than the calculated weight for cemented plates, 4.9% less for non cemented.[11] Since the Admiralty rolling thickness tolerances were 0% over, 2% under, that provided a hidden weight 'saving'.

**Provisional price *Prince of Wales* 14in mountings, Barrow Contracts 316G and 317G**

| Item | | Quad £ | Twin £ |
|---|---|---|---|
| 1 | Turntable, working chamber, trunks, roller paths | 64,350 | 45,700 |
| 2 | Training gear, racks, engines, gearing, shafts & brackets | 24,000 | 15,600 |
| 3 | Gun slides, cradles & elevating gear | 69,500 | 33,800 |
| 4 | Hydraulic pumping engines (Elswick) | 40,000 | 20,000 |
| 5 | Pipes, valves, etc on revolving structure | 15,100 | 9,400 |
| 6 | Shell room machinery | 30,000 | 17,200 |
| 7 | Magazine machinery and flashtight doors | 18,400 | 15,800 |
| 8 | Transporting gear, tilting trays, cordite containers etc | 20,300 | 11,300 |
| 9 | Shell and cordite hoists | 68,200 | 37,900 |
| 10 | Pipes, valves etc on fixed structure | 11,400 | 2,600 |
| 11–12 | Miscellaneous | 22,550 | 14,200 |
| | Sub total 1–12 | 383,800 | 223,500 |
| | Armour | 58,500 | 41,430 |
| | Delivery and erection | 90,000 | 45,000 |
| | Drawings, patterns, jigs, tools etc (half to each quad) | 48,400 | 14,890 |
| | Total | 580,700 | 324,820 |

Total per ship: £1,486,220. These costs exclude guns, director firing gear and new pits. The Elswick accounts for 1941 show a sales value for *King George V*'s three mountings as £1,336,334 with a profit of £60,713.[12] However these Elswick figures were based on the provisional prices and appear to exclude gunhouse armour. The prices were later revised in the light of actual costs, which were subject to wartime inflation of wages and material prices, the many modifications and the later reduced armour prices. The Elswick accounts for years 1942–46 suggest that the final cost to the Admiralty of *Anson*'s three mountings including an assumed £73,000 for armour came to £1.48M excluding guns, say £580,000 for a quad. The profit margin was a meagre 3.7% of cost well below the 20% or so typical of cruiser mountings pre-war. For *Prince of Wales* the final price in 1948 after adjustments for was £1.614M, say £500 per ton excluding guns; assuming about £22,000 per gun, total for the ship £1.834M.[13] *Howe*'s from Barrow were cheaper as there were no costs for jigs etc and they had the benefit of the learning curve: estimated £575,000 including guns for each quad and £350,000 for the twin, so total for ship £1.5M, 21% of ship cost.

But Elswick's contribution to the 14in programme was not limited to design, manufacture and installation of guns and mountings. In addition, they built a wooden mock-up of a 14in gunhouse, they proofed each gun at their Ridsdale range (about £700 each), they made 3pdr and 6pdr sub-calibre guns (3pdr £152 each for 5.25in guns), they carried out inspections, they made spare and replacement parts, they made modifications to mountings in service, they overhauled mountings at major refits (*Duke of York* in 1945 at Cammell Laird £36,677), they relined guns (14in about £4100 each) and they made empty shells.

## PERFORMANCE

The 14in mounting handbook claimed 'The performance of the mountings in their early life has given good reason to expect that the design will prove to be thoroughly sound and reliable.'[14] But that had been written in 1942 over a year after the first mountings entered service. Given time, the tools and the skills of the ordnance artificers, most of the minor faults could be overcome, but failures could cause missed salvoes which could be critical in action. The importance of regular tests, drills and lubrication was emphasised without which mechanical components could get misaligned or damaged, particularly on the quad mounting which was more complex than the twin, which latter was considered more reliable.

The first real test in action was *Prince of Wales*' action with *Bismarck* on 24 May 1941. She had been rushed into service without time to tune the mountings and train the gun crews properly. It is disappointing that more of the problems had not been ironed out during the building period, probably due to the rush to complete. The purpose of erecting a complete mounting in a pit was to ensure the exact fit and smooth working of all parts especially in hoists, loading gear and interlocks, where faults could be corrected under workshop conditions. While disassembly and re-erection in the ships could cause some disruption to well oiled components, there had been six months between installation of the last mounting (B in September 1940) and acceptance after trials in March 1941. V-A technicians had remained on board when she left Birkenhead in January for completion at Rosyth, and one was still on board in each turret when she sailed to meet *Bismarck*.[15]

A turret was the most affected during the first action in which *Hood* was lost. Over the eight salvoes fired, only 55 guns fired out of a possible 74, but at least at a rate of 1.9 rounds per gun per minute, nearly the same as that designed, namely 2. Y turret, which initially could not bear on *Bismarck*, also had some failures but B – the simplest – performed well. Problems encountered included shell ring rammers jamming, hoists loaded only with shell not with charges owing to interlock failures, flashtight doors not opening fully and overshoots in traversers with shells jamming. In A gunhouse, heavy seas breaking over the low freeboard bow entered the gunports, flooding not only the gunhouse but draining down to the shell rooms, where the crews had to work with water slopping around owing to inadequate drainage, slowing them down. The spray also made use of the turret rangefinder difficult.

[11] Gun Mounting Contracts 316 & 317G. BAE Systems Barrow shipyard archives
[12] Tyne & Wear Archives DS.VA/3/5/14
[13] Gun Mounting Contracts 316 & 317G. *Ibid*

[14] ADM 234/271, *op cit*
[15] Garzke, William H, Dulin, Robert, Jurens, William. *Battleship Bismarck*. Naval Institute Press, 2019, p221

Her sister *King George V* engaged *Bismarck* on 27 May, firing a total of 339 14in APC shells, an average of 34 per gun. Her mountings suffered intermittent failures with A out of action for 30 minutes after firing about 23 rounds per gun due to a jam between the fixed and revolving structure in the shell room. Four guns in A and B mountings had to cease firing after about 30 rounds owing to jams. There were also a few failures from misfires and drill errors, not surprising from new largely untried mountings and an as yet inexperienced crew.[16] Both Y and B missed a few salvoes but by the end of the action which saw *Bismarck* sinking they were all 'working in perfect order'.[17]

Radar gave *Duke of York* a great advantage against *Scharnhorst* in the darkness of 26 December 1943 but her 14in output was only 68% of that possible, owing both to minor mechanical failures and to faulty drill by some of the crew. Compared with *King George V* against *Bismarck,* the report noted fewer breakdowns.[18] But three guns in Y turret failed together so missing 17 broadsides over 15 minutes. Poor ventilation in the magazines resulted in stale air and a fatigued crew handling the 84½lb quarter charges. Indeed, the *KGV* class's ventilation overall was poor when they joined the British Pacific Fleet in 1945, compared with US battleships with air conditioning.

*King George V* undertook shore bombardments of Japan with her 14in, now fitted with RP.40 remote control. The first was at the Hitachi works on 14 July 1945 with ten rounds of HE per gun, but 15 were missed owing to failures in flashtight doors and interlocks. Fifteen days later she fired at Hamamatsu making aircraft parts, ten rounds per gun with reduced charges as the range was only 20,000 yards. 265 were fired out of the planned 300 with some failures due to misfires in the lock firing mechanisms.[19]

The four remaining *KGV*s were all broken up in 1957–58, their 14in guns and mountings being scrapped with the ships.

| **14in BL Mark VII Gun** | | |
|---|---|---|
| Calibres, bore | 45 | |
| Length overall, in | 650.85 | |
| Weight incl BM, tons | 79.55 | |
| Muzzle velocity nominal, ft/sec | 2475 | |
| Shell, lb | 1590 | |
| Charge SC300, lb | 338 | |
| Life, efc | 340 | |

| **14in Mounting** | **Quad Mark III** | **Twin Mark II** |
|---|---|---|
| Revolving weight, tons | 1550 | 898 |
| Max elevation, degrees | 40 | 40 |
| Max depression, degrees | -3 | -3 |
| Loading angle, degrees | 5 | 5 |
| Training angle, degrees | A 286, Y 270 | 270 |
| Training speed, deg/sec | 2 | 2 |
| Elevating speed, deg/sec | 8 | 8 |
| Recoil, in | 45 | 45 |
| Rounds per gun per minute | 2 | 2 |
| Max range, yards | 37,100 | 37,100 |
| Mounting crew | 112 | 62 |

[16] Pursuit and Destruction of German Battleship *Bismarck*. 1941–42. ADM 199/1188 & Roberts, John in *Warship* 28

[17] Garzke. *Op cit*, p425

[18] Sinking of *Scharnhorst*. 1943. ADM199/913

[19] National Archives ADM 1/18485

The end of a quadruple 14in mounting, as *King George V*'s Y is demolished at Dalmuir on 21 May 1958. The thickness of the gunhouse armour is apparent, front 520lb (12.74in), side rear 280lb. The semi-circular splinter screen (mantlet) also helped keep blast and spray out. *(T W Ferrers-Walker)*

CHAPTER SEVENTEEN

# The 5.25-inch Twin Mountings

## INTRODUCTION

The early 1930s saw many design studies for battleships with main armament ranging from 12in to 16in. Instead of separate low angle and high angle secondary armament (*eg* 6in LA and 4.7in HA as in *Nelson* and *Rodney*), most designs featured dual-purpose HA/LA guns, which made better use of the available topside space. Initially these were power operated 4.5in BD (between decks) as adopted in the modernised *Queen Elizabeth* and *Renown,* then followed by 5.25in. Vickers-Armstrongs, now the only major gun mounting manufacturer, were asked to prepare a design. Since there would still be some manual handling, the Admiralty considered that a shell weight of 80lb could be managed. In May 1936, they were given an order to design and manufacture a twin 5.25in mounting with 70° elevation and a rate of fire of 10–12 rpg. This was to be delivered by December 1937 and fitted in the partially disarmed battleship *Iron Duke,* and then planned to be fitted in the 1936 Programme battleships. Contract 319G at Barrow included £35,000 for the design and £50,000 for the actual mounting. As well as being suitable for battleships, it could also be mounted in smaller cruisers replacing both the normal 6in LA and 4in HA. The resulting *Dido* class had five Mark II twin 5.25in mountings all on the centreline, three forward (A, B & Q) and two aft (X & Y), with revolving long trunk ammunition hoists reaching right down to deep magazines holding 360 rpg. This permitted a smaller crew, unlike battleships where wing short trunk mountings had to be fitted amidships but the deep magazines with 400 rpg had to be fore and aft of the machinery spaces. Separate hoists were provided for HA shell, LA shell and cartridges, with a complex ammunition supply for the four superimposed mountings with horizontal transfer chutes on the lower deck to feed hoists to the 5.25in casemates beneath the actual mounting. In the hoists between working chamber and gunhouse, the cartridges and LA shells were moved vertically, but HA shells were horizontal for ease of transfer to the fuze setting machine. The guns and gunhouses were similar inside, protected by 1in D.1 (40lb) quality high tensile steel in the Mark Is and 0.5in in Mark IIs at £28.75 per ton. In the battleships, the casemates under each mountings had 1½in NC (non cemented). In both Marks hydraulic mechanisms were used in the gunhouses for elevating, training and ramming, supplied by an electrically powered 80hp Newton-Derby pump beneath. The working medium was mineral oil at 1000lb/sq in, which was less prone to leakage than the water used earlier.[1] *(See colour plate on page 172)*

A 5.25in Mark I gun at the Explosion Museum, Gosport. Engraved on the breech is N Q.F. 5.25 IN MK I VA 1941 No. S/367 L. *(Author)*

The 50-calibre Mark I guns were a new design, weighing 4.3 tons with breech mechanism, jacket and a replaceable 1.65-ton loose barrel; the gun life was estimated to be only 750 equivalent full charges. With gun cradle and balance weight (to keep the centre of gravity close to the trunnions) the elevating mass was 11 tons, maximum 70°. The revolving weight of each battleship mounting, gunhouse and working chamber, was 80 tons. For the cruisers, revolving weight included the hoist trunk, varying from 84 tons for the shortest Y to 95 tons for the longest Q.

Plans were drawn up to rearm the cruisers *Hawkins* and *Frobisher* with six 5.25in mountings replacing their 7.5in LA guns. Together with 16 Mark Is, 35 Mark IIs for the first seven *Didos* and ten spares, the order book had reached 73 mountings by December 1937. But it was

[1] *Handbook for 5.25-inch QF Mark I Gun on HA/LA Twin Marks I and II Mountings. 1941.* (CB.4108/BR.915) ADM 234/183

soon realised that the rundown armament industry could not supply anything like that number by the required 1940 date, as well as the 14in for battleships and 6in for cruisers. With WW1 battleship modernisation also under way, the dockyards had no spare refit capacity to re-arm the cruisers, so the plans were cancelled. It was noted on 23 November 1938 that armament supply would limit new construction for cruisers with 4.5in or 5.25in mountings to only four ships a year.[2]

Vickers-Armstrongs had received the first production order on 20 November 1936 with eight Mark I at Barrow (Contract 340G at a price of £33,000 each excluding guns) plus eight at Elswick for the two 1936 Programme battleships (Contract TM.121). On the same date 51 Mark II were ordered at Barrow (328G) with delivery between March and December 1939, although 33 were soon transferred to Elswick. The Barrow order was later cut back to just ten, for the first two *Dido* class cruisers. Their final price after Admiralty adjustments came to £50,950 each excluding guns – their more complex hoists adding to the Mark I price. Barrow received further orders in November 1937 and in mid-1939. Further Mark II orders were placed with the newly refurbished Scotswood works and also for spare hoist trunks. Most of Elswick's orders were later transferred to Scotswood.

Harland & Wolff's newly reinstated ordnance works at Scotstoun received the order for 26 Mark I mountings for the three 1937 Programme battleships and two spares. Nine Mark I* were ordered from Barrow early in 1941 for *Vanguard,* one being a spare (386G). These were designed for remote power control from the start, although manufacture did not start until 1943. They also had thicker gunhouse protection (90lb NC sides, 60lb roof) so were heavier at 95 tons – 60lb was reckoned to be proof against most bomb fragments.

Orders for the corresponding QF Mark I guns included 70 for Elswick in early 1937 at final cost of £2312 for the gun plus £527 for the breech mechanism, plus Beardmore in Glasgow and the Royal Gun Factory at Woolwich. Campbell gives a total of 267 guns made including those for the Army.[3] Ninety-nine 3pdr sub-calibre guns were ordered from Elswick at £152 each.[4]

Despite a large amount of sub-contracting of components such as gunhouse plating and ammunition hoists, Vickers-Armstrong struggled to complete not only the sixteen 5.25in Mark I mountings for *King George V* and *Prince of Wales* but also the twenty-five 5.25in Mark II mountings for the five 1937 Programme *Dido* class, delivered over a year late. The hulls for the latter had all been launched by July 1939 by which time the pilot Mark I mounting had only just been installed in *Iron Duke,* removed later that year. The V-A production programme of February 1939 showed the first *King George V* mounting due to be delivered in February 1940. It also showed sixteen for *Lion* and *Temeraire* due to be delivered from H&W from August 1941. In June 1939, DNO requested V-A to deliver the *Dido* mountings in the following order to suit the shipbuilders: Hawthorn Leslie Y, X, Q, B, A but the other shipyards A, B, Q, X, Y.[5]

Elswick had already transferred fifteen of their *Dido* class orders to nearby Scotswood to concentrate on the heavier mountings, but that inexperienced works struggled with these new mountings. Erection of the first started only in the third quarter of 1939 but it was only in March 1940 that Elswick completed the first of their three for *Naiad* (Scotswood two, with some parts from Elswick), by which time Barrow had also completed four for *Bonaventure.* The result was that only *Naiad* completed with all five mountings in July 1940, although a problem arose with A mounting (the last to be fitted) jamming due to lack of rigidity and circularity in the supporting hull structure. The first *Dido* to complete was *Bonaventure* from Scotts at Greenock in May 1940, but with only four mountings. In place of X was a single 4in for star shells. By the time her X mounting had been completed (from Scotswood), she had been torpedoed in March 1941 in the Mediterranean, so it (Reg Nº 5) was re-allocated to the delayed *Sirius* at Portsmouth as B.

*Dido* completed at Cammell Laird at Birkenhead in September 1940, also with only four mountings, with a single 4in in place of Q. She received a Q mounting at Brooklyn Navy Yard about October 1941,

One of *Sirius'* 5.25in Mark II mountings at Scotswood ready for shipment to Portsmouth in 1941. The ammunition trunks were made separately by sub-contractors so no pits were needed for erection. *(Vickers)*

[2] National Archives. ADM 265/2
[3] *Naval Weapons of World War Two*, p44
[4] Tyne & Wear Archives. DS.VA/3/5/13
[5] 5.25in Erection Correspondence. Tyne & Wear Archives. DS.VA/6/36/13/18

where she was being repaired after being damaged off Crete in May 1941. She actually got Nº 37 planned as *Sirius'* B as the latter was not yet ready to receive it, while *Dido*'s Q (Nº 8) was switched to become *Sirius'* Q early in 1942. *Sirius'* construction at Portsmouth had been delayed by bomb damage in the dockyard in May 1941, so delivery of three mountings already completed at Barrow was held up. *Phoebe* was completed by Fairfield of Glasgow in September 1940 also with only four mountings, three from Scotswood (Nºs 21–23) and X from Barrow (Nº 24) so lacking Q. Nº 38 planned for *Sirius* Q was earmarked for despatch abroad, probably intended to become *Phoebe*'s Q at Brooklyn Navy Yard early in 1942 where she was repairing after being torpedoed in the Mediterranean in August 1941. However, it was not installed and probably became a spare, as *Phoebe* got a quadruple 2pdr in Q position instead, while in 1943 after another refit at Brooklyn she received a quadruple 40mm Bofors there. The price of the Barrow mountings had now fallen to an average of £38,000 each, benefiting from the learning curve; erection of all the mountings in each ship costing about £10,000.

Such was the shortage of mountings that *Scylla* (planned for Scotswood) and *Charybdis* were redesigned to take four twin 4.5in Mark III UD mountings, little more than the armament of a large destroyer. However, with delays to the cruisers, the War Office could be

*Argonaut* was torpedoed aft in the Mediterranean in December 1942 and repaired at Philadelphia Navy Yard. Both X and Y mountings were removed for overhaul. Y is awaiting re-installation, while her stern has yet to be rebuilt. *(Author's collection)*

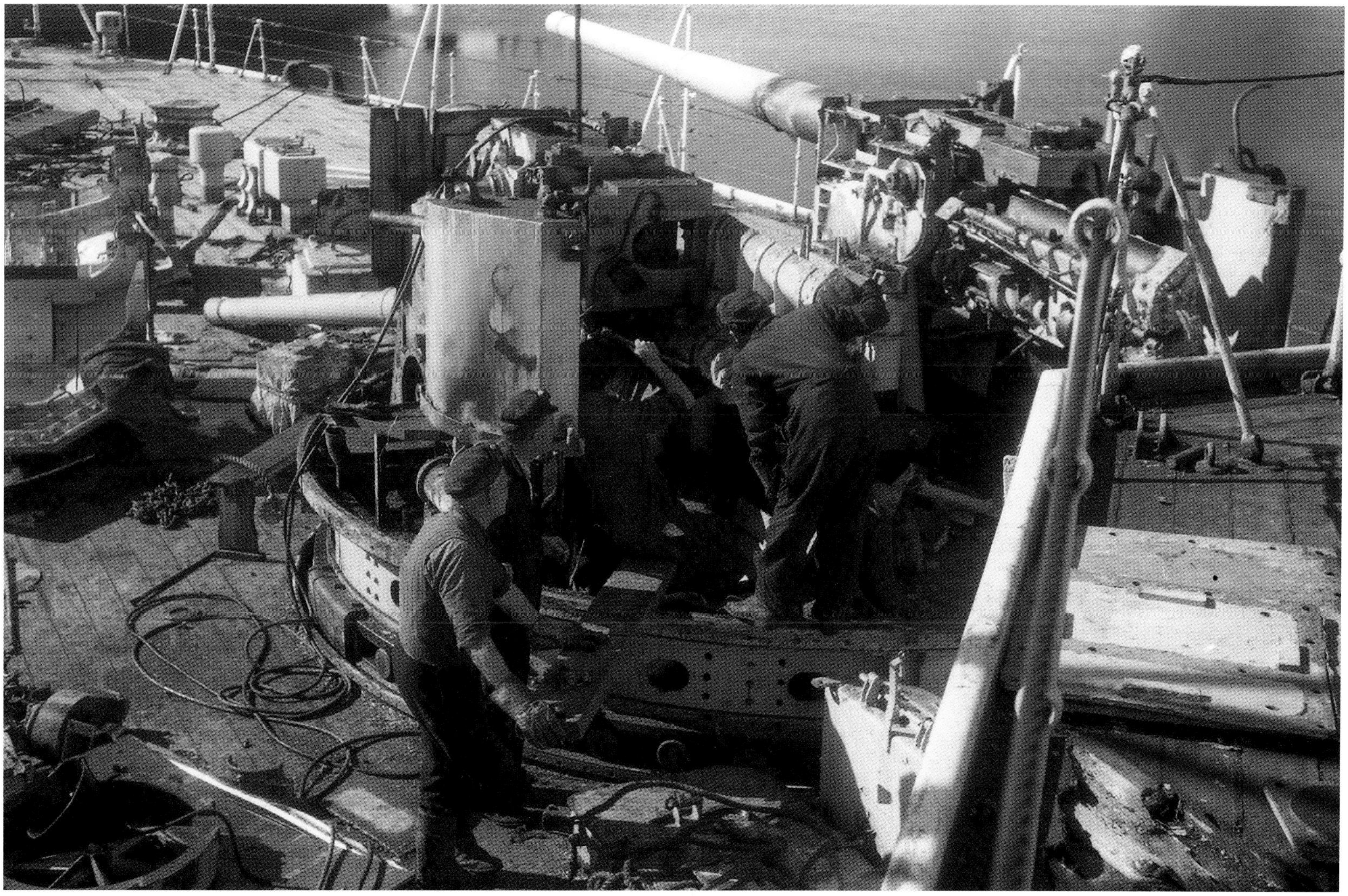

**Below:** An unusual view of a 5.25in mounting: *Dido*'s Y in course of demolition at Barrow in 1958. The left gun has been cut off, the right cut through. The tank at left is oil supply for the electrically driven hydraulic pump. The two loading trays and some of the 48 rollers are visible. *(Ken Royall)*

The interior of a 5.25in Mark II mounting looking forward. Both guns are elevated. The vertical shell on the left is for low angle fire, on the right is a shell and cartridge on the loading tray. High angle shells were hoisted horizontally next to the fuze setting machine. Trainer and sightsetter were both seated on the centre line. *(HMS* Excellent*)*

allocated three twin 5.25in Mark II from Scotswood (one being Nº 25 planned as *Phoebe*'s Q), which were erected near London for anti-aircraft defence in 1942, one on Primrose Hill, one at Wimbledon, one at Eltham.[6] Reducing the main armament of the last five *Dido*s to four mountings apiece also helped deliveries, even allowing at least two spare mountings to be completed, possibly shipped to overseas dockyards. From 1942 remote power control was introduced, *Argonaut*'s X and Y being completed with RP.10 equipment. This not only doubled elevating and training speed but allowed the fire control system to control the pointing of the guns directly, rather than by the gunlayers and trainer following pointers. The gear was made by Metropolitan-Vickers at Manchester.[7] The last five (Modified *Didos*) had all their four mountings fitted with RP.10. During her refit at John Brown October 1943 to August 1944, *Euryalus* had X and Y mountings removed to Barrow and replaced with two new ones, spares converted to RP.10. A and B were also modified to RP.10 but Q was removed. Her two old X and Y were converted to RP.10 at Barrow then stored at Rosyth in 1947. In 1944, *Cleopatra* and *Argonaut* had Q removed and replaced by a quad 40mm or 2pdr, which increased close range fire ahead and reduced topweight. Scotswood produced sixteen RPC conversion kits in 1947, possibly for two *KGV* battleships.

### 5.25-INCH MARK II MOUNTING ALLOCATION

The whole 5.25in mounting production programme from 1937 to 1943 suffered many delays, postponements and re-allocations as ship-

[6] National Archives. ADM 1/17228

[7] Addendum to BR.915 (RP.10) 1944. ADM 234/184

### 5.25in Mark II Mounting Allocation

| Reg Nº | Works | Allocation | Mtg | Shipped | Notes |
|---|---|---|---|---|---|
| 1 | Barrow | *Bonaventure* | | 2/40 | Originally planned for Scotswood. Scotts, Greenock |
| 2 | Barrow | *Bonaventure* | | 3/40 | |
| 3 | Barrow | *Bonaventure* | | 3/40 | |
| 4 | Barrow | *Bonaventure* | | 3/40 | |
| 5 | Scotswood | *Sirius* | B | 4Q/41 | Planned for *Bonaventure* X |
| 6 | Barrow | *Dido* | A | 6/40 | Cammell Laird |
| 7 | Barrow | *Dido* | B | 7/40 | |
| 8 | Barrow | *Sirius* | Q | 1/42 | Planned for *Dido* Q |
| 9 | Barrow | *Dido* | X | 8/40 | |
| 10 | Barrow | *Dido* | Y | 7/40 | |
| 11 | Elswick | *Naiad* | Y | 3/40 | Hawthorn Leslie |
| 12 | Elswick | *Naiad* | X | 4/40 | |
| 13 | Elswick | *Naiad* | Q | 4/40 | |
| 14 | Scotswood | *Naiad* | B | 4/40 | |
| 15 | Scotswood | *Naiad* | A | 6/40 | |
| 16 | Scotswood | *Euryalus* | Q | 10/40 | Chatham Dockyard |
| 17 | Scotswood | *Euryalus* | B | 11/40 | |
| 18 | Scotswood | *Euryalus* | A | 1/41 | |
| 19 | Scotswood | *Euryalus* | Y | 2/41 | |
| 20 | Scotswood | *Euryalus* | X | 3/41 | |
| 21 | Scotswood | *Phoebe* | Y | 3Q/40 | Fairfield |
| 22 | Scotswood | *Phoebe* | A | 3Q/40 | |
| 23 | Scotswood | *Phoebe* | B | 3Q/40 | |
| 24 | Barrow | *Phoebe* | X | 3Q/40 | |
| 25 | Scotswood | War Office | | 3Q/41 | Planned for *Phoebe* Q |
| 26 | Barrow | *Hermione* | A | 12/40 | Stephen |
| 27 | Scotswood | *Hermione* | B | 4Q/40 | |
| 28 | Barrow | *Hermione* | Q | 10/40 | |
| 29 | Elswlck | *Hermlone* | X | 4Q/40 | |
| 30 | Scotswood | *Hermione* | Y | 4Q/40 | |
| 31 | Scotswood | *Cleopatra* | Q | 3/41 | Hawthorn Leslie |
| 32 | Scotswood | *Cleopatra* | Y? | 3/41 | |
| 33 | Scotswood | *Cleopatra* | X? | 3/41 | |
| 34 | Scotswood | *Cleopatra* | A | 4/41 | |
| 35 | Scotswood | *Cleopatra* | B | 6/41 | |
| 36 | Barrow | Spare? | | | Planned for *Sirius* A |
| 37 | Barrow | *Dido* | Q | 4Q/41 | Planned for *Sirius* B. Fitted at Brooklyn Navy Yard |
| 38 | Barrow | Spare | | | Planned for *Sirius* Q |
| 39 | Barrow | *Sirius* | X | 11/41 | Portsmouth Dockyard |
| 40 | Barrow | *Sirius* | Y | 10/41 | |
| 41 | Scotswood | War Office | | 3Q/41 | Planned for *Royalist* A |
| 42 | Scotswood | War Office | | 3Q/41 | Planned for *Royalist* B |
| 43 | Scotswood | *Sirius* | A | 4Q/41 | Planned for *Royalist* Q |
| 44 | Scotswood | *Royalist* | X | 1Q/43 | Scotts. RPC |
| 45 | Scotswood | *Royalist* | Y | 1D/43 | RPC |
| 46 | Barrow | *Argonaut* | A | 1Q/42 | Cammell Laird |
| 47 | Barrow | *Argonaut* | B | 1Q/42 | |
| 48 | Barrow | *Argonaut* | Q | 2Q/42 | |
| 49 | Barrow | *Argonaut* | X | 6/42 | RPC |
| 50 | Barrow | *Argonaut* | Y | 5/42 | RPC |
| 51 | Scotswood | *Bellona* | A | 2Q/43 | Fairfield. RPC |
| 52 | Scotswood | *Bellona* | B | 2Q/43 | RPC |
| 53 | Scotswood | *Royalist* | A | 2Q/43 | Planned for *Bellona* Q. RPC |
| 54 | Scotswood | *Bellona* | X | 3Q/43 | RPC |
| 55 | Scotswood | *Bellona* | Y | 3Q/43 | RPC |
| 56 | Barrow | *Euryalus* | X | 2Q/44 | Planned as spare. RPC replacing orig X |
| 57 | Barrow | *Euryalus* | Y | 2Q/44 | Planned as spare. RPC replacing orig Y |
| 58 | Barrow | *Black Prince* | A | 3Q/43 | Harland & Wolff. RPC |
| 59 | Barrow | *Black Prince* | B | 3Q/43 | RPC |
| 60 | Barrow | Spare | | | Planned for *Black Prince* Q. |
| 61 | Barrow | *Black Prince* | X | 3Q/43 | RPC |
| 62 | Barrow | *Black Prince* | Y | 3Q/43 | RPC |
| 63 | Scotswood | *Diadem* | | 8/43 | Hawthorn Leslie. RPC |
| 64 | Scotswood | *Diadem* | | 9/43 | RPC |
| 65 | Scotswood | *Hoyalist* | B | 2Q/43 | Planned for *Diadem* Q. RPC |
| 66 | Scotswood | *Diadem* | | 9/43 | RPC |
| 67 | Scotswood | *Diadem* | | 9/43 | RPC |
| 68 | Barrow | *Spartan* | A | 1Q/43 | RPC |
| 69 | Barrow | *Spare* | | 4/43 | Planned for *Spartan* Q. |
| 70 | Barrow | *Spartan* | B | 4/43 | RPC |
| 71 | Barrow | *Spartan* | X | 2Q/43 | RPC |
| 72 | Barrow | *Spartan* | Y | 2Q/43 | RPC |

*Royalist* received four modernised RPC mountings at her major refit 1954–56, probably converted spares.

building priorities changed and due to shortages of materials and labour. This was most acute for the Mark II mountings for cruisers, with some mountings diverted to the Army for London anti-aircraft defences. No original Admiralty gun mounting registers have been found which would have recorded manufacturer, dates, destinations, cancellations, modifications and disposal for every mounting. It has been necessary to piece together information from ship gunnery trial reports (*eg Dido* class Ships Covers), Vickers-Armstrongs works quarterly reports from Barrow, Elswick and Scotswood,[8] Barrow gun mounting contract ledgers and even floating crane logs. Enough Registered Numbers and shipment dates have been found to compile what is considered the most likely record of 72 mountings. It appears that *Dido* class 5.25in orders were placed in fives, even for the later ships which actually only received four, with the fifth Q later being reallocated. No Reg Nºs appear to have been allocated to *Scylla* and *Charybdis,* which suggests that as early as 1939 deliveries were expected to be late, hence the decision to arm them with 4.5in.

### BATTLESHIP MOUNTINGS

Production delays also affected Mark I deliveries for battleships. Only four of Elswick's eight were ready in August 1940 to be shipped in *King George V* (Nºs 10–13) so the first four (foremost) from Barrow

[8] Cambridge University Library, Manuscripts. Vickers Archive 192–210

A twin 5.25in Mark I mounting being loaded from Harland & Wolff's Scotstoun works into a coaster, probably *Citrine* for transport to Birkenhead for *Prince of Wales* early in 1941. *(Author's collection)*

intended for *Prince of Wales* (Nos 2–5) had to be sent to the Tyne from June to August 1940 – she left the shipyard on 17 October for drydocking at Rosyth. *Prince of Wales* got the next four from Barrow in October and November 1940 (Nos 6–9) plus the first four from Harland & Wolff at Scotstoun (Nos 18–21). *Duke of York* got all eight from nearby H&W between February and June 1941 (Nos 22–29), as did *Howe* early in 1942 (Nos 34–41) at a ship erection cost each of about £16,000. *Anson* got four (foremost) from Scotswood (Nos 14–17, having been switched from Elswick) and four from Scotstoun (Nos 30–33). Each Scotswood mounting, weighing complete with gunbay and guns 84 tons, was loaded on to a barge and

**5.25in QF Mark I Gun**

| | |
|---|---|
| Calibres, bore | 50 |
| Length overall, in | 275.5 |
| Weight incl BM, tons | 4.29 |
| Shell weight, lb | 80 |
| Charge, cordite SC140, lb | 18.6 |
| Cartridge, lb | 38 |
| Muzzle velocity, nominal, ft/sec | 2690 |
| Life, efc | 750 |

**5.25in Twin Mounting Marks I & II**

| | |
|---|---|
| Revolving weight, tons | 80-95 |
| Max elevation, degrees | 70 |
| Max depression, degrees | -5 |
| Training speed, deg/sec | 10 |
| Elevating speed, deg/sec | 10 |
| Recoil, in | 24 |
| Max range at 45° yards | 24,700 |
| AA ceiling, ft | 46,500 |
| Mounting crew | 48 |

towed downstream to Swan Hunter & Wigham Richardson's shipyard at Wallsend where it was lifted on board by the *Titan II* floating crane as S1, S2, P1, P2. The 150-ton crane was jointly owned by Swan Hunter, Hawthorn Leslie and Wallsend Slipway, charging £1 per ton for such a load. The first mounting was lifted on 25 May 1941, then 30 May, 27 June and 29 August. The four from Scotstoun were each loaded on to the new 871 grt coaster *Empire Gat,* which took about ten days to sail northabout from the Clyde to the Tyne, having to wait at Methil until a suitable east coast convoy was available. Each was then lifted aboard *Anson* by *Titan II* on the following dates: 9 September, 27 September, 7 November and 23 December 1941. The recorded weight was only 65 tons, which suggests that the guns were not included.[9]

Sub-contractors were used to make less critical components wherever possible – by 1943 Barrow works had used over ninety such companies for work across all their products valued at £2 million. Elswick brought in members of the Richardsons Westgarth group of engineers. North Eastern Marine, normally making marine engines at Wallsend, fabricated eight Mark I gunhouses at £3350 each, completed by December 1940 and ten of the thinner Mark II at £2510 each by April 1941. The D.1 steel plates were a free issue from Vickers-Armstrongs. The contract value totalled £53,371 yielding a profit of £5179. Marine engineer George Clark at Sunderland made 41 sets of Mark II ammunition trunks, ordered from February 1939 and delivered between August 1940 and mid-1942. Richardsons Westgarth at Hartlepool made twenty-two Mark II gunhouses and fifty 14in shell cages at £280 each.[10]

*Anson*'s Mark I mountings were removed at her Devonport refit starting in June 1944, and converted to RP.10 before she went to the Pacific in April 1945; three at Scotswood, three at Barrow and presumably two at Scotstoun. She also got the earliest deliveries of the four associated Mark VI directors, Reg Nos 2–5. At least one hundred single 5.25in anti-aircraft mountings with Mark II guns were made at Elswick and Barrow for the War Office to a different design between 1942 and 1944.[11]

*Vanguard* received her eight 5.25in mountings from Barrow, Mark I* with RP.10, delivered in early 1946 by coaster *Empire Jack* from Barrow, plus one spare. If the *Lion* class had been built, they also would have had eight twin 5.25in mountings. In 1943 Elswick were asked to design and build two new 80° 5.25in gun mountings, one all-electric and one RP.40, but these appear not to have been completed. Post-war a Mark III twin mounting with a new Mark IV gun was designed for the Admiralty, and cruiser designs prepared but none went into service. Anti-aircraft 5.25in singles were made for the Army. Most of the 5.25in mountings were scrapped with each ship, but a Mark II mounting from the Pakistani *Babur* (ex-*Diadem*) was preserved as a gate guardian outside a naval establishment at Karachi in the 1980s.

When delivered there were all the usual teething troubles of a new mounting. The designed rate of fire in the battleships could not be

[9] *Titan II* Log 1941. Marine Technology Special Collection, 01250/23
[10] Marine Technology Special Collection 01363, 01364
[11] Tyne & Wear Archives. DS.VA/3/5/14-16

**Above:** *Vanguard* nearly complete about April 1946 at Clydebank, showing some of her twin 5.25in Mark I* mountings. Also visible is one the four US Mark 37 directors which controlled them and a 6-barrelled 40mm Mark VI mounting and its black topped director. *(UCS 118-567-416)*

**Right:** After the Pakistan Navy's *Babur* (ex-*Diadem*) was scrapped, one of her 5.25in turrets became a gate guardian at a Navy shore base. *(Author's photo 15 December 1988 at Karachi)*

achieved owing to the complex ammunition supply with separate projectile and cartridge case, only about 8 rather than the 10–12 rpg expected. The gunhouse was cramped with thirteen men. Raven and Roberts noted: 'In general the mountings were not liked because they were complex and difficult to maintain. Their high angle fire moreover was not up to expectations because, apart from the reduced rate of fire, their training speed of 10° to 11° per second was not considered adequate.'[12] They were also vulnerable to any loss of electrical power.

Particularly in the cruisers, the 5.25in was used mainly in its HA role. But *King George V* used hers against *Bismarck* on 27 May 1941 with repeated hits on her superstructure at ranges of about 11,000 yards. 660 rounds were fired, with more from her starboard battery than port. There were very few material failures, apart from occasional problems with hoists.[13]

Estimated production including spares and 3 War Office:

| *Works* | *Mark I* | *Mark I** | *Mark II* | *Total* |
|---|---|---|---|---|
| Barrow | 8 | 8+1 | 30+4 | 51 |
| Elswick | 4 | | 4 | 8 |
| Scotswood | 4 | | 34 | 38 |
| Scotstoun | 24+2 | | | 26 |
| Total | 42 | 9 | 72 | 123 |

[12] *British Battleships of World War Two*, p286
[13] Pursuit and Destruction of German Battleship *Bismarck*. 1941-42. ADM 199/1188

An unusual view of a 5.25in mounting: one of *King George V*'s being demolished by Arnott Young at Dalmuir on 21 May 1958. Even though the guns are off, it looks out of reach of and heavier than the 50-ton dockside crane could lift, so how did it get into this unhelpful position? *(T W Ferrers-Walker)*

*Royalist* in July 1956 after modernisation at Devonport Dockyard. Her 5.25in mountings are now Mark 2* fitted with RP.10 remote power control. Her new fire control system includes two Mark 6M directors with Type 275P radar, Gunnery Direction System Mark 2* linked to Type 293Q target indicating radar and Flyplane System Mark 5. Three twin 40mm STAAG Mark 2M replace her quadruple 2pdr. *(Admiralty)*

CHAPTER EIGHTEEN

# The 4.7-inch Twin Mounting Mark XX

FOLLOWING THE INTRODUCTION of the transferable twin 4.7in CP Mark XIX in the 'Tribal's, 'J' and 'K' class destroyers, the Admiralty wanted a power worked twin 4.7in but with higher elevation and a powered central hoist feeding directly into an enclosed weather-proof gunhouse. The result was the large impressive looking Mark XX, armed with a new QF Mark XI gun. Three mountings were fitted in the 'L' and 'M' class destroyers, slightly larger ships than the 'Tribal's. The gunhouse rotated about a fixed hoist on its centre line, with two shell and two cartridge hoists, which could deliver 10 rpg. The elevation was increased from 40° to 50°, though not really enough for fully effective anti-aircraft fire. Partly for this reason but mainly because of production delays, four of the eight 'L's were completed with four twin 4in Mark XIX HA (*Lance, Legion, Lively* and *Gurkha* which were all war losses).

Full power operation with fewer transfer points meant that a heavier shell could be fired, 62lb vs 50lb. An off-mounting 45hp electrically

'M' class destroyer *Martin* newly completed by Naval Yard in April 1942 with her three twin 4.7in Mark XX mountings. Owing to production delays, four sister 'L' class had to fit four twin 4in instead. *(Author's collection)*

A twin 4.7in Mark XX without its shield in the Barrow works, probably the first to be made for the destroyer *Lightning* under contract 345G. The balance weights are prominent. *(Barrow Dock Museum 5000)*

driven VSG pump powered the hoists, training and ramming but elevation remained manual, which made it difficult to track aircraft. The Mark IV director sent training and elevating signals to receivers in the gunhouse, where the trainer and gunlayers matched their pointers to correspond. Two fuze setting machines at the top of the twin hoists set the predicted time to the shell, which was transferred manually from the central hoist onto a tilting tray and then slid to the gun loading tray for ramming.[1] *(See colour plates on page 173)*

The guns were in separate cradles so required two gunlayers. The rest of the gunhouse crew comprised a trainer, one sight setter, one fuze follower and one captain of the turret (in a lookout) and for each gun

[1] *Handbook for 4.7-inch Mark XI Gun on the Twin Mark XX Mounting, 1942.* ADM 234/185

**4.7in QF Mark XI Gun**

| | |
|---|---|
| Calibres, bore | 50 |
| Length incl BM, in | 247.7 |
| Weight incl BM, tons | 3.35 |
| Nominal muzzle velocity, ft/s | 2540 |
| HE shell, lb | 62 |
| Charge SC122, lb | 12.8 |
| Weight of cartridge, lb | 32.6 |
| Life, efc | 850 |

**4.7in Twin Mounting Mark XX**

| | |
|---|---|
| Rotating weight, tons | 37.6 |
| Total incl on fixed structure, tons | 48.8 |
| Max elevation, degrees | 50 |
| Max depression, degrees | -10 |
| Training speed, deg/sec | 10 |
| Rounds per minute | 10 |
| Max range, yards | 21,200 |
| Max height, ft | 21,000 |

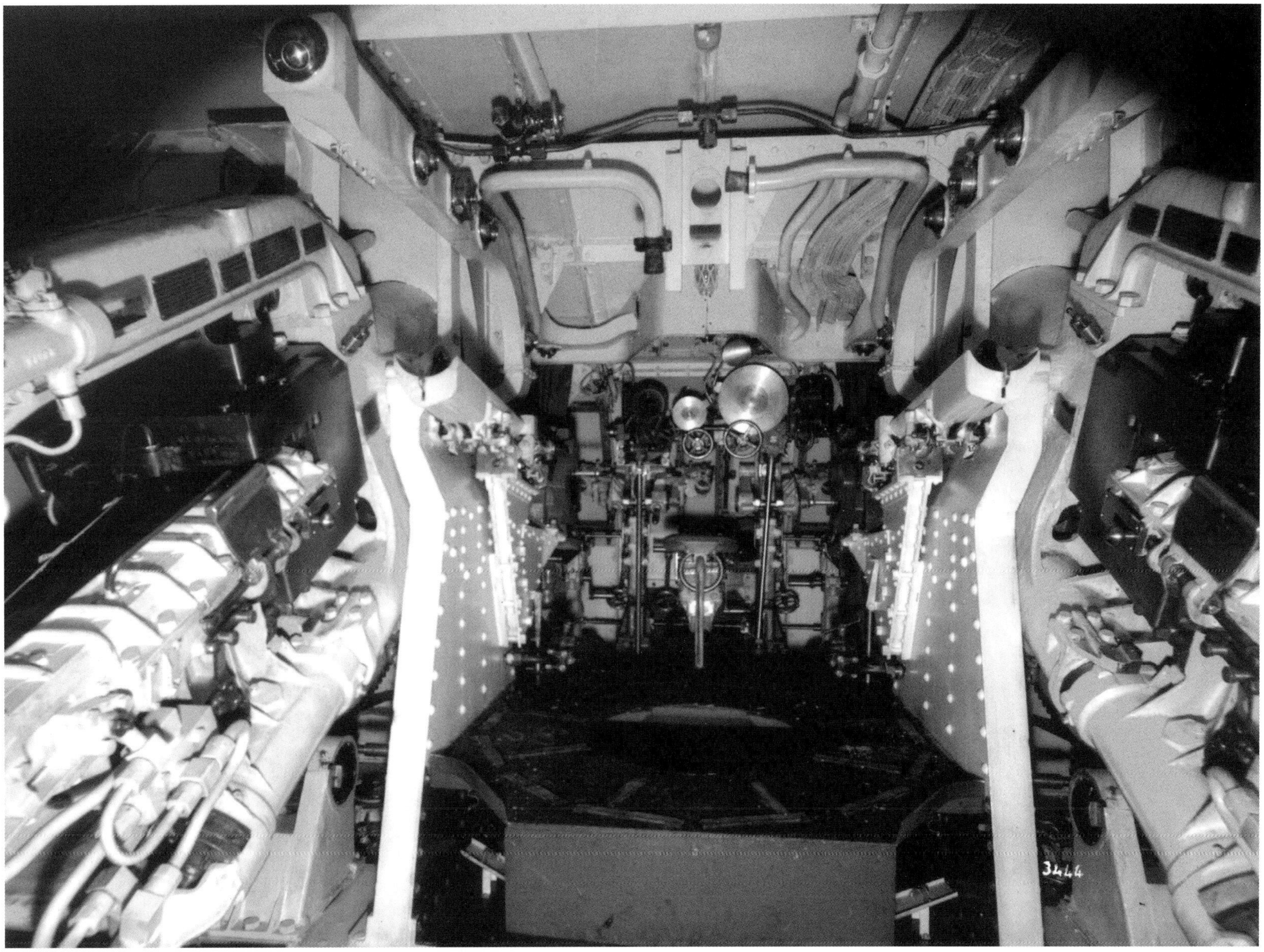


**Above:** Inside a twin 4.7in Mark XX looking forward with the top of the ammunition hoist in the centre and loading trays left and right. *(Barrow Dock Museum 4996)*

**Right:** Not a new gunboat for the Navy but how Vickers-Armstrongs delivered 38-ton twin 4.7in Mark XX mountings from their Scotswood works to their Tyneside shipyard. *(Tyne & Wear Archives)*

one fuze setter, two ammunition supply numbers and one tray worker; a total of fourteen.

The Mark XI gun itself was of 50 calibres, with a replaceable barrel weighing 1.4 tons and a jacket 63% of the length. Total weight per gun including breech mechanism and balance weight was 3.35 tons. Campbell reports 87 made, which implies 15 spares, as twelve ships had six each.[2] The new shell was a 5/10 crh, both HE and SAP.

The enclosed gunhouse required its own ventilation system, both for fresh atmosphere and for clearing fumes. Used cartridge cases were pushed out of holes at the rear of the gunhouse. The shield was only 0.5in thick, which helped keep the rotating mass down to 37.6 tons.

Barrow received an order for 28 mountings on 29 October 1937 worth £30,000 each, excluding guns, due for delivery in March 1940. This was five months before the hulls were ordered, indicating the lead time needed for the detailed design of the new mounting. Elswick received an order for 21, but they were actually built at Scotswood; only twelve were completed. Barrow delivered its first to *Lightning* at Hawthorn Leslie in December 1940 and its last to *Mahratta* (ex-*Marksman*) to Scotts in February 1943. Registered numbers were 1–40, *ie* 36 in twelve ships plus four spares. One each of the latter went to Portsmouth, Devonport and Coventry Gun Mounting Store, while Nº 25 replaced *Marne*'s damaged Y, repaired by Swan Hunter from February 1943 to February 1944.

The mounting was regarded as a better LA weapon than HA, and no further classes were fitted with the Mark XX. Four of the 'M's were sold to Turkey in 1957, surviving until 1972.

A large number of destroyers had transferable 4.7in mountings, which are not included.

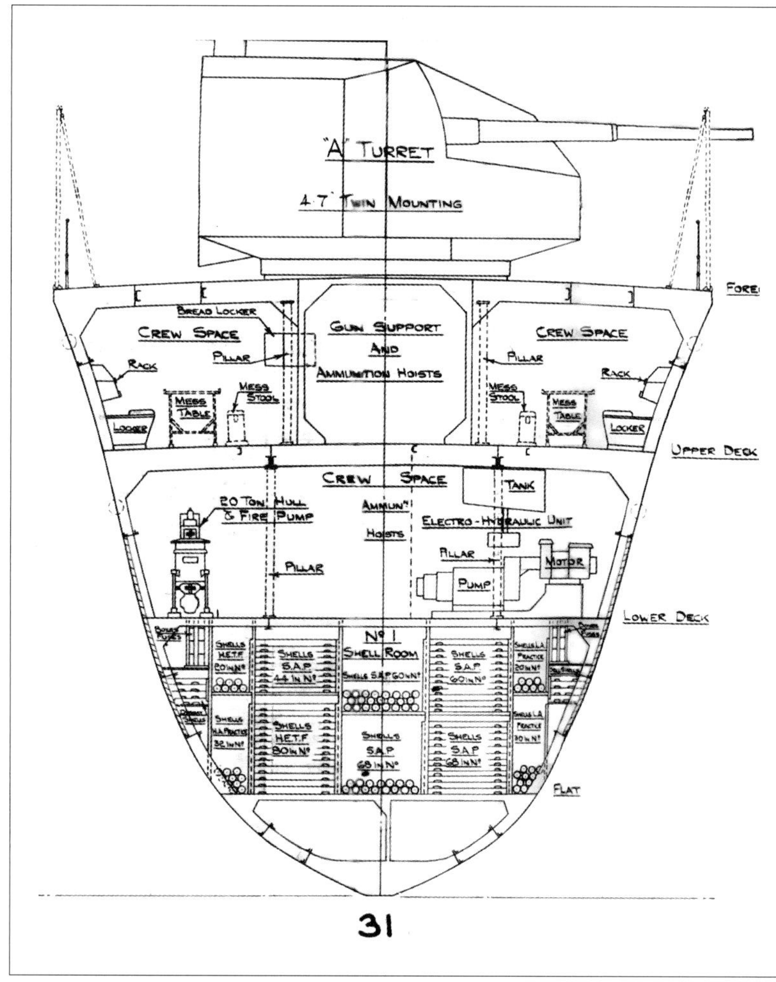


The cross section at Frame 31 through *Musketeer*'s A mounting shows how it dominated the design, not just the large gunhouse but the hoists and shell rooms. The electrically driven hydraulic pump was fitted off the mounting to reduce the rotating weight. *(General Arrangement, Marine Technology Special Collection 04418/2)*

[2] Campbell, John. *Naval Weapons of World War Two*, p47

Four 'M' class destroyers laid up at Penarth in August 1956 prior to their transfer to Turkey, still in their wartime guise. From left to right, *Marne* (lacking director), *Matchless, Milne, Meteor. (Author)*

CHAPTER NINETEEN

# 4.5-inch Gun Mountings

## THE 4.5-INCH TWIN MOUNTING MARK II

By the mid-1930s, the threat to warships from aircraft was all too apparent. *Nelson* and *Rodney* had only six single 4.7in HA, while the *Queen Elizabeth*s and 'R' class generally had by that time four to eight 4in, single or twin but hand operated, retaining their 6in LA batteries. When reconstruction of *Queen Elizabeth* and *Valiant* was being planned in 1936, it was recognised that their 6in and 4in should all be removed and replaced with dual-purpose HA/LA mountings. These would need a high elevation and a heavier shell than the 4in. While a twin 4.7in transferable mounting was being designed for the 'Tribal' class destroyers, it could only elevate to 40°. However, 4.5in was already an Army calibre, so a new 45-calibre QF was designed, the Mark I and III being essentially similar, the actual bore being 4.45in. (The Mark II was an Army AA gun.) This could fire a slightly heavier shell (55lb) than the 4.7in (50lb) with improved ballistics. The full fixed round (shell plus cartridge) at 85lb and 50in long was reckoned the largest handleable at sea. A one-ton loose barrel was fitted, although the gun had to be removed from the mounting to change it. Apart from five experimental guns, Campbell reports 46 Mark I made and 524 Mark III.[1] *(See colour plate on page 176)*

Vickers-Armstrongs were tasked to design a twin power operated between-decks mounting. Most of the circular gunhouse and the mechanisms were below the deck level, with only 5ft protruding above. The gun cradles were combined and could elevate to 80°. Fixed ammunition was expected to speed up loading to give a rate of about 12 rpg. The mounting was designated 4.5in Mark II BD and weighed 38 tons with 0.5in thick gunhouse. A 46hp electric motor powered the hydraulic pump for the various motions. It would be suitable both for rebuilt battleships with five on each beam and the new armoured aircraft carriers with eight mountings. In the latter, two mountings were placed on sponsons at each 'corner', with the gun bay for ammunition transfer alongside, and able to fire across the flight deck. The 4.5in magazines could not be sited directly beneath the mountings but six decks below the gunbay, requiring the endless chain hoists to be angled both fore and aft and athwartships.

A prototype Mark II was ordered from Barrow in January 1935 for installation in *Iron Duke* in September 1936. The production order for ten for *Renown* plus two spares (Contract 310G Reg N^os^ 2–13, although 2–7 were actually made at Scotswood) was placed with

The production of twin 4.5in Mark II BD mountings was shared between Barrow and Scotswood. One of the latter's first mountings is shown in N^o^ 12 Shop in October 1939, possibly for *Valiant*. Only 5ft protruded above deck level. Fixed ammunition was used, with three rounds loaded from the gunbay by hand onto the two small platforms before transfer into the mounting. *(Tyne & Wear Archives)*

[1] 'British Naval Guns 1880-1945' Part 16. *Warship* 34, 1985

Barrow in April 1936, costing about £19,000 each excluding guns. Twenty more for *Queen Elizabeth* and *Valiant* were ordered later that year (330G N$^{os}$ 14–33) but actually made at Scotswood, although those already made for *Queen Elizabeth* were diverted to complete *Formidable* first. In November 1936 a further sixteen planned for carriers *Illustrious* and *Victorious* were ordered from Barrow two months before the hulls were ordered, although the latter's eight were actually fitted in the Barrow-built *Indomitable*. Elswick then got an order for a further sixteen for *Formidable* and *Victorious* but they were actually made at Scotswood with *Formidable*'s being used for *Queen Elizabeth*. All such changes were made to get hulls into service as soon as possible.

In parallel with the BD mounting, an upper deck (UD) Mark III was also designed, transferable in being bolted to the deck with nearby ammunition hoists. The gun was the same, with a gunshield open at the rear to enable loading. Its weight was less at 30 tons and slightly cheaper at £16,000 each. Eight were fitted in *Ark Royal* in 1938, while four were fitted to each of the new submarine and destroyer depot ships, *Maidstone, Forth, Tyne, Hecla* and *Adamant* for LA fire. Although some production was switched from Barrow to Scotswood, there were still delays such that *Maidstone* left Clydebank in May 1938 without hers which were later fitted at Malta. Although planned to rearm 'D' class cruisers and *Repulse*, there was simply not the capacity to undertake that extra work. However, enough had been completed by 1941 to arm cruisers *Scylla* and *Charybdis* whose 5.25in mountings were not yet ready. Four twin 4.5in was a poor substitute for five twin 5.25in, yet neither ever received their planned main armament.

The four 1937 carriers did get their planned mountings, which served them in good stead during air attacks in the Mediterranean in 1941. Although there is no complete record for the Mark II and III mountings, it is known that several spares were made, at least one replacing a damaged mounting in *Renown* in 1941 and one sent to the US in 1942, possibly for *Victorious* before her temporary transfer to the US Navy.

By the time the two bigger carriers *Implacable* and *Indefatigable* were

The fleet carriers each received eight twin 4.5in Mark II BD mountings. Here one of *Indefatigable*'s port forward mountings is about to be installed at Clydebank on 18 October 1943, now fitted with RP.10 as Mark II**. *(UCS)*

| **4.5in QF Mark III Gun** | |
|---|---|
| Bore, in | 4.45 |
| Calibres | 45 |
| Length overall, in | 212 |
| Weight incl BM, tons | 2.81 |
| Muzzle velocity ft/sec | 2350 |
| HE shell, lb | 55 |
| Charge, lb | 13.6 |
| Cartridge, lb | 33.5 |
| Life, efc | 750 |

| **4.5in Mark II Mounting** | |
|---|---|
| Rotating weight, tons | 38 |
| Max elevation, degrees | 80 |
| Max depression, degrees | -5 |
| Elevating, deg/sec | 10 |
| Training speed, deg/sec | 15 |
| Rounds per minute | 12 |
| Max range, yards | 20,000 |
| Max height, ft | 39,000 |

ready to ship their sixteen in 1943, remote power control had become available so theirs were Mark II** with RP.10. Elevating and training was thereby increased to 20° per second, with weight increasing to 43 tons. Not all were new builds, the two spares ordered with *Renown* being modified as were two removed from *Indomitable* in 1942 for repair, most of the rest coming from Scotswood. *Implacable*'s general arrangement plan showed stowage of an average of 385 HE time fuzed rounds per gun:

| *Magazine* | *HETF* | *Star* | *Practice* | *Total* |
|---|---|---|---|---|
| A forward starboard | 1654 | 50 | 145 | 1849 |
| B forward port | 1551 | 44 | 491 | 2086 |
| X aft starboard | 1480 | 100 | 331 | 1911 |
| Y aft port | 1471 | 50 | 428 | 1949 |
| Total | 6156 | 244 | 1395 | 7795 |

Although ordered in 1942, *Eagle* and *Ark Royal* were not completed until the 1950s, so got the modified Mark II*** (although now designated Mark 2) with a flat top to the gunhouse flush with the flight deck. Each received eight (weight 49 tons), while others for their planned two sisters were cancelled. To control the mountings each also got four US Mark 37 directors with British Type 275 radar, whereas *Illustrious* got the RN Mark IV HA director. In later refits, the mountings were progressively removed to reduce weight (the eight mountings alone weighing 400 tons) and increase space until by 1970 *Ark* had lost all of hers.

## THE 4.5-INCH TWIN MOUNTING MARK IV

By the end of 1941 it was clear that a destroyer with better AA armament than the Emergency destroyers was required with better fire control, and with a longer endurance now that Japan had entered the war. None of the twin 4.7in mountings was really a HA weapon, but

A 4.5in Mark II BD showing the interior, with the two guns in a single cradle elevating as one, elevating position at left, training at right. *(Author's collection)*

the twin 4.5in Mark II BD was. This could be adapted as an above-deck mounting with gunbay below, not only for the hoist mechanisms but providing space for recoil at 80° elevation and used cartridge cases. The circular gunhouse could be extended at the rear to include a gun captain lookout (useful for HA barrage fire) and full ventilation to extract fumes. There was a spare Mark II aircraft carrier mounting completing at Barrow which could be converted so it became the prototype Mark IV. This was mounted in A position in the 'S' class destroyer *Savage* whose two single mounts aft were fitted with a 4.5in gun in place of the normal 4.7in to simplify ammunition supply. She was completed in June 1943. *(See colour plates on page 175)*

**Above:** One of the first twin 4.5in Mark IV mountings nearly complete at Scotswood on 11 April 1944, probably for the first 'Battle' class destroyer *Barfleur.* The compressed air bottle is at the front; at the same level is the support ring to be bolted to the deck and behind it the training rack and rollers. *(Vickers)*

**Right:** A 4.5in Mark III gun has been lifted off *Lagos'* Mark IV mounting, being demolished by P & W MacLellan at Bo'ness in June 1967. Visible are the loading tray, balance weight, recoil and recuperator cylinders. *(Patrick Munro)*

Captain (E) G C de Jersey in the *Journal of Naval Engineering*[2] describes the rapid development from February 1942 to produce this mounting, which was the first designed from the start for RPC (RP.10). Fixed ammunition in the Marks II and III had been found difficult to handle at sea, so the Mark IV was designed for separate ammunition. The gun was the Mark III, the bodies typically costing about £1550, breech mechanism £480 and replaceable loose barrel £440.[3] Meanwhile the first eight-ship flotilla had been ordered to become the 'Battle' class with two twin 4.5in forward, a single 4in for star shell abaft the funnel and four twin 40mm Hazemeyer mountings. Sixteen 4.5in mountings were ordered from Barrow in mid-1942 with more from Scotswood. Unfortunately, no details have been found of the number or allocation of the mountings made. A pair of Denny-Brown roll stabilisers were planned to be fitted to provide a steadier gun platform. Only *Camperdown* and *Finisterre* received them, as it was reckoned that RPC could compensate for most of the ship motions.

The first delivery was made in May 1944 with floating crane *Titan II* lifting two from Scotswood into *Barfleur* at Wallsend. But progress on the new Mark VI HA director with its Type 275 radar was frustratingly slow, *Barfleur* completing in September without hers, which was not lifted aboard until 29 November, made by V-A at Crayford.[4] Although English Electric at Bradford had been brought in to make Mark VIs, they had only completed four by the end of 1945.[5] At least five 'Battle's and 'C' class destroyers were completed without a director, severely limiting their effectiveness and deployment overseas. Not until mid-1945 had enough directors been fitted (£22,000 each) to enable the first 'Battle's to deploy to the Far East, but too late to participate in the war against Japan. Even cruiser *Ontario* completed in May 1945 with only one of her three HA directors. Reportedly the Mark IV mounting was heavy on maintenance.

The Third flotilla of eight was completed with two Mark IV* mountings with RP.10, with a single 4.5in Mark V* mounting abaft the funnel. Orders for the fourth and fifth flotillas of 'Battle's had been placed in April 1943. Mounting orders will have been placed also but details are lacking, but if all had been completed nearly one hundred will have been required. Twenty-four 'Battle's were completed by 1947, plus two built in Australia but with 4.5in Mark 6 mountings (see below). Two Mark 6 mountings had also been planned for the Fifth flotilla, the same number as the cancelled *Gallant* class would have had. The others were cancelled in late 1945. Six hulls had been launched, three more were taken to launching stage and then towed away for scrap (*Trincomalee, Belle Isle* and *Poictiers*), the rest were broken up on the slip or had yet to be laid down. Of those already launched, most had their machinery already installed but no armament. Three were allocated for trials by the new Naval Construction Research Establishment, both structural and explosion tests. The opportunity was taken to use *Albuera* for structural tests at Rosyth in 1948–50. The bare hull was

This aerial view of *Saintes* at speed in 1956 shows off the gunhouse shape well, with the Mark VI director above the bridge. *(World Ship Society)*

[2] 'The Development of Destroyer Main Armament 1941–45'. *JNE* Vol 7, N^os 1 January 1954

[3] Tyne & Wear Archives. DS.VA/3/5/12-14

[4] *Titan II* Log 1945. Marine Technology Special Collection, 01250/26

[5] CAB 102/526

Three twin 4.5in Mark IV were fitted in three Venezuelan destroyers built at Barrow. Here *Nueva Esparta*'s A mounting is lifted aboard in 1953 by the 250-ton Devonshire Dock crane. *(Barrow Dock Museum 0716)*

Demolition often revealed aspects of gun mountings not normally seen. At Sunderland in December 1974, *Agincourt*'s 4.5in gunshields have been removed, while *Corunna*'s 4.5in mountings are still intact; she was later broken up at Blyth. *(Author)*

supported in drydock, known loads applied and bending moments and deflections measured. The results were re-assuring, the measurements confirming the design calculations and assumptions.

Ammunition nominal capacity was 400 rounds per gun. *St James* shell rooms and magazines had as built:[6]

| | *A* | *B* | *Total* |
|---|---|---|---|
| High explosive shell | 310 | 410 | 720 |
| Semi armour piercing | 240 | 240 | 480 |
| HA practice | 25 | 91 | 116 |
| LA practice | 80 | 80 | 160 |
| Star shell | 50 | 50 | 100 |
| Drill | 10 | 10 | 20 |
| Total shells | 715 | 881 | 1596 |
| Cartridges | 715 | 877 | 1592 |

Oil-rich Venezuela had for some years been discussing with Vickers-Armstrongs new warships for its navy, including cruisers. In 1950 they finally settled on two destroyers based on the *Daring* class and armed with three twin 4.5in mountings, but Mark IV with RP.10. The advantage to V-A was that it supplied not only hull and machinery but also armament and ammunition. The latter comprised one outfit of 60 rounds of 4.5in SAP plus 60 of HE per gun stowed on board plus two more outfits sent as reserves to Venezuela. Six loose barrels were also supplied, *ie* one spare per mounting. Each vessel, *Nueva Esparta* and *Zulia,* was worth about £2.5M. Another was ordered three years later, as *Aragua.*

**4.5in Mark IV RP.10 Mounting**

| | |
|---|---|
| Rotating weight, tons | 46 |
| Weight on fixed structure, tons | 6.2 |
| Max elevation, degrees | 80 |
| Max depression, degrees | -5 |
| Training angle, degrees | 540 |
| Elevating & training speed, deg/sec | 20 |
| Rounds per minute | 20 |
| Max range, yards | 20,000 |
| Max height, ft | 39,000 |

The four RN 'Battle's converted to radar picket destroyers in 1962 kept their two mountings, although the old gun sighting arrangements through ports in the front of the gunhouse were no longer needed.

## THE 4.5-INCH TWIN MOUNTING MARK 6

### *Development and Orders*

The Mark 6 was the first post-war 'heavy' gun mounting with its asso-

[6] *St James* General arrangement plan. Marine Technology Special Collection 04419/5

'Battle' class destroyer *Tobruk* received the first two production 4.5in Mark VI mountings made by Elswick, seen here in 1963. *(World Ship Society)*

ciated ammunition hoists which were integral below the gunhouse. The Mark IV in the 'Battle's had been a relatively successful short term solution to improved HA fire in destroyers but was essentially an interim design, whereas the Mark 6 was designed from scratch to incorporate all the wartime lessons. Arrangements were made for relatively easy replacement, so it was not strictly 'non transferable' in the old nomenclature. It remained in service for six decades, albeit with updates and new fire control systems. The first vessel to ship it was the Australian 'Battle' class destroyer *Tobruk*, completed at Cockatoo Dockyard in 1950. While she had been intended to receive two mountings made at the Government Ordnance Factory at Bendigo, Victoria, these were not ready in time, so she got from Elswick the first two ship production mountings shipped in August 1948; her sister *Anzac* received two Bendigo mountings. Although the last Mark 6 mounting was completed in 1969, there were enough available to arm two Chilean *Leander* class frigates launched in 1972 and six Indian-built *Leander*s, the last *Vindhyagiri*, not completing until 1981. *(See colour plates on page 176)*

Records of the development of a gun mounting are rare but an account of that of the dual-purpose twin 4.5in Mark 6 (called Mark VI until 1950) was given by de Jersey.[7] By September 1942, enough expe-

[7] *JNE, op cit*

The twin 4.5in Mark 6 was widely fitted in frigates. *Whitby* was the first Type 12 to get one in 1956. Fitting out alongside at Cammell Laird's yard at Birkenhead is *Tenby*. *(Marine Technology Special Collection)*

rience had been obtained to consider a totally new design of high angle/low angle gun mounting for destroyers. The twin 4.5in Mark II in carriers and the related Mark IV in 'Battle' class destroyers were essentially of pre-war design. Requirements for a new Mark 6 included:

A higher rate of fire, at least 12 rounds per minute
All-angle loading
Higher elevating and training speeds
Remote power control *eg* RP.40
Quick change of type of shell being fired
Short 'dead time' between fuze setting and firing
Continuous all-round training.

Vickers-Armstrongs at Elswick was to be the designer with DNO staff on site to advise and approve drawings and manufacture. Both guns would be in the same cradle, *ie* elevate together. Separate ammunition (shell and cartridge) would be easier to handle and there would be two shell hoists into the gunhouse, one direct to the fuze setting machine, one for other types of round. New features included a loose liner for the QF Mark 5 gun, which when worn could be easily replaced. While the jacket and breech ring remained on the cradle, the gun barrel could be withdrawn through the two vent doors in the rear of the gunhouse (see drawing on page 209). Although the whole gun weighed 3.3 tons, the barrel itself was only 0.79 tons. The large breech ring incorporated a vertically sliding breech with one recoil cylinder above the breech and two run-out cylinders below. A wooden mock-up was built to try out positioning of all the components.

The order for a pilot mounting for trials was placed in May 1943 planned to be fitted in *Jutland* in February 1945. At much the same time, preliminary design of a twin 4.5in Mark VII mounting for the *Malta* class aircraft carriers was started, sharing many of the same components, apart from a larger diameter roller path.

The faster elevating, training and ramming mechanisms were all hydraulic, powered by three separate electrically driven pumps on the mounting. The peak power was nearly 50% higher than the Mark IV, while it was found that using 220V direct current required larger motors than 440V alternating current. This was one reason why AC was adopted for four of the *Daring*s, also to gain experience of a technology already used by the US Navy.

Proof firings of the guns and cradle took place in March 1945 at Ridsdale, Elswick's proving ground north west of Newcastle. But with the end of the war in Europe followed by the surrender of Japan, there was less urgency to complete the new ordnance quickly. It was not until October1946 that the first mounting was completed at V-A's Elswick works in Newcastle. By that time, *Jutland* had been cancelled at Hawthorn Leslie, though the hull had been launched and was used for post-war tests, and the name transferred to *Malplaquet* building at Stephen on the Clyde, but having the 'Battle's Mark IV mounting. Therefore, the pilot mounting was sent to the Fraser gunnery range at Eastney, Portsmouth for trials. By now 15 mountings had been ordered from Elswick and 23 at Barrow. With 16 *Daring*s with three mountings each and eight *Gallant*s with two each and the last flotilla of 'Battle's

A twin 4.5in Mark 6 turret plus gunbay hoist mechanism were used at gunnery school HMS *Excellent* for instructional purposes. Both were moved to the Explosion Museum at Gosport, the mounting outside, the hoist inside. Shells and cartridges from the magazine below were delivered by fixed hoists (not seen here) into the gunbay, then transferred manually to the upper circular trough (shells) and lower trough (cartridges) which rotated with the mounting above. The selected shells (HE, SAP, star etc) were then transferred manually to each gun's hoists, cartridge at right, two shell per gun at centre. *(Author)*

with two each, at least 80 would be required, making with spares probably about 90 required in 1945. Further orders were planned for Harland & Wolff's Scotstoun works in Glasgow, while Marshall Brothers (Successors) Ltd at Gainsborough were also to be used to achieve the planned but optimistic output of three mountings per month. But with all but eight of the *Daring*s cancelled in December 1945, some orders including ten at Barrow were cancelled.

The first production mounting was fitted in B position in the new 'Battle' class destroyer *Saintes* at Chatham Dockyard early in 1947, although none of her close range weapons were fitted. By the end of 1947 Elswick had fifteen on order, Barrow ten plus possibly six at H&W but full details are lacking. In early 1951, with new fast frigates planned, Elswick received an order for six at an estimated cost of £350,000, although not completed until 1957.

The first British set of three 44-ton (as fitted) mountings was installed in *Daring* in November 1950 at Swan Hunter, supplied from Elswick upriver from the Wallsend yard. Elswick then supplied the mounts for *Diamond*, *Duchess* (January 1952) and *Decoy* (August 1952). Two of *Diamond*'s mountings were shipped in coaster *Seatern* from Elswick on 18 May 1951, arriving at Clydebank on 23 May. Barrow supplied three each to *Dainty* and *Delight* in 1951, then *Diana*. Scotstoun supplied three to *Defender* building nearby at Stephen about 1951, presumably their first. Four ships, *Daring*, *Dainty*, *Delight* and *Defender*, had DC electrics, with the mounting powered electro-hydraulically and designated Mark 6, while the other four had AC, so powered electrically designated Mark 6*. Both Marks had RP.41 remote power control.

*Blackpool's* magazine held over 400 cartridges per gun with the fixed hoist in the centre (left); her shell room was basically similar (right). *(Marine Technology Special Collection)*

Australian production appears to have been sixteen mountings: *Anzac* (two), *Voyager, Vendetta, Vampire* (three each, the fourth *Daring* having been cancelled). The four 'River' class frigates *Derwent, Parramatta, Stuart* and *Yarra* got one Mod 1 mounting each; these were later modified to Mod 3 by Bendigo. There was one for training at West Head gunnery range near Melbourne. *Torrens* and *Swan's* Mod 3 version were refurbished mountings from Elswick in 1968.

### *The Mounting*

The whole gun mounting with attached hoists was designed to be transported and shipped as a single unit, weighing with its transport cage 46 tons, 21ft high, 14ft wide and 27ft long from muzzle to gunhouse rear. The 10ft diameter 2.6-ton training base and rack would already have been fitted to the upper deck of the receiving ship. The mounting was also designed to be shipped in smaller units:

| | |
|---|---|
| Gunhouse, turntable, training and elevating gear | 24.5 tons |
| Two guns, breech mechanism and cradle | 12.2 |
| Revolving lower platform, pumps, shell and cartridge chutes | 2.4 |
| Shell and cartridge hoists | 3.6 |
| Central trunk (also for spent cartridges) | 1.6 |
| Total revolving weight | 44.3 |

The circular gunbay with its cantilever beams supporting the training base, rollers and gunhouse was designed to resist a firing force of 54 tons upwards or downwards. The mounting and supply cables had a training arc of 400° right or left, *ie* a complete rotation when tracking a target. The gunhouse itself was a welded structure of D.1 high tensile steel, originally planned to be ½in but reduced to ⅜in to save weight. It was somewhat akin to an upturned biscuit tin, the turntable floor forming the lid, with sides and roof a single weldment. Although small panels were fitted for access to internal fittings, replacement of the guns could be done without having to remove the gunshield itself. Maximum elevation was 80°, depression 15°.[8]

The operation of the mounting was somewhat akin to that of the short trunk 6in mountings. The magazine held both shells (in bins) and cartridges (in bottle racks) stowed horizontally which were transferred manually to fixed hoists, two for shells and two for cartridges both carried vertically to the gunbay. Here they were transferred manually to circular loading troughs (fixed) surrounding the revolving platform, the upper holding shells, the lower cartridges. These were transferred manually to a tilting tray at the bottom of the cartridge hoist on the revolving portion, which delivered to the rear of the gun, whence it was manually transferred to the loading tray. There were two shell loading hoists per gun, one delivering time fuzed shells vertically to the fuze setting machine to the side of each gun, the second adjacent for other types of shell. The gunbay could hold 80 shells but only 45 of the longer cartridges.

The gunhouse crew was nine: the officer of the quarters whose position included a lookout hood, the captain of the gunhouse who could select the type of ammunition (*eg* SAP, DA, TM, VT, star), a gunlayer, a trainer, a sightsetter, and one shell loader and one cartridge loader per gun. In the gunbay were a further eight and a few more in the magazine. These numbers were probably slightly different in later Mods. Electric power fed through flexible cables powered four VSG pumps: in the AC ships 18hp for the training motor, 7hp for the elevating gear

**4.5in Mark 6 Twin Mounting**

| | |
|---|---|
| Rotating weight, tons | 45 |
| Max elevation, degrees | 80 |
| Max depression, degrees | -15 |
| Elevating & training speed, deg/sec | 20 |
| Rounds per minute per gun | 16 |

[8] *Handbook for QF 4.5-inch Mark 5 gun on RP.41 Twin Mark 6 Mounting, 1961.* ADM 234/477

A twin 4.5in Mark 6 on display at Explosion Museum in 2016 showing at the rear the access door and the two circular holes through which gun barrels could be replaced. *(Author)*

and 18hp for general services including hoists and rammers, plus a similar one for powering the fixed hoists. These could train the turret at 20°/sec through 400° left or right. Air pressure bottles at 4000lb/sq in supplied air blast (to clear residues from the gun) and the recuperator (to run the gun out after firing).

### *The Gun*

The new Mark 5 QF gun had a performance broadly similar to the previous marks but with the breech end adapted for faster loading. The actual bore was 4.45in but always referred to as 4.5in. The jacket held a replaceable loose barrel, whose life was over 1000 rpg. A 2pdr sub-calibre gun was provided for the left gun for practice. In 1949 Elswick received an order for 14 gun bodies at £3000 each and 14 breech mechanisms at £1400 each. Campbell suggests about 300 guns were made.[9]

Theoretically 24 rounds per minute per gun was possible, but in practice a normal rate was about 16. The ammunition outfit was nominally 300 rounds per gun in the *Daring*s and 400 in the frigates. The actual magazine capacity for the three turrets in *Daring* as completed was 540 SAP, 1260 HE, 200 star and 420 practice.[10] Type 12

### 4.5in Mark 5 QF Gun

| | |
|---|---|
| Mark 5 QF gun bore | 4.45 |
| Calibres | 45 |
| Length bore, in | 200.25 |
| Length incl breech ring, in | 241.25 |
| Weight, tons | 3.38 |
| Shell, lb | 56 |
| Cartridge, lb | 33.4 |
| Charge, NFS 168-048, lb | 13.6 |
| Muzzle velocity, ft/sec | 2350 |
| Max range, yards | 18,000 |
| Time of flight, sec | 48 |
| AA ceiling, ft | 39,000 |
| Life, efc | 1200 |

9 *Warship* 34, *op cit*

10 HMS *Daring*: Gun Weights, Bolting, Ammunition and other Details. Marine Technology Special Collection 04690/21

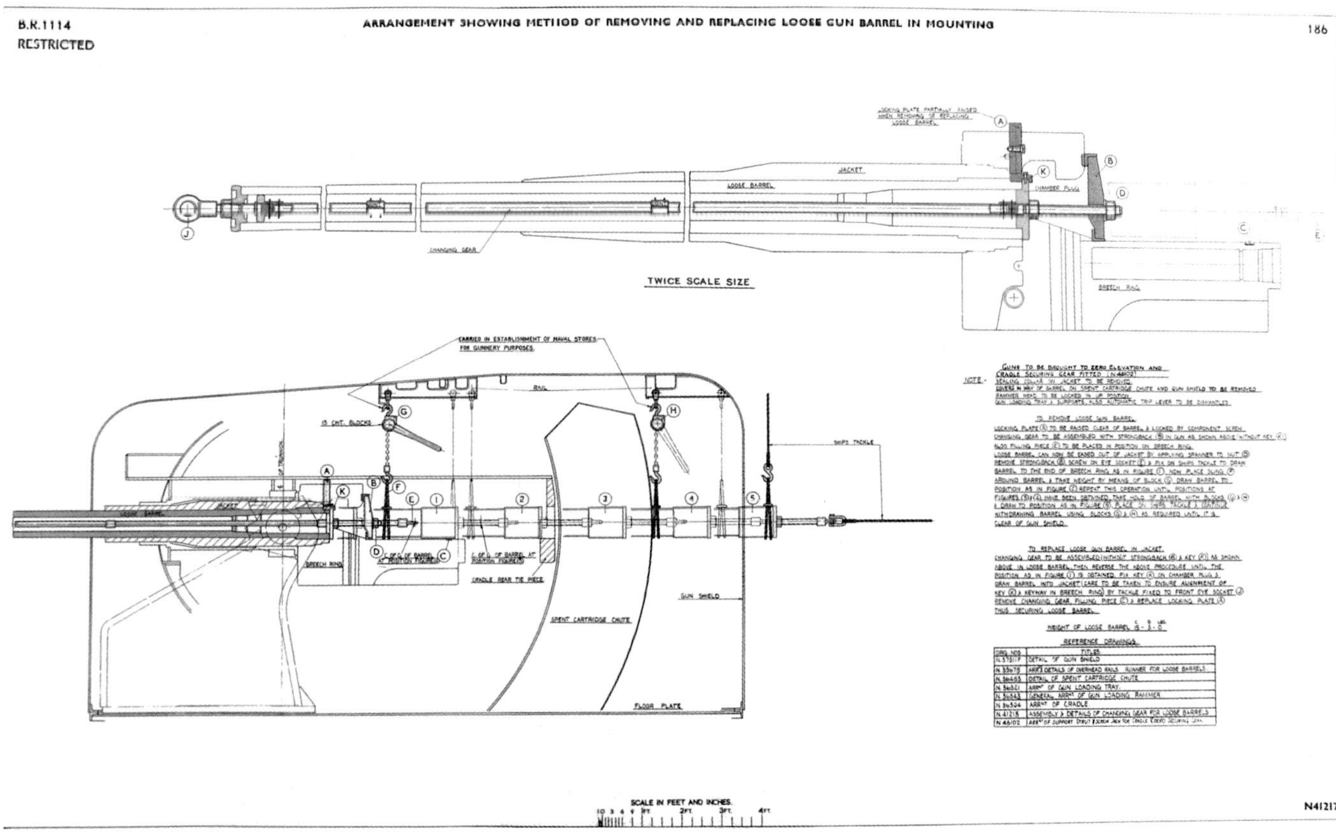


The Mark 5 gun had a replaceable barrel which could be withdrawn out of its jacket through circular doors in the rear plate. (BR.1114)

*Scarborough* had 818 and 135 practice for her single turret. Type 41 *Puma* (two turrets) had in her X magazine 292 HEVC, 215 HE AA practice, 48 HE fuzed, 61 star, 162 surface practice and 16 Res X (radar echo?) making 794 total. The practice allowance per turret in the 1970s was about 60 per month, when a shell was costed at about £80.

### *Production of the Mountings*

Unfortunately, no complete record of Mark 6 gun mountings has been found nor any gun mounting registers. Mounting production has had to be estimated from fragmentary records from Barrow and Elswick and individual ships and passing references in other sources. The only complete record is a list of the 54 mountings made by Elswick with delivery dates and Registered Numbers, about 36% of the estimated total of 150 made. The table on p212 shows the estimated production. There appear to be two batches of numbers, from 1 up to about 121, and from 200 up. No numbers have been found for the sixteen believed built at Bendigo. The last number found was 234 but that does not mean that 234 were made; apart from blanks in the order series there would have been cancellations, especially in the early years. Both Elswick (for India) and Scotstoun (for spare) completed their last in 1969. Vickers (as the company was now called) then concentrated gun mounting production at Barrow, so probably made the last five for Indian *Leander*s there, unless some were made in India under licence.

The RN had at least two mountings for training, one at HMS *Excellent* gunnery school Portsmouth, one at HMS *Cambridge* at Plymouth and possibly *Caledonia* at Rosyth, as well as the Eastney trials mounting. A 1965 document noted at least seven RN spare and training mountings. The pool of spare mountings could be drawn on when the original was lost or damaged or removed for refurbishment. At least one was stored cocooned at H&W Scotstoun works in the 1960s.

Unlike the 5.25in and heavier mountings, they were not 'made to order' for a specific ship but ordered en bloc to be allocated by the Admiralty to each ship as construction progressed. Nor was there the same tendency to use Elswick to supply east coast shipyards and Barrow west coast, as transport was not a major problem. The mountings for frigates were generally sent to the shipyard three to nine months after

The frigate *Jaguar*'s two Mark 6 mountings were delivered before the ship was ready to receive them, so they were stored cocooned under Denny's 50-ton fitting out crane, seen in September 1958. *(Author)*

launch, although in the case of *Jaguar* at Dumbarton her two arrived (probably from Scotstoun) in June 1957, one month before her launch. They were stored next to the 50-ton fitting out crane until installed in October 1958.

The Mark 6 mounting went through at least three modifications, with Mod 1 fitted new in the Type 12, 41 and 61 frigates from the mid-1950s, incorporating all the modifications issued to date. Mod 2 updated the *Darings*' Mod 0s. Mod 3 was the most extensive involving improved power drives and RPC electronics. It was introduced about 1960 and fitted in all but the first two 'County' class and all but the first four *Leanders*. It was also retrofitted to the earlier frigates as shown in the table below. *Whitby*'s was despatched (Reg Nº 83) from Elswick on 8 January 1962 for fitting during her refit at Rosyth. Her original mounting was probably sent for refurbishment to Mod 3 standard, either at a dockyard gun mounting workshop or at one of the maker's works. *Plymouth* at the Falklands had Mod 3 Nº 213, which would have been made in the mid-1960s, *ie* not her original mounting from 1959. The table on p213 shows which versions went into each ship, derived from various editions of the official 'Particulars of War Vessels'.[11]

Vickers-Armstrongs annual accounts show financial results of some Barrow and Elswick Mark 6 contracts. Early ones were cost plus, but had become fixed price by the mid-1950s, albeit with adjustments as work progressed and modifications made.[12]

| Works | Number | Contract | Delivered | Sales Value £ | Profit £ | Price each £ | Profit % on cost |
|---|---|---|---|---|---|---|---|
| Elswick | 6 | TM.1495 | 1954–57 | 593,440 | 41,834 | 98,906 | 7.6 |
| Barrow | 10 | 11/412 | 1952–57 | 970,863 | 128,702 | 97,086 | 15.3 |
| Elswick | 4 | TM.1774 | 1959–60 | 448,473 | 35,970 | 112,118 | 8.7 |
| Barrow | 6 | 11/450 | 1958–62 | 774,133 | 91,036 | 129,022 | 18.8 |
| Elswick | 4 | TM.1826 | 1960–62 | 434,593 | 51,249 | 108,648 | 13.3 |

The last two suggest that Elswick costs were now less than Barrow, albeit rather less profitable. In 1961 Elswick received an order for fourteen 4.5in Mark 5 guns and eighteen loose barrel liners worth £102,808, which equates to about £7343 for one complete gun, breech mechanism presumably included. That would make the price of a complete mounting with two guns in the early 1960s about £130,000, or only about 3% of the price of a *Leander* class frigate, although of course the ammunition hoists, close range weapons, fire control systems and

[11] For example, 1959 ADM 259/550, 1973 ADM 239/833

[12] Cambridge University Library, Manuscripts, Vickers Archive 1704–1714

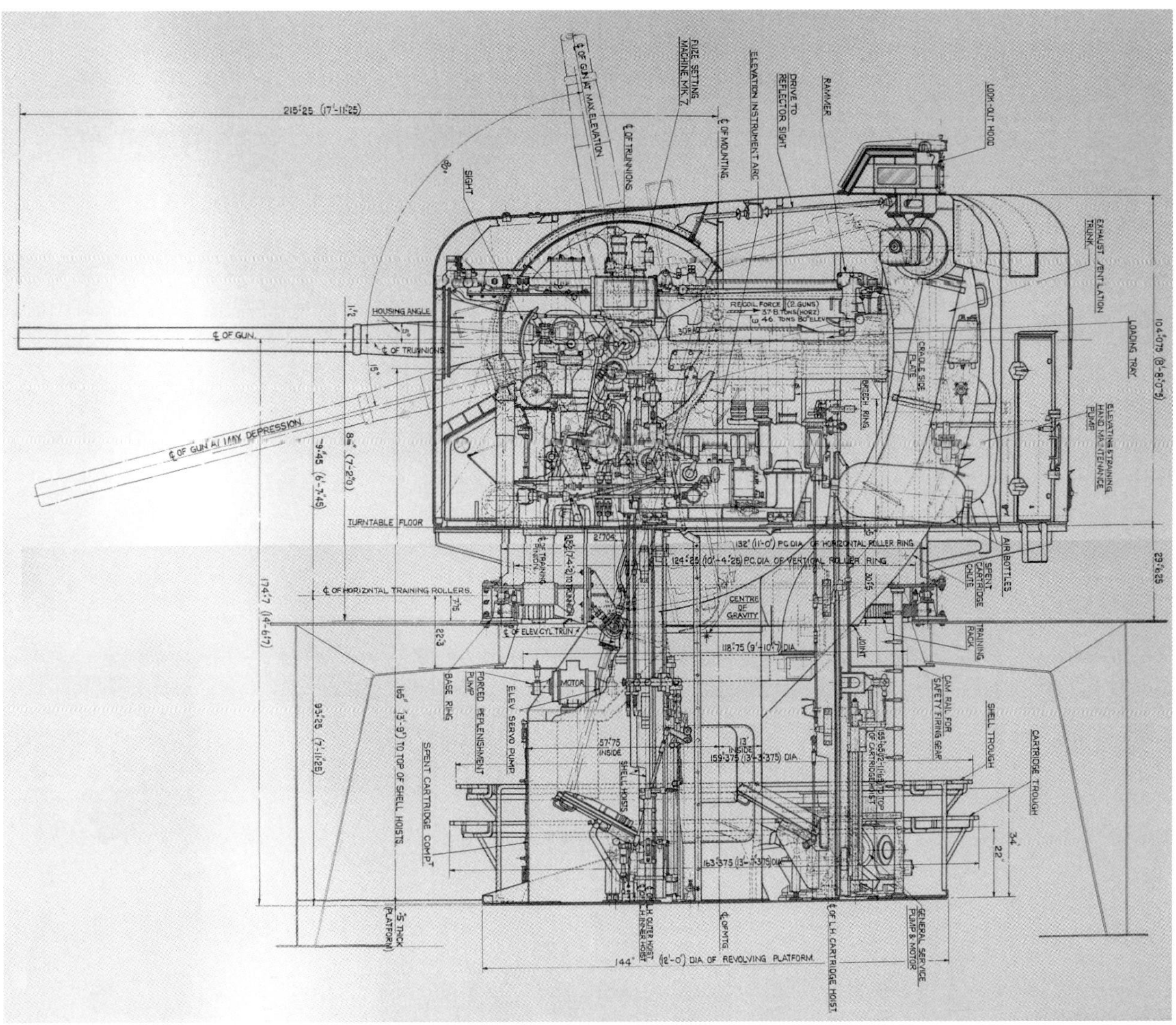


Nearly 300 large coloured dimensioned drawings of the 4.5in Mark 6 were supplied by the Naval Ordnance Department for use by the ship's officers. The detailed profile of both turret and gunbay is shown. It notes a recoil force for both guns of 46 tons at 60° elevation and roller ring diameter of 11 ft. (*4.5inch Twin RP.41 Mountings Marks 6 and 6*. Ships' Officers' Drawings, 1953.* BR.1114)

### 4.5in Mark 6 Registered Numbers

| Reg No | Maker | Contract | Supplied to | Date sent | Notes |
|---|---|---|---|---|---|
| (1) | Elswick | TM.943 | Fraser, Portsmouth | 3.10.46 | Pilot Mark 6 |
| 2 | Elswick | TM.953 | *Saintes* B | 13.1.47 | To Chatham. Then spare? |
| 3 | Elswick | TM.953 | Elswick | 3.5.48 | Research |
| 4 | Elswick | TM.953 | *Tobruk* | 26.8.48 | To Cockatoo |
| 5 | Elswick | TM.953 | *Tobruk* | 26.8.48 | To Cockatoo |
| 6 | Elswick | TM.953 | *Daring* | 28.11.50 | To Swan Hunter |
| 7 | Elswick | TM.953 | *Daring* | 29.11.50 | To Swan Hunter |
| 8 | Elswick | TM.953 | *Daring* | 18.12.50 | To Swan Hunter |
| 9 | Elswick | TM.953 | *Diamond* | 18.5.51 | First Mark 6*. To John Brown |
| 10 | Elswick | TM.953 | *Diamond* | 18.5.51 | To John Brown |
| 11 | Elswick | TM.953 | *Diamond* | 23.6.51 | To John Brown |
| 12 | Elswick | TM.953 | *Duchess* | 16.1.52 | To Thornycroft |
| 13 | Elswick | TM.953 | *Duchess* | 16.1.52 | To Thornycroft |
| 14 | Elswick | TM.953 | *Duchess* | 16.1.52 | To Thornycroft |
| 15 | Elswick | TM.953 | *Decoy* | 5.8.52 | To Yarrow |
| 16-18? | Barrow? | 11/412 | *Dainty* | 1951? | To J S White |
| 19-21? | Barrow? | 11/412 | *Delight* | 1952? | To Fairfield |
| 22-24? | Barrow? | 11/412 | *Diana* | 1953? | To Yarrow |
| 25 | Barrow? | 11/412 | *Salisbury*? | 1954? | To Devonport. Or spare? |
| 26-28? | Scotstoun? | | *Defender* | 1952? | To Stephen |
| 29? | Scotstoun? | | *Dainty* A | | Original *Dainty* A to *Excellent* |
| 30-31? | Scotstoun? | | Spares? | | |
| 32 | Elswick | TM.1359 | *Decoy* | 5.8.52 | To Yarrow |
| 33 | Elswick | TM.1359 | *Decoy* | 5.8.52 | To Yarrow |
| 34? | Barrow | 11/441 | *Whitby*? | 1955? | To Cammell Laird |
| 35? | Barrow | 11/441 | *Tenby*? | 1956? | To Cammell Laird |
| 36? | Barrow | 11/441 | *Eastbourne*? | 1957? | To Vickers-Armstrongs, Barrow |
| 37? | Barrow | 11/441 | *Blackpool*? | 1958? | To Harland & Wolff, Belfast |
| 38 | Elswick | TM.1495 | *Torquay* | 15.11.54 | To Harland & Wolff |
| 39 | Elswick | TM.1495 | *Scarborough* | 7.3.56 | To Vickers-Armstrongs, Newcastle |
| 40 | Elswick | TM.1495 | *Leopard* | 13.11.57 | To Portsmouth |
| 41 | Elswick | TM.1495 | *Leopard* | 13.11.57 | To Portsmouth |
| 42 | Elswick | TM.1495 | *Llandaff* | 13.9.56 | To Hawthorn Leslie |
| 43 | Elswick | TM.1495 | *Rothesay* | 27.11.56 | To Yarrow |
| 67 | Elswick | TM.1774 | *Betwa* | 18.2.60 | To Vickers-Armstrongs, Newcastle |
| 68 | Elswick | TM.1774 | *Londonderry* | 23.11.59 | To J S White |
| 69 | Elswick | TM.1774 | *Betwa* | 18.3.60 | To Vickers-Armstrongs, Newcastle |
| 70 | Elswick | TM.1774 | *Cambridge* | 23.11.59 | Gunnery school, Plymouth. Also No 47? |
| 81 | | | *Ajax* | 1963? | To Cammell Laird |
| 82 | | | | | |
| 83 | Elswick | TM.1826 | *Whitby* | 8.1.62 | Mod 3. Fitted at Rosyth |
| 84 | Elswick | TM.1826 | *Falmouth* | 25.9.60 | To Swan Hunter |
| 85 | Elswick | TM.1826 | *Devonshire* | 21.10.60 | To Cammell Laird |
| 86 | Elswick | TM.1826 | *Devonshire* | 21.10.60 | To Cammell Laird |
| 87 | Elswick | TM.2019 | *Hampshire* | 2.10.61 | To John Brown |
| 88 | Elswick | TM.2019 | *London* | 22.2.62 | To Swan Hunter |
| 89 | Elswick | TM.2019 | *London* | 16.3.62 | To Swan Hunter |
| 90 | Elswick | TM.2019 | *Dido* | 2.7.62 | To Yarrow |
| 91 | Elswick | TM.2019 | | 17.2.63 | To Bombay. Spare for Indian Type 12? |
| 92 | Elswick | TM.2019 | *President Pretorius* | 15.5.62 | To Yarrow |
| 93 | Elswick | TM.2119 | *Penelope* | 27.2.63 | To Vickers-Armstrongs, Newcastle |
| 94 | Elswick | TM.2119 | *Aurora* | 16.5.63 | To John Brown |
| 95 | Elswick | TM.2119 | *Euryalus* | 22.7.63 | To Scotts |
| 96 | Elswick | TM.2119 | *Galatea* | 8.7.63 | To Swan Hunter |
| 97 | Elswick | TM.2119 | Portsmouth | 7.4.64 | Spare? |
| 98 | | | | | |
| 99 | | | | | |
| 100 | Bendigo? | | | | |
| 120 | Elswick | TM.1307 | *Swan* | | Refurbished mounting. To Williamstown |
| 121 | Elswick | TM.1307 | *Torrens* | | Refurbished mounting. To Cockatoo |
| 200 | | | | | |
| 208 | Elswick | TM.1034 | *Van Speijk* | 15.2.65 | To Netherlands |
| 209 | Elswick | TM.1034 | *Van Galen* | 15.2.65 | To Netherlands |
| 210 | Elswick | TM.1034 | *Tjerk Hiddes* | 26.8.65 | To Netherlands |
| 211 | Elswick | TM.1034 | *Van Nes* | 26.8.65 | To Netherlands |
| 217 | Elswick | TM.1194 | *Isaac Sweers* | 10.2.66 | To Netherlands |
| 218 | Elswick | TM.1194 | *Evertsen* | 10.2.66 | To Netherlands |
| 224 | Scotstoun | | | 11.69 | To UCS, Glasgow. Spare? |
| 225 | Elswick | TM.1259 | *Hermione* | 7.9.67 | To Stephen |
| 226 | Elswick | TM.1259 | *Antrim* | 31.1.68 | To UCS, Govan |
| 227 | Elswick | TM.1259 | *Antrim* | 11.4.68 | To UCS, Govan |
| 231 | Scotstoun | | | 11.69 | To Vickers, Barrow. Spare? |
| 234 | Elswick | TM.1278 | | 24.9.69 | To Bombay. For *Nilgiri*? |

Elswick mountings Tyne & Wear Archives DS.VA/9/PH/15/9

magazine fittings would roughly double that percentage for 'armament'.

Mountings were removed in the modernised *Leander*s from 1970 to 1980, replacing them with an Ikara A/S torpedo launcher or four Exocet missiles or a Seawolf launcher. The six Dutch *Leander*s had each of theirs replaced by a 76mm OTO-Melara gun in the late 1970s, before going to Indonesia. The last four 'Counties' had their B mounting replaced by four Exocets in the mid-1970s.

There were four ships mounting the Mark 6 at the Falklands. The advantages of having naval gunfire support included mobility, availability and surprise. The Mark 6s fired some 3500 rounds, with *Glamorgan* firing over 1200.

The RN disposed of its last operational ship which mounted the Mark 6 in the 1990s, but others survived into the present century in overseas navies. The Bangladesh Navy ex-RN Type 41 and 61 *Abu Bakr, Ali Haider, Umar Farooq* probably kept theirs until decommissioned about 2014. Also into the 2010s, were the Indian *Leander*s *Dunagiri* and *Taragiri*, not thought to have been rearmed. But the longevity record must be the Ecuadorian ex-Chilean *Leander*s *Presidente Eloy Alfaro* and *Moran Valverde* shown on Google Earth in August 2022 still sporting their 4.5in Mark 6, 72 years after it was first at sea, albeit these particular mountings were probably built in the late 1960s.

Some ships had had their mountings removed before the hulls were

**4.5in Mark 6 gun mountings**

| Mark | | Period fitted | Ships Fitted |
|---|---|---|---|
| 6 | As built | Early 1950s | *Daring, Defender, Dainty, Delight, Anzac*, Tobruk* |
| 6* | As built | Early 1950s | *Diamond, Duchess, Decoy, Diana, Vendetta*, Vampire*, Voyager** |
| 6* Mod 1 | As built | Mid-1950s to early 1960s | *Whitby, Torquay, Scarborough, Tenby, Eastbourne, Blackpool, Leopard, Lynx, Puma, Jaguar, Salisbury, Chichester, Llandaff, Lincoln, Brahmaputra, Beas, Betwa, Talwar, Trishul, Parramatta*, Yarra*, Derwent*, Stuart*, Rothesay, Rhyl, Yarmouth, Londonderry, Otago, Taranaki, Plymouth, Lowestoft, Falmouth, Brighton, Berwick, Devonshire, Hampshire, President Kruger, President Steyn, President Pretorius, Leander, Ajax, Dido, Aurora* |
| 6 Mod 2 | Modernised | Early 1960s | *Daring, Defender, Dainty, Delight* |
| 6* Mod 2 | Modernised | Early 1960s | *Diamond, Duchess?, Decoy, Diana?* |
| 6* Mod 3 | As built | Early 1960s to early 1970s | *Kent, London, Penelope, Euryalus, Galatea, Naiad, Arethusa, Sirius, Phoebe, Minerva, Cleopatra, Glamorgan, Fife, Antrim, Norfolk, Swan, Torrens, Juno, Danae, Argonaut, Andromeda, Jupiter, Hermione, Charybdis, Bacchante, Scylla, Diomede, Achilles, Ariadne, Apollo, Waikato, Canterbury,* 6 Indian-built *Leanders*, 6 Netherlands-built *Leanders*, *Condell, Lynch* |
| 6* Mod 3 | Modernised | Mid-1960s | *Whitby, Torquay, Scarborough, Tenby, Eastbourne, Blackpool, Leopard, Lynx, Jaguar, Salisbury, Chichester, Lincoln, Llandaff, Rothesay, Rhyl, Yarmouth, Londonderry, Plymouth, Lowestoft, Falmouth, Brighton, Berwick, Devonshire, Hampshire, Duchess* A & B, *Vendetta, Vampire, Parramatta, Yarra, Derwent, Stuart, Otago, Taranaki* |
| 6* Mod 4 | Modernised | Mid-1960s | *Puma.* Reconverted to Mod 3 *c*1969. |

* Made at Bendigo

Redundant 4.5in mountings were sold off for scrap, this one yet to be broken up at Pounds' scrapyard at Portsmouth in May 2005. *(Richard Holme)*

broken up, *eg Delight* scrapped at Inverkeithing in 1970, *Puma* scrapped at Blyth in 1976, *Duchess* scrapped in the Far East in 1980 and *Vendetta*. Others had theirs scrapped with the ship, *eg Dainty* at Cairnryan in 1971, *Scarborough* at Blyth in 1977, *Hampshire* at Briton Ferry in 1979 and *Diamond* on the Medway in 1981. There were a goodly number stored at naval depots in the 1970s from which they would have been probably sold for scrap in the 1980s. There was one mounting still at Pounds scrapyard at Portsmouth in 2011 in very poor condition, probably now scrapped. At least nine have been preserved, most in Australia: one at Priddys Hard (ex-*Excellent*), three on HMAS *Vampire* preserved at Sydney, one from *Parramatta* at Spectacle island, former armament depot in Sydney harbour, one from *Stuart* at Garden Island west near Fremantle, one at West Head gunnery range and one at HMAS *Cerberus* both near Melbourne, one at Albany WA, possibly one in India.

### THE 4.5-INCH SINGLE MOUNTING MARK 8

The mid-1960s saw the need for a new medium calibre gun and mounting to replace the WW2 designs – the 4.5in single Mark 5 and twin Mark 6. Guided missiles were now the primary anti-air weapon so multiple gun barrels were less necessary, but surface fire against fast attack craft and for shore bombardment remained important. A more compact, lighter, longer range automatic 4.5in system could meet these requirements and could be fitted in smaller vessels.

RARDE (Royal Armament Research and Development Establishment) had produced the 105mm Abbot self-propelled gun for the British Army which could be redesigned into a 55-calibre 113mm (4.45in) gun designated Mark 8. Its muzzle brake would reduce recoil length allowing a smaller gunhouse, while a fume extractor halfway along the barrel would eliminate air blast. Better barrel material would avoid water cooling for modest bursts of fire. A new shell and propellant would extend range to 24,000 yards and increase barrel life.

One of the early single 4.5in Mark 8 mountings in the erection shop at Barrow in about 1970. *(Vickers)*

For the mounting, a desirable feature was the fixed round (*ie* shell plus cartridge) which could be stowed vertically and use a simplified hoist mechanism. A feed ring holding sixteen rounds in the gun bay allowed indexing, *ie* moving the required shell to the single hoist on the rotating section. This would allow a choice of High Effect direct fuzed for surface targets, HE proximity VT time fuzed for air targets, illuminating (star) shell, radar echo or practice to be fired automatically. The round would be delivered at the top of the hoist into a tray which swung up to align with the breech for ramming. Fewer ammunition transfer points would improve reliability and reduce crew number, and still achieve 25 rounds per minute.[13]

Vickers Armament Division at Barrow developed this new Mark 8 mounting, manufactured in the same workshops long used for gun mountings. With a new glass reinforced plastic curved gunshield, the rotating mass was only 15 tons; including equipment and training motor on the fixed structure 26 tons. Remote power control for training and elevating was electrical using thyristors, although the hoist was hydraulically worked using an electrically driven pump sited below the gunhouse. The mounting could train and elevate at 40° per second, requiring a peak AC power of 178 kVA or about 110 kW.

With ammunition already loaded onto the ring and an unmanned turret, fire could be opened from the Operations Room in 15 seconds. In the new Type 42 air defence destroyers being designed at the same time (also with Sea Dart missiles), the Type 996 radar would indicate a target with its range, height and speed, for the GSA 1 fire control system to predict its future position and point the gun automatically, tracked by the same Type 909 radar that controlled the Sea Dart. Local control from the mounting was also possible with a crew of four.[14]

The guns were made at the Royal Ordnance Factory at Nottingham (later taken over by BAE) and the mountings at Barrow, Elswick gun mounting shops being now closed. The existing shops at Barrow were used, although not requiring pits as the mounting and the gunbay mechanisms beneath could be erected at floor level. A prototype was completed in 1966, followed by an order for one for HMS *Bristol.* Early export orders were obtained, including four for Iranian frigates, the first of which, for *Zaal*, was shipped in 1971. Some thought was given to fitting one in place of the twin Mark 6 in the *Leander*s, but there were insufficient advantages to outweigh the cost. The first batch of six Type 42 *Sheffield* class destroyers had one each, as did the eight Vosper designed Type 21 frigates (GSA 4 fire control). The Mark 8 was also fitted in six Vosper-built Brazilian frigates from 1974, the two general purpose vessels having two mountings each, and five corvettes. Only the four Type 22 frigates in Batch 3 received one. By the mid-1980s, Vickers were reporting 54 mountings completed or on order, 22 of which were for five overseas navies, 32 for the RN, which included some spares. The original plan for the Type 23 frigates omitted the gun, but one was fitted in the final design, sixteen entering service from 1990. Barrow delivered the first to Yarrow for *Norfolk* in 1988. Following the order for four mountings for the first Type 23s, Barrow received an order early in 1988 for another four worth over £10M for delivery between end 1990 and mid-1992 including those for *Iron Duke, Monmouth* and *Montrose.*[15] There was room for only about 260 rounds, barely half that of the Type 42s.

Thirteen vessels with Mark 8s took part in the Falklands War, with *Sheffield, Coventry, Ardent* and *Antelope* being lost. The primary use was for naval gunfire support as the guns could range ten miles inland. Many of the rounds were VT fuzed to burst 80ft above the ground. Consequently when *Ardent* sailed south, her CO had stowed 650 4.5in rounds, more than her official capacity.[16] On 28 May 1982, *Arrow* fired 22 rounds of star shell and 150 of HE against Argentine positions near Goose Green.[17] *Alacrity* reported hers as worn out after 900 rounds,

This 1990 view of Type 42 destroyer *Southampton* shows off well both her 4.5in Mark 8 and her underwater form. She was being repaired in Hebburn drydock after a collision with container ship *Tor Bay* in the Gulf in 1988. *(Swan Hunter)*

A 4.5in Mark 8 at Explosion Museum in 2016, with used cartridge ejection chute at front. The muzzle brake and fume extractor are evident. *(Author)*

[13] VSEL Armaments Division brochures

[14] Marland, Peter. 'Postwar Weapons in the Royal Navy'. *Warship* 2015. Also 'Postwar Fire Control in the Royal Navy'. *Warship* 2014

[15] VSEL Link, March 1988. Marine Technology Special Collection, 01167/4

[16] Brown, Paul. *Abandon Ship*. Osprey Publishing, 2021, p85

[17] *Ibid*, p245

The Mark 8 mounting was modernised from 2001 to Mod 1 with an all-electric hoist and new gunshield. *Dauntless* was open to the public on 24 February 2013. The small hatch is for striking down ammunition to the magazine. *(Author)*

which was well under the claimed 3300 round life, and had to be ordered home. *Arrow* and *Avenger* were reportedly forced to cease fire on several occasions due to faults. Microswitches could fail during sustained firing. *Antelope*'s had a history of unreliability.[18]

The Mark 8 had been planned for a thirty-year life, so by 1998 an upgrade was needed, to be provided by BAE's Royal Ordnance Defence division. The Mod 1 design featured a new faceted gunshield to reduce radar echo, was lighter with an all-electric hoist (reducing fire risk), required less maintenance, had extended range ammunition which could reach 30,000 yards but no longer had an anti-air capability. It is not clear if any were new builds, rather than converted Mod 0s. The first ship to receive the Mod 1 was the earliest Type 23 *Norfolk* in 2001. Type 22 *Cumberland* and Type 42s *York* and *Edinburgh* were the only older ships so fitted. All of the Type 23s got one by about 2009, except *Grafton* sold to Chile in 2005. A second batch of about twelve mountings were refurbished by Babcock's Devonport Weapons Workshop, including *Sutherland* in 2008. The mountings had been removed from the Type 42s when they were scrapped from 2000, mostly in Turkey. Possibly some of these were refurbished, as well as mountings removed while ships were refitting. The six new Type 45 destroyers all received one each from 2007 (with GSA 9), as did training establishment *Collingwood* at Gosport.

This mounting has the dubious distinction of being the last British designed and built medium or large calibre naval mounting, as the Type 26 will have one US 5in 62-calibre gun on the Mark 45 Mod 4 mounting manufactured in Minneapolis, albeit by the British company BAE Systems at a price of about £40M each.

[18] *Ibid*, p256

**4.5in QF Mark 8 Gun**

| | |
|---|---|
| Bore, in | 4.45 |
| Calibres | 55 |
| Length, in | 245 |
| Weight incl BM, tons | 2.4 |
| Muzzle velocity ft/sec | 2850 |
| HE shell, lb | 46 |
| Burster, lb | 5.5 |
| Charge, lb | 16 |
| Weight of round, lb | 81 |
| Length of round, in | 48.7 |

**4.5in Mark 8 Mounting**

| | |
|---|---|
| Rotating weight, tons | 15.1 |
| Weight on fixed structure, tons | 10.9 |
| Total mounting, tons | 26 |
| Max elevation, degrees | 55 |
| Max depression, degrees | -10 |
| Training angle, degrees | 330 |
| Elevating & training speed, deg/sec | 40 |
| Rounds per minute | 25 |
| Max range, yards | 24,000 |
| Max height, ft | 29,500 |

A view inside the gunhouse of *Daring*'s 4.5in in 2009, showing the open breech. *(Author)*

APPENDIX ONE

# Barrow Gun Mounting Output

OVER A PERIOD OF A CENTURY, Vickers' Barrow works produced gun mountings mostly for the Admiralty but also for the War Office and overseas navies. Fortunately, the company gave each major contract a G series number, over 450 of them. While no complete record of these has been found, it has been possible to put together a nearly complete list from financial records. Each year the Vickers directors were presented with a full set of accounts for the Barrow works. Among the 200 or so handwritten pages were lists of Work in Progress and Completed Contracts by G number, including sales value, cost and profit. BAE Systems company Barrow shipyard archive has a run from 1897 to 1929 by which time the company had become Vickers-Armstrongs; the Marine Technology Special Collection at Newcastle

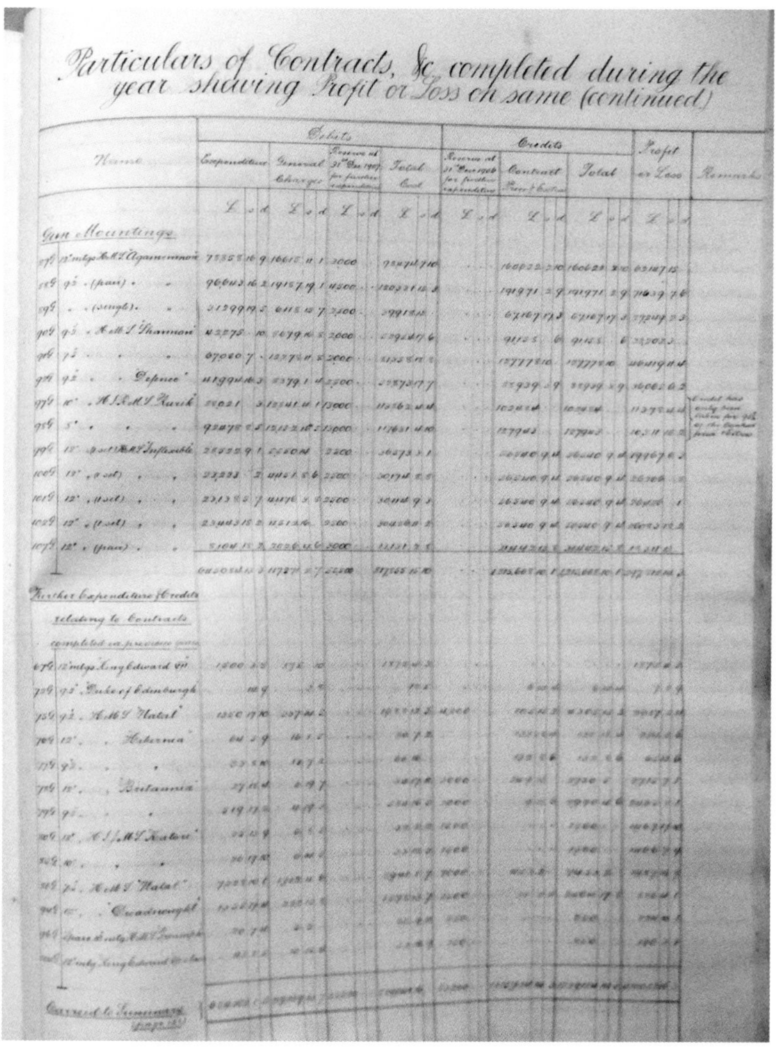

Particulars of Contracts, &c completed during the year showing Profit or Loss on same (continued)

A page from the Vickers, Sons & Maxim Barrow works annual accounts for 1907. Mountings for five ships were completed that year: *Agamemnon, Shannon, Defence,* Russian *Rurik* and *Inflexible.* The columns show: Name, Debits (*ie* costs): Expenditure, General Charges (overheads), Reserve at 31 December 1907 for further expenditure, Total Cost. Credits (*ie* income): Reserve at 31 December 1906 for further expenditure, Contract Price & Extras, Total. Profit or Loss. Remarks. Lower down the page are amounts relating to contracts completed in previous years, *eg Dreadnought.* The bottom line carried to a Summary page reads: £654,302-1s-5d, 119,239-14-7, 55,500 - -, 829,041-16 - -, 23,200 - -, 1,215,924-14-3, 1,239,124-14-3, 410,082-18-3. This latter represents the gun mounting department's profit for the year, *ie* 49.46% on costs.

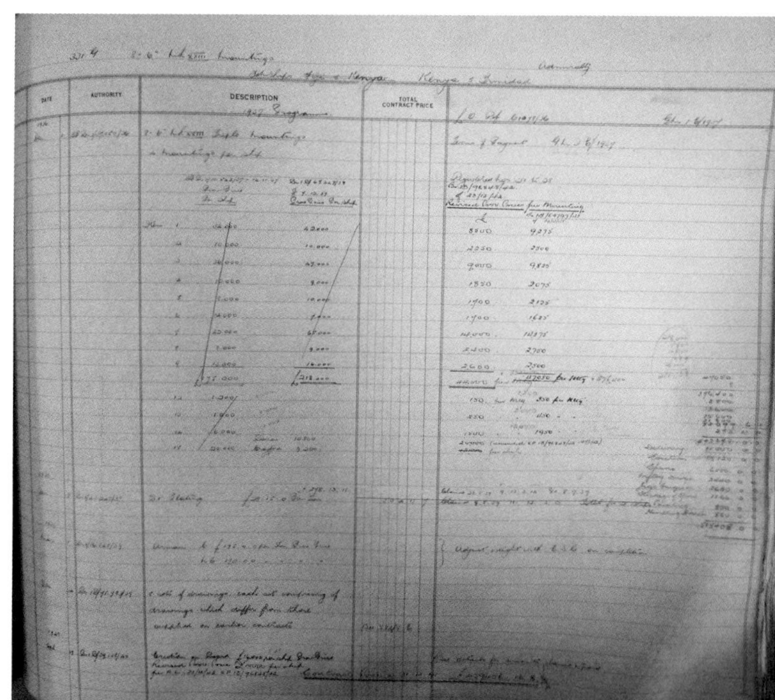

DATE | AUTHORITY | DESCRIPTION | TOTAL CONTRACT PRICE

A page from a Barrow gun mounting contracts payments ledger, showing for 331G the order of 1 December 1936 for eight 6in Mk XXIII triple mountings for *Kenya* and *Trinidad.* There were three further pages of detailed claims made and when paid plus Admiralty additions to the contract, *eg* for spare parts, drawings etc. The final price agreed in October 1944 for the two ships was £451,521-6-6. That included £52,099 for the gunhouse armour that ESC supplied.

A picture postcard in 1907 of Barrow's gun mounting Bay 12 looking north showing four of *Shannon*'s single 7.5in Mark II mountings at maximum depression (Contract 91G). On top of each is a 12pdr; in the distance twin 12in. *(Author's collection)*

University also has a selection. The BAE archive also contains looseleaf ledgers with payment details of major gun mounting contracts from the mid-1920s to mid-1940s. These give details of when the contract was placed, prices for the dozen or so component items (sometimes provisional), a record of claims made and when paid (usually by the Admiralty) and often the Registered Number and delivery date of each. For the post WW2 period, comparable detail has not been found but summary information can sometimes be extracted from the Vickers Armstrongs Barrow works accounts in Cambridge University Library.

Although the bulk of costs is recorded in the year of contract completion, further costs and income sometimes stretched over several later years, partly due to 'extras and credits', usually at the customer's request, and partly from renegotiation of prices. Most of the contracts were based on an agreed tender price, but during both world wars, 'cost plus' contracts came in where it was production at almost any price, which covered labour, materials, establishment charges (overheads) and an agreed profit rate, typically 5-10% of cost. Export orders generally comprised guns (from Sheffield or after the V-A merger Elswick), mountings, directors, ammunition as well as gunhouse armour.

Each Admiralty mounting received a unique Registered Number in a separate series for each Mark of mounting. Separate series of Registered Numbers were allocated for each Mark of gun, though Barrow did not make such. For the larger mountings, a number was given to each gun and cradle (hence two for twin 12in) but a single number for the smaller mountings, usually shipped as a complete unit. After 1950, Mark numbers were no longer quoted in Roman numerals. Up to about 1921, the Reg N^os^ were listed by ship or shore establishment in the Quarterly Appropriation Lists of Gun Mountings.

The list here includes all major contracts but excludes minor ones such as for small overseas contracts, spare parts or where no information has been found. It provides a unique and comprehensive picture of the range of output – second only to Armstrong at Elswick – and often their profitability, which made a significant contribution to the company's viability over a long period. Indeed, profit rates on gun mountings varied between 40 and 80% of cost for most years up to WW1. A record of the minor contracts omitted can be found in the World Ship Society's *Warships* Nos 192–195 (2018–19).

Four Argentine warships fitting out in Devonshire Dock at Barrow in November 1937: three destroyers *Buenos Aires, Entre Rios* and *Corrientes* (Yard Nos 719–721, 4.7in mountings 308G) with cruiser *La Argentina* (No 711, 6in mountings 293G) with triple A mounting installed. The 1908 150-ton crane is still in use but the tower for a new 250-ton crane is being erected nearby. *(Author's collection)*

### Barrow Gun Mounting Output

| Contract Number | Mountings | Ship | Customer | Ordered | Completed | Price excluding guns £ | Cost | Profit | Average price per mounting | Notes | Registered Numbers |
|---|---|---|---|---|---|---|---|---|---|---|---|
| 1G | Two twin 12in B.V | Vengeance | Admiralty | Mar 1898 | 1901 | 63455 | 99367 | -35912 | 31728 | For delivery Jun 1899. Ship delayed by dock collapse. Spare No 75 £5000 | 71-74 |
| 2G | Two twin 12in B.VII | Irresistible | Admiralty | Mar 1898 | 1901 | 64739 | 84659 | -19920 | 32370 | Spare No 90 | 86-89 |
| 3G | Two single 9.2in Mk VI | Hogue | Admiralty | Apr 1898 | 1902 | 43279 | 67048 | -23769 | 21640 | Excluding 6in gunshields. Ship delayed by dock collapse | 7, 8 |
| 4G | Two single 9.2in Mk VI | Sutlej | Admiralty | Apr 1898 | 1902 | 43067 | 47784 | -4717 | 21534 | Excluding 6in gunshields. | 9, 10 |
| 5G | Two twin 12in B.VII | Venerable | Admiralty | Aug 1898 | 1902 | 68623 | 80341 | -11718 | 34312 | For delivery Mar 1900. Spare No 105 | 101-104 |
| 6G | Two single 9.2in Mk VI | Euryalus | Admiralty | Nov 1898 | 1902 | 43740 | 47525 | -3785 | 21870 | Excluding 6in gunshield. Erection £3500. | 15, 16 |
| 7G | Two single 9.2in Mk VI | Bacchante | Admiralty | Nov 1898 | 1902 | 43463 | 44191 | -728 | 21732 | Excluding 6in gunshield. Includes spares nos 17, 18 | 13, 14 |
| 8G | Two single 9.2in Mk VI | King Alfred | Admiralty | Jun 1899 | 1902 | 47520 | 47379 | 141 | 23760 | Excluding 6in gunshield. Spare No 30 £12000 | 28, 29 |
| 9G | Two single 9.2in, twelve single 6in. Re-armed | Messudieh | Turkey | 1899 | 1902 | 26987 | 27804 | -817 | | Modernised at Ansaldo | |
| 10G | Two twin 12in B.VII | Exmouth | Admiralty | Jul 1899 | May 1902 | 86063 | 82716 | 3347 | 43032 | Erection £5470. Spare No 125 | 121-124 |
| 11G | Two twin 12in B.VII | Albemarle | Admiralty | End 1899 | Nov 1903 | 86263 | 70096 | 16167 | 43132 | Spare No 130 to drill turret at Portsmouth | 126-129 |
| 12G | Two twin 6in Mk I | Monmouth | Admiralty | Nov 1899 | 1903 | 22500 | 32900 | -10400 | 11250 | Includes gunshield armour. 1905 spare mounting No 10 £13161 | 1, 2 |
| 13G | Two twin 6in Mk I | Bedford | Admiralty | Nov 1899 | 1903 | 22500 | 18610 | 3890 | 11250 | Includes gunshield armour. 1905 spare mounting No 11 £12898 | 3, 4 |
| 14G | Two twin 6in Mk I | Kent | Admiralty | Nov 1899 | 1903 | 22500 | 18251 | 4249 | 11250 | Includes gunshield armour. 1905 spare mounting No 12 £12898 | 5, 6 |
| 15G | Two twin 6in Mk I | Essex | Admiralty | Nov 1899 | 1903 | 22500 | 17159 | 5341 | 11250 | Includes gunshield armour. 1905 spare mounting No 9 £12905 | 7, 8 |
| 16G | Boat hoisting machinery | King Alfred | Admiralty | 1900 | 1902 | 4064 | 8430 | -4366 | | | |
| 17G | Boat hoisting machinery | Exmouth | Admiralty | 1900 | 1903 | 4063 | 4649 | -586 | | | |
| 18G, 19G | Two single 9.2in coast defence mountings | | War Office | 1900 | 1902 | 14254 | 24769 | -10515 | 7127 | | |
| 20-25G | Six single 9.2in Mk V CD garrison mountings | | War Office | 1900 | 1902 | 44366 | 39242 | 5124 | 7394 | For 9.2in Mk.X gun, approx price £5500 each | |
| 26-44G | Nineteen single 9.2in Mk V CD garrison mountings | | War Office | Sep 1900 | 1903 | 144256 | 101370 | 42886 | 7592 | | |
| 48G | 9.2in pedestals and holdfasts | | War Office | 1900 | 1903 | 23365 | 16328 | 7037 | | | |
| 52G, 53G | Two single 9.2in CD mountings | | War Office | 1900 | 1903 | 14605 | 10152 | 4453 | 7303 | | |
| 56G | Two twin 6in Mk I | Cornwall | Admiralty | Dec 1900 | 1903 | 24750 | 13511 | 11239 | 12375 | Includes gunshield armour. 1905 spare mounting No 15 £13952 | 13, 14 |
| 57G | Two twin 6in Mk I | Suffolk | Admiralty | Dec 1900 | 1903 | 24750 | 12710 | 12040 | 12375 | Includes gunshield armour. 1905 spare mounting No 18 £13952 | 16, 17 |
| 58G | Two twin 12in B.VII | Prince of Wales | Admiralty | Jan 1901 | End 1903 | 97270 | 72994 | 24276 | 48635 | Delivery at Chatham in 24 months. Erection £12000 | 141-144 |
| 61-63G | Three single 9.2in CD mountings | | War Office | 1901 | 1904 | 22752 | 15494 | 7258 | 7584 | | |
| 66G | Two twin 10in B.V | Triumph | Admiralty | Mar 1902 | End 1903 | | 52458 | | | Ordered for Chile (Libertad). Armament price included with ship | |
| 67G | Two twin 12in B.VII S | King Edward VII | Admiralty | Jul 1902 | Mid-1904 | 115715 | 80418 | 35297 | 57858 | Delivery Devonport. Erection £8830 | 154, 155, 157, 160 |
| 68G | Four single 9.2in Mk V S | Dominion | Admiralty | Mid-1902 | 1904 | 89310 | 53536 | 35774 | 22328 | Strengthened mounting | 35-38 |
| 69G | Two twin 12in B.VII S | Dominion | Admiralty | Mid-1902 | 1904 | 131170 | 67455 | 63715 | 65585 | Strengthened mounting | 150-153 |
| 70G | Two twin 12in B.VII S | New Zealand | Admiralty | Mid-1902 | 1904 | 111276 | 57540 | 53879 | 55638 | Later Zealandia | 166-169 |
| 71G | Four single 9.2in Mk V S | New Zealand | Admiralty | Nov 1902 | 1904 | 80789 | 42840 | 37949 | 20197 | Includes spare mounting £13900 | 51-54 |
| 72G | Six single 9.2in Mk V S | Duke of Edinburgh | Admiralty | Jan 1903 | 1905 | 120993 | 66239 | 54754 | 20166 | Includes gunshield armour | 63-68 |
| 73G | Fourteen single 7.5in Mk IV | Triumph | Admiralty | Mar 1902 | 1904 | | | | | Ordered for Chile. Armament price included with ship | 5503-5516 |
| 75G | Six single 9.2in Mk V S | Natal | Admiralty | Nov 1903 | 1906 | 134515 | 66319 | 68196 | 22419 | Strengthened mounting. Includes gunshield armour approx £25000, erection £8000 & spare mtg £8150 | 106-111 |
| 76G | Two twin 12in B.VII S | Hibernia | Admiralty | Jan 1904 | 1905 | 109413 | 64970 | 44443 | 54707 | Includes gunshield armour | 179-182 |
| 77G | Four single 9.2in Mk V S | Hibernia | Admiralty | Jan 1904 | 1905 | 73156 | 38527 | 34629 | 18289 | Includes gunshield armour. Plus spare mtgs Nos 104-5 | 100-103 |
| 78G | Two twin 12in B.VII S | Britannia | Admiralty | Jan 1904 | 1905 | 111873 | 65922 | 45951 | 55937 | Includes gunshield armour | 175-178 |
| 79G | Four single 9.2in Mk V S | Britannia | Admiralty | Jan 1904 | 1905 | 76026 | 42085 | 33941 | 19007 | Includes gunshield armour | 96-99 |
| 80G | Two twin 12in 45cal | Katori | Japan | Early 1904 | May 1906 | 153720 | 110690 | 43030 | 76860 | Built by Barrow | |
| 81G | Four single 10in 45cal | Katori | Japan | Early 1904 | May 1906 | 118770 | 85526 | 33244 | 29693 | | |
| 85G | Four twin 7.5in | Pisa? | Orlando Co, Italy | 1904 | 1908 | 45400 | 42427 | 2973 | 11350 | | |
| 86G | Four single 7.5in Mk.II | Natal | Admiralty | 1904 | 1906 | 71418 | 52124 | 19294 | 17855 | Includes erection £4000 & spare mtg del Portsmouth £6400 | 16-19 |
| 87G | Two twin 12in B.VIII | Agamemnon | Admiralty | Dec 1904 | 1907 | 163947 | 101347 | 62600 | 81974 | Includes gunshield armour | 1-4 |
| 88G | Four twin 9.2in Mk.VII | Agamemnon | Admiralty | Dec 1904 | 1907 | 197213 | 124631 | 72582 | 49303 | Includes gunshield armour | 1-8 |
| 89G | Two single 9.2in Mk.VIII | Agamemnon | Admiralty | Dec 1904 | 1907 | 69995 | 41475 | 28490 | 34998 | Includes gunshield armour | 1-2 |
| 90G | Two twin 9.2in Mk.VII | Shannon | Admiralty | Mar 1905 | 1907 | 93490 | 53841 | 39649 | 46745 | Includes gunshield armour | 15-18 |
| 91G | Ten single 7.5in Mk.II | Shannon | Admiralty | Mar 1905 | 1907 | 130414 | 81445 | 48969 | 13041 | | 31-40 |
| 92G | Two twin 9.2in Mk.VII | Defence | Admiralty | Mar 1905 | Late 1907 | 92620 | 54404 | 38216 | 46310 | Includes gunshield armour | 10-13 |
| 93G | Ten single 7.5in Mk.II | Defence | Admiralty | Mar 1905 | Early 1908 | 127582 | 83738 | 43844 | 12758 | | 21-30 |
| 94G | Two twin 12in B.VIII (X & Y) | Dreadnought | Admiralty | Mid-1905 | Mid-1906 | 137713 | 80511 | 57202 | 68857 | Parts taken from 87G. Includes delivery & erection Portsmouth £11000. Elswick A, P & Q Nos 14-19 | 10-13 |
| 95G | Two twin 12in B.IX (A & Y) | Invincible | Admiralty | End 1905 | Mid-1908 | 148371 | 132795 | 15573 | 74186 | Electric. Elswick P & Q Nos 6-9 | 1-4 |
| 97G | Two twin 10in | Rurik | Russia | Mid-1905 | Early 1908 | 141857 | 160611 | -18754 | 70929 | Excludes guns from Sheffield at £7000 each excl BM | |
| 98G | Four twin 8in | Rurik | Russia | Mid-1905 | Early 1908 | 167677 | 168697 | -1020 | 41919 | Excludes guns from Sheffield at £3900 each excl BM | |
| 99G | One twin 12in B.VIII | Inflexible | Admiralty | Nov 1905 | Jul 1907 | 56613 | 36793 | 19820 | 56613 | | 31-32 |
| 100G | One twin 12in B.VIII | Inflexible | Admiralty | Nov 1905 | Oct 1907 | 56613 | 30191 | 26422 | 56613 | | 33-34 |
| 101G | One twin 12in B.VIII | Inflexible | Admiralty | Nov 1905 | Nov 1907 | 56613 | 30118 | 26495 | 56613 | | 35-36 |
| 102G | One twin 12in B.VIII | Inflexible | Admiralty | Nov 1905 | Dec 1907 | 56613 | 30460 | 26153 | 56613 | | 37-38 |
| 104G | Spare 12in cradles (2?) | King Edward VII class | Admiralty | 1905 | 1906 | 10680 | 4542 | 6138 | | | |
| 107G | Spare hydraulic pumping engines | Inflexible | Admiralty | Jan 1906 | 1907 | 31442 | 12043 | 19399 | | | |
| 108G | Five twin 12in B.VIII* | Temeraire | Admiralty | Mid-1906 | Mid-1908 | 277014 | 179497 | 97517 | 55403 | | 51-60 |
| 109G | Five twin 12in B.VIII* | Bellerophon | Admiralty | Mid-1906 | Mid-1908 | 268473 | 168717 | 99756 | 53695 | | 40-49 |
| 110G | Five twin 12in B.XI | Vanguard | Admiralty | Dec 1907 | End 1909 | 300449 | 205283 | 95166 | 60090 | | 95-104 |
| 111G | Five twin 12in B.XI | Neptune | Admiralty | End 1908 | End 1910 | 289909 | 191105 | 98804 | 57982 | | 63-72 |
| 112G | Five twin 12in B.XI | Colossus | Admiralty | Early 1909 | End 1910 | 320772 | 212095 | 108677 | 64154 | | 52-61 |
| 114G | Two twin 12in 50cal | España | Spain | 1909 | 1912 | 146115 | 128421 | 17694 | 73058 | Two mountings Elswick. Gun trials Eskmeals early 1913. Ship trials Jun 1914 | |
| 115G | Five twin 13.5in Mk II | Orion | Admiralty | Aug 1909 | Dec 1911 | 330306 | 223085 | 107221 | 66061 | Mk.V gun (£12850 each). Includes delivery to Portsmouth & erection £23000 | 1-10 |
| 116G | Two twin 12in 50cal | Settsu | Japan | 1909 | End 1911 | 142116 | 120512 | 21604 | 71058 | | |
| 119G | Four 13.5in Mk II | Princess Royal | Admiralty | Dec 1909 | Mid-1912 | 260607 | 170668 | 89939 | 65152 | | 32-39 |
| 120G | Three 13.5in Mk II | Thunderer | Admiralty | Dec 1909 | Early 1912 | 191471 | 125875 | 65596 | 63824 | Two mountings Elswick Nos 47-50 | 41-46 |
| 121G | Four twin 12in B.VIII* | Australia | Australia | Mar 1910 | Jan 1913 | 250061 | 162723 | 87338 | 62515 | | AN.1-8 |
| 123G | Spares for 13.5in | Orion class | Admiralty | 1910 | 1912 | 27115 | 17803 | 9312 | | | |
| 125G | Five twin 13.5in Mk II* | King George V | Admiralty | End 1910 | Nov 1912 | 323917 | 227827 | 96090 | 64783 | | 1-10 |
| 126G | Two twin 12in 50cal | Alfonso XIII | Spain | End 1910 | 1913 | 128135 | 99790 | 28345 | 64068 | Two mountings Elswick. Gun trial Eskmeals Mid-1913 | |
| 127G | Four twin 14in 45cal Type 43 | Kongo | Japan | End 1910 | Mid-1913 | 354379 | 293183 | 61196 | 88595 | Gun trials Eskmeals to early 1913 | |
| 129G | Five twin 13.5in Mk II* | Audacious | Admiralty | Early 1911 | Oct 1913 | 401906 | 303151 | 98755 | 80381 | | 22-32 |
| 130G | Three triple 12in 46cal | Leonardo da Vinci | Italy | 1911 | 1912 | 286708 | 264773 | 21935 | | Large parts of mountings. Rest from La Spezia | |
| 131G | Two twin 12in 46cal | Leonardo da Vinci | Italy | 1911 | 1912 | | | | | | |
| 133G | Two twin 13.5in Vickers | Erin (X & Y) | Admiralty | Mid-1911 | Early 1914 | 155396 | 126044 | 29352 | 77698 | Ordered for Turkish Reshadieh. Three mountings Elswick (A, B, Q) | |
| 134G | Three triple 12in 46cal | Andrea Doria | Italy | 1911 | 1914 | 199996 | 191148 | 8848 | 66665 | Large parts of mountings. Rest from La Spezia | |
| 135G | Five twin 13.5in Mk II** | Emperor of India | Admiralty | Early 1912 | Oct 1914 | 329665 | 251699 | 77966 | 65933 | | 1-10 |

| Contract Number | Mountings | Ship | Customer | Ordered | Completed | Price excluding guns £ | Cost | Profit | Average price per mounting | Notes | Registered Numbers |
|---|---|---|---|---|---|---|---|---|---|---|---|
| 136-138G | Nine hydraulic pumping engines | | Japan | 1912 | 1912 | 38209 | 22279 | 15930 | | For three battlecruisers? | |
| 139G | Four twin 13.5in Mk II** | Tiger | Admiralty | Early 1912 | Sep 1914 | 313527 | 249999 | 63528 | 78382 | | 13-20 |
| 140G | Two twin 12in 50cal | Jaime I | Spain | 1912 | 1914 | 140395 | 122443 | 17952 | 70198 | Two mountings Elswick | |
| 142G | One twin 6in Vickers | Javary | Brazil | Early 1912 | End 1913 | 13795 | 15782 | -1987 | 13795 | 1914 British Humber. Gun trials Eskmeals Aug 1913 #1420A & 1421A. Presume includes all design & tooling costs | |
| 143G | One twin 6in Vickers | Solimões | Brazil | Early 1912 | End 1913 | 14097 | 10640 | 3457 | 14097 | 1914 British Severn | |
| 144G | One twin 6in Vickers | Madeira | Brazil | Early 1912 | End 1913 | 14396 | 10711 | 3685 | 14396 | 1914 British Mersey | |
| 145G | Four twin 15in Mk I | Valiant | Admiralty | End 1912 | End 1915 | 469433 | 437126 | 32307 | 117358 | Orig contract £359638 + extras & acceleration | 28-35 |
| 146G | Four twin 15in Mk I | Barham | Admiralty | End 1912 | Mar 1915 | 506767 | 432843 | 73924 | 126692 | Includes c£25000 erection at Clydebank | 37-44 |
| 147G | Conversion of two twin 12in to hydraulic | Invincible | Admiralty | 1913 | Aug 1914 | 52336 | 38317 | 14019 | | From electric. A & Y. P & Q converted by Elswick | |
| 148G | Spares for 13.5in | Iron Duke class | Admiralty | 1913 | 1914 | 27021 | 18368 | 8653 | | | |
| 150G | Four twin 15in Mk I | Ramillies | Admiralty | Sep 1913 | Mid-1916 | 440173 | 377839 | 62334 | 110043 | 1 mounting diverted to Marshal Ney, 1 to Marshal Soult, 2 to Glorious (see also 163G & 167G) | |
| 151G | Four twin 15in Mk I | Revenge | Admiralty | Sep 1913 | Early 1916 | 431697 | 375394 | 56303 | 107924 | | 73-80 |
| 152G | Eight single 8in | | Russia | | 1915 | 30927 | 14910 | 16017 | 3866 | | |
| 154G | Four twin 15in Mk I | Ramillies | Admiralty | Mid-1914 | End 1916 | 311297 | 262214 | 49083 | 77824 | Original order for Resistance (cancelled). Replacement for diverted 150G early 1915 | 82-89 |
| 155G | Three twin 13.5in mountings | Fatih | Turkey | Apr 1914 | | | | | | Hull & machinery from Vickers. Cancelled Nov 1914. Costs incurred by 1920 £10945 (design work?). Two mountings Elswick | |
| 156G | Spares for ten 15in mountings | 5 'QE' and 5 'R' class | Admiralty | May 1914 | 1916 | 99893 | 88425 | 11468 | | | |
| 157G | Two spare 13.5in cradles and parts | Erin | Admiralty | Oct 1914 | 1916 | 25869 | 21673 | 4196 | | | |
| 158G | Conversion & erection of twin 12in B.II | Sir John Moore | Admiralty | Jan 1915 | Mid-1915 | 10716 | 10240 | 476 | | To monitor with 30° elevation ex-Mars | |
| 159G | Conversion & erection of twin 12in B.II | Earl of Peterborough | Admiralty | Jan 1915 | End 1915 | 10711 | 9944 | 767 | | To monitor with 30° elevation ex-Mars | |
| 160G | Conversion & erection of twin 12in B.II | Sir Thomas Picton | Admiralty | Jan 1915 | End 1915 | 11823 | 11562 | 261 | | To monitor with 30° elevation ex-Hannibal | |
| 161G | Naval truck mounting for 12in gun Mk IX | | Admiralty | Mar 1915 | 1915 | 17938 | 16613 | 1325 | | Railway gun for land service | |
| 162G | Spares for 15in | Renown & Repulse | Admiralty | Mar 1915 | 1916 | 23308 | 17971 | 5337 | | Three original Repulse Elswick, Renown Coventry Ordnance Works | |
| 163G | 2 trunks & 2 shields for 15in Mark I* | Glorious | Admiralty | 1915 | End 1916 | 50444 | 45000 | 5444 | | Part in 150G | 19-22 |
| 164-166G | Three twin 15in mountings | | Admiralty | May 1915 | | | | | | Probably for cancelled monitors M.34, 35, 37 | |
| 167G | Two sets of 15in gun machinery | Glorious | Admiralty | Jul 1915 | End 1916 | 35019 | 30606 | 4413 | | Part in 150G | |
| 168G | Spares for 13.5in | Erin | Admiralty | 1915 | 1916 | 49780 | 21878 | 27902 | | | |
| 169G | Single 12in open mounting for Mk IX gun | Sub M.1 | Admiralty | Mar 1916 | Mid-1918 | 66926 | 54254 | 12672 | 66926 | | |
| 170G | Four twin 15in Mk II | Hood | Admiralty | Mid-1916 | End 1919 | 816534 | 685710 | 130824 | 204134 | | |
| 171G | Single 12in open mounting for Mk IX gun | Sub M.2 | Admiralty | 1916 | 1919 | 37445 | 31674 | 5771 | 37445 | All design and tooling presumably charged to 169G | |
| 172G | Four twin 15in Mk II | Howe | Admiralty | Mid-1916 | | 181662 | 167814 | 13848 | | Cancelled 1919 after considerable work done | |
| 173-175G | Conversion of three naval 9.2in to land service mounting | | Admiralty | 1916 | 1917 | 12445 | 10527 | 1918 | | Probably using surplus Mark X guns after loss of ships converted for use in France | |
| 176-177G | Two railway bogie trolleys for 12in Mk IX gun | | Min of Munitions | Nov 1916 | 1918 | 44051 | 39990 | 4061 | | | C2826 |
| 179G | Seventy 11in naval howitzers | | Admiralty | Mar 1917 | 1919 | 128413 | 108096 | 20317 | 1834 | Anti-submarine bomb thrower. 210 total ordered. Not all | |
| 180G | Conversion of two reserve 13.5in mountings | Orion class | Admiralty | 1917 | 1918 | 9001 | 6874 | 2127 | | Possibly to 30° elevation | |
| 181G | Conversion of two 12in mountings | Commonwealth | Admiralty | Aug 1917 | 1918 | 19204 | 17457 | 1747 | | For 30° elevation. Conversion at Portsmouth | |
| 182G | Conversion of two 12in mountings | Zealandia | Admiralty | 1917 | 1918 | 17224 | 15652 | 1572 | | For 30° elevation | |
| 183G | One 12in railway mounting | | Min of Munitions | Oct 1917 | 1919 | 142951 | 114882 | 28069 | 142951 | For land service. Five others ordered. Only one completed? | 100 |
| 187G | Conversion of thirty-one 13.5in mountings to pneumatic run-out | Orion class | Admiralty | Feb 1918 | 1922 | 35083 | 27399 | 7684 | | Original price £2635 each. Not all completed. Final price suggests about thirteen. | |
| 188G | Conversion of 16in proof mounting | | Admiralty | 1918 | 1921 | 8138 | 6728 | 1410 | | | |
| 189G | Spares for six railway bogie mountings | | Min of Munitions | 1918 | 1921 | 16592 | 14182 | 2410 | | | |
| 191G | Spares for 15in Mk II | Hood | Admiralty | Nov 1918 | 1920 | 17402 | 19945 | -2543 | | | |
| 192G | Three triple 16in mountings | 'G3' class | Admiralty | End 1921 | | | | | | Cancelled early 1922. £98000 paid Oct 1921 (for design work and includes 193G?) | |
| 193G | Three triple 16in mountings | 'G3' class | Admiralty | End 1921 | | | | | | Cancelled 1922 | |
| 194G | Six twin 6in Mk XVIII | Nelson | Admiralty | Dec 1922 | Mid-1926 | 237465 | 228412 | 9053 | 39578 | Includes erection £3600 per mounting and £14758 design costs. Shipped in coaster Sagenite Barrow-Tyne | |
| 195G | Six twin 6in Mk XVIII | Rodney | Admiralty | Dec 1922 | 1926 | 222093 | 179131 | 42962 | 37016 | Includes erection at Birkenhead £3600 per mounting. | |
| 196G | Two spare twin 6in Mk XVIII | Nelson & Rodney | Admiralty | Dec 1922 | 1926 | 59428 | 51512 | 7916 | 29714 | Price at Dec 1926. Later costs £2176. Probably not the complete mounting, as lower price than 195G | |
| 197G | One twin 6in Mk XVII | Enterprise | Admiralty | Sep 1921 | 1925 | 73579 | 75451 | -1872 | 73579 | | |
| 198G | Two twin 5.2in Mk I | Sub X.1 | Admiralty | Jul 1922 | 1924 | 70315 | 64466 | 5849 | 35158 | Erection at Chatham | |
| 200G | Reconditioning of 971 4in CP.I mountings | | Admiralty | 1923 | 1924 | 51996 | 28432 | 23564 | 54 | | |
| 201G | Reconditioning of 346 4in P.IX mountings | | Admiralty | Nov 1923 | Apr 1924 | 33220 | 22785 | 10435 | 96 | | |
| 202G | Reconditioning of 157 4in P.IV, VI, VII mountings | | Admiralty | 1923 | 1924 | 10964 | 8600 | 2364 | 70 | | |
| 203G | Reconditioning of 300 4.7in P.V & P.X mountings | | Admiralty | 1923 | 1924 | 10704 | 9843 | 861 | 36 | | |
| 204G | Reconditioning of 91 12pdr P.I mountings | | Admiralty | 1923 | 1924 | 1856 | 1647 | 209 | 20 | | |
| 205G | Four twin 8in Mk I | Cumberland | Admiralty | Jun 1924 | 3 qtr 1927 | 589066 | 441910 | 147156 | 147267 | £19600 alterations to pits at Barrow, £48750 jigs and gauges, £34000 erection at Barrow. Excludes parts from Erith. | 15-18 |
| 206G | Four twin 8in Mk I | Kent | Admiralty | Jun 1924 | 4 qtr 1927 | 532505 | 338637 | 193818 | 133126 | Includes erection at Chatham £36000. Berwick 4, Suffolk 4 at Elswick. Cornwall 4 at Openshaw | |
| 207G | Reconditioning of 123 4in P.II mountings | | Admiralty | 1924 | 1925 | 9545 | 8681 | 684 | 78 | Fitted in light cruisers, sloops and DAMS in WW1 | |
| 208G | Reconditioning of 67 12pdr HA Mk VIII mountings | | Admiralty | 1924 | 1925 | 4655 | 3269 | 1386 | 69 | | |
| 209G | Four twin 8in Mk I | Canberra | Australia | May 1925 | 1 qtr 1928 | 481304 | | 30000 | 120326 | Includes director firing gear c£50k. Approx profit after costs of c£450k. Australia two at Elswick, two Openshaw | AN5-8 |
| 210G | Four single 15in CD | | Spain | Sep 1925 | Early 1929 | 292400 | | 40000 | 73100 | Delivery at English port 1st 18 months, 4th 24 months. For Ferrol Campelo (2) & Prior (2) batteries. Trials at Ferrol Sep 1933 with guns #2012A-2015A. Approx profit is after costs of c£250k. Contract c£415k including guns | |
| 211G | Four twin 8in Mk I* | Devonshire | Admiralty | Dec 1925 | 3 qtr 1928 | 472251 | | 73000 | 118063 | Approx profit after costs of c£400k. Sussex four at Elswick. | 33-36 |
| 212G | Four twin 8in Mk I* | Shropshire | Admiralty | Dec 1925 | Early 1929 | 472490 | 390000 | 82490 | 118123 | Includes erection at Dalmuir £33500. Excludes guns c£7500 each. Final handover 28.8.29 | 37-40 |
| 213G | One 6in HA CD mounting | | War Office | 1926 | 1929 | 28467 | 28907 | -440 | 28467 | Probably not complete mounting | |
| 214G | Four single 15in CD | | Spain | Jul 1926 | End 1929 | 269400 | | 115000 | 67350 | Delivery at English port 1st 26 months, 4th 32 months. 2 for Ferrol (Priorino battery), 2 for La Coruña (San Pedro now museum). Firing trials end 1933. Guns from Elswick. Approx profit is after costs of c£155k | |
| 215G | Four twin 8in Mk II | Norfolk | Admiralty | Oct 1926 | Early 1930 | 495793 | | 55000 | 123948 | Final handover 24.4.30. Dorsetshire 4 at Elswick | |
| 216G | Two spare twin 8in Mk II | | Admiralty | Oct 1926 | Mid-1930 | 174766 | | | 87383 | To Portsmouth | |
| 217G | Ten single 15in CD | | Spain | Jul 1927 | 1934 | 650000 | | | 65000 | Four completed end 1929 price £252k, profit £111k. Other six completed early 1931, not delivered then as sites not ready. 4 for Cartagena (Las Cenizas and Castillitos batteries), 6 for Mahon (La Mola, Llucalary & Favaritx batteries 1934-36). Trials on last four had not been completed by 1939 - due to Civil War? | |
| 218G | One twin 4in HA | Resolution | Admiralty | | Sep 1930 | | | | | | |
| 219G | Twelve 8 barrel 2pdr M Mark V | Battleships | Admiralty | Aug 1928 | 1932 | 103689 | | | 8641 | 2 each Hood, Royal Sovereign, Furious. 1 each Nelson, Rodney, Revenge, Renown, Valiant, Malaya. Prototype in Valiant 1930 (No 1 made at Erith?) | 2-13 |

| Contract Number | Mountings | Ship | Customer | Ordered | Completed | Price excluding guns £ | Cost | Profit | Average price per mounting | Notes | Registered Numbers |
|---|---|---|---|---|---|---|---|---|---|---|---|
| 221G | Twenty-four single 4.7in CP XIV | 'B' class destroyers | Admiralty | Dec 1928 | End 1930 | 73629 | | | 3068 | Includes four for Canada ordered Mar 1929 (Saguenay?) | 45-64 |
| 222G | One twin 8in 50cal mounting with guns | Canarias | Spain | Feb 1929 | 1931 | 180786 | | | 180786 | Delivery Ferrol in 22 months. Price includes drawings £25000 and guns from Elswick c£7000 each. Three mtgs built Spain. Installed as A mounting at Ferrol 1933 | |
| 223G | Four twin 8in Mk II | Northumberland | Admiralty | Feb 1929 | | | | | | Cancelled Feb 1930. £120k claimed from Admiralty (Contract £430k). Four at Elswick for Surrey cancelled. | |
| 225G | Two single 4.7in Upper Deck mountings and guns | Araucano | Chile | Jul 1929 | 1930 | 13875 | | | 6938 | Submarine depot ship. Includes guns from Elswick £2860 | 31763-4 |
| 226G | Four single 4.7in UD mountings | Maipo & Rancagua | Chile | Jul 1929 | 1930 | 15853 | | | 3963 | Navy oilers. Excludes guns | 31765-8 |
| 227G | One single 4in HA Mk XVI | | Admiralty | Sep 1929 | 1932 | 11648 | | | 11648 | | |
| 228G | Repair of twin 8in mounting | Devonshire | Admiralty | Dec 1929 | Early 1931 | 30432 | | | | Explosion in X mounting 26.7.29 | |
| 229G | Ten single 4.7in CP XIV | Cygnet & Crescent | Admiralty | Mar 1930 | Dec 1931 | 30404 | | | 3040 | Includes 2 spares? | |
| 230G | Four twin 6in Mk XXI | Achilles | Admiralty | Dec 1930 | Early 1933 | 255109 | | | 63777 | Includes erection at Cammell Laird £17800. Four each at Elswick: Leander, Neptune, Orion | 14-17? |
| 231G | Two spare 6in Mk XXI mountings | | Admiralty | Dec 1930 | Mid-1933 | 88880 | | | 44440 | Delivered to Devonport. Probably not the complete mounting | |
| 232G | Twenty single 4.7in CP XIV | 'D' class destroyers | Admiralty | Dec 1930 | Dec 1932 | 46937 | | | 2347 | Dainty 104-7, Duncan 108-11, Defender 112-5, Diamond 116 9, Delight 120-3 | 104-123 |
| 235G | Thirty-six single 4.7in mountings + spares | Destroyers & sloops | Portugal | Jul 1931 | Mid-1934 | 88233 | | | 2451 | 4 each in Tejo, Douro, Vouga, Lima, Dao, Alf de Albuquerque, Bart Dias, 3 each in Gonc Velho, Gonc Zarco, 2 in Ped Nunes. Guns from Elswick. Fire control & ammunition from Crayford and Dartford | 32554-89 |
| 236G | Two single 4in + spares | Delfim, Espadarte | Portugal | Jul 1931 | 1934 | 5974 | | | 2987 | Submarines | 32590-1 |
| 239G | Twenty 4 barrel 0.5in M Mk I HA | | Admiralty | Sep 1930 | 1932-33 | 32253 | | | 1613 | 2 Barham, 4 Leander, 2 Berwick, 2 Dainty, 2 Decoy, 2 Delight, 2 Duchess, 2 Duncan. Price excludes guns from Crayford £200 each | 2-21 |
| 240G | Six 8 barrel 2pdr Mk VI | | Admiralty | Nov 1931 | End 1933 | 61126 | | | 10188 | Barham 1-2, 2 in Ramillies, 1 Nelson, 1 Rodney | 1-6 |
| 242G | Two twin 4in BD HA Mk XVIII | Repulse | Admiralty | Feb 1932 | Mid-1934 | 54112 | | | 27056 | Trials 1934 then delivered to ship Dec 1936 | |
| 243G | Ten 40mm | | Japan | Early 1932 | | 45000 | | | 4500 | Includes ammunition | |
| 245G | Four twin 6in Mk XXI | Ajax | Admiralty | May 1932 | End 1934 | 234374 | | | 58594 | Plus four for Amphion/Perth at Elswick (20-23) | 24-27? |
| 246G | Three twin 6in Mk XXI | Arethusa | Admiralty | May 1932 | End 1934 | 178417 | | | 59472 | Includes erection by Elswick at Chatham £15000, armour £12000, director £12000. Excludes guns £2900 each | 28-30? |
| 247G | Fifteen 4 barrel 0.5in Mk I HA | | Admiralty | Sep 1932 | Oct 1933 | 16902 | | | 1127 | 3 each in Achilles, Orion, Neptune. 6 Portsmouth | 23, 31-2, 35-46 |
| 249G | Twenty-seven single 4.7in CP XVII | 'E' class destroyers | Admiralty | Oct 1932 | End 1934 | 66769 | | | 2473 | Esk 1-4, Electra 5-8, Express 9-12, Encounter 13-16, Escapade 17-20, Exmouth 21, Escort 22-25, Eclipse 26-7. Plus RGF Woolwich: 3 Exmouth, 4 Echo (28-34?) | 1-27 |
| 250G | Six 8 barrel 2pdr Mk VI | | Admiralty | Nov 1932 | End 1934 | 59326 | | | 9888 | Courageous 7-8, Glorious 9-10, Warspite 11-12. Price excludes guns from Crayford, £1090 each | 7-12 |
| 251G | Three twin 6in Mk XXI | Galatea | Admiralty | Dec 1932 | Early 1935 | 177607 | | | 59202 | Includes erection at Scotts £14500 | 37-39 |
| 252G | Two twin 6in Mk XXI | Apollo/Hobart | Admiralty | Dec 1932 | Mid-1935 | 119456 | | | 59728 | Elswick other two (42-43) plus four Phaeton/Sydney (33-36). Directors from Barrow & Crayford £18500. | 40-41 |
| 255G | Forty single 4.7in CP XVII | 'F' class destroyers | Admiralty | Mar 1933 | Early 1935 | 92717 | | | 2318 | Eclipse 36-7, Fearless 44-7, Foresight 48-51, Faulknor 52-56, Forester 57-60, Fame 61-64, Fortune 65-8, Firedrake 69-72, Foxhound 73-6, Fury 77-80. Spare 83 | 36-7, 44-80, 83 |
| 256G | One single 4in | Golfinho | Portugal | Mar 1933 | End 1934 | 2535 | | | 2535 | Submarine | 33154 |
| 257G | Two 4in UD | Almirante Saldanha | Brazil | May 1933 | Mar 1934 | 11439 | | | 5720 | Training ship built Barrow | 33151-2 |
| 258G | One 3in | Almirante Saldanha | Brazil | May 1933 | Mar 1934 | | | | | | |
| 259G | Sixteen 4 barrel 0.5in Mk I | 'E' class destroyers | Admiralty | May 1933 | Jun 1934 | 17600 | | | 1100 | 4 each Esk, Electra and two 'E' class | 63-78 |
| 260G | Two single 15in CD Mk II | Singapore | War Office | Jul 1933 | End 1936 | 274846 | | | 137423 | For Johore battery. V-A Barrow 60%, Elswick 40%. (Elswick had received an order for one mounting in Oct 1928) | |
| 261G | One 4 barrel 2pdr Mk VII | | Admiralty | Aug 1933 | 1935 | 10164 | | | 10164 | Prototype | 1 |
| 262G | Fifty-three 4 barrel 0.5in Mk II | | Admiralty | Oct 1933 | Mid-1935 | 62658 | | | 1182 | 5 in Ajax, 3 in Amphion, Apollo, Sydney, 2 in Valiant, Courageous, Glorious, Furious, Norfolk, Australia, Galatea, Arethusa, Woolwich, eight F destroyers, 1 in Indus, spares | 1-45, AN1-8 |
| 263G | Six 8 barrel 2pdr Mk VI | Cruisers | Admiralty | Nov 1933 | End 1935 | 60076 | | | 10013 | Curlew 14-15, Coventry 16-17, Norfolk 18, 1 experimental 13 | 13-18 |
| 264G | Three twin 6in Mk XXI | Penelope | Admiralty | Dec 1933 | Mid-1936 | 174278 | | | 58093 | | 44-46 |
| 265G | Four triple 6in Mk XXII | Southampton | Admiralty | Dec 1933 | Sep 1936 | 320287 | | | 80072 | Four for Newcastle at Elswick plus one prototype/spare | 6-9? |
| 266G | Eight single 4.7in | Douro, Tejo | Portugal | Mar 1934 | End 1935 | 24505 | | | 3063 | Replacement mountings for destroyers sold to Colombia | 33499-33506 |
| 267G | Six 40mm HA | Douro, Tejo | Portugal | Mar 1934 | Mid-1936 | 6732 | | | 1122 | 266 & 267G part of armament contract for guns, ammunition & fire control £192260 | |
| 268G | Thirty-nine single 4.7in CP XVII | 'G' class destroyers | Admiralty | Feb 1934 | End 1935 | 90090 | | | 2310 | Greyhound 84-7, Gallant 88-91, Glowworm 92-5, Garland 96-9, Griffin 100-3, Gipsy 104-7, Grenade 108-11, Grafton 112-5, Grenville 116-20, two spare | 84-122 |
| 269G | Ten single 3in CP V | 'S' class submarines | Admiralty | May 1934 | End 1935 | 23689 | | | 2369 | Snapper, Starfish, Seahorse, Seawolf, Swordfish, Sturgeon, Shark, Sealion, Salmon, spare | 1-10 |
| 270G | Ten single 3in 40 cal AA | Fuket class torpedo boats | Siam | Aug 1934 | End 1935 | 17244 | | | 1724 | Delivered to Bangkok. Ships built at Trieste. Mounting price excludes guns and ammunition. Total £31106 | 33602-11 |
| 271G | Six 8 barrel 2pdr Mk VI | Battleships | Admiralty | Sep 1934 | Oct 1936 | 58857 | | | 9810 | Repulse 19, Queen Elizabeth 20-1, Royal Oak 22-3, Malaya 24 | 19-24 |
| 272G | Fifteen 4 barrel 0.5in M Mk II* | | Admiralty | Oct 1934 | 1936 | 17550 | | | 1170 | 2 each in Rodney, Repulse, Courageous, Gallant, Greyhound, Glowworm, Griffin, 1 in Furious | 46-60 |
| 273G | Twelve twin 4in HA Mk XIX | | Admiralty | Oct 1934 | End 1936 | 81878 | | | 6823 | Royal Oak 6-7, Penelope 8-9, Suffolk 10-11, Malaya 12-15, Fleetwood 16-17. Plus 4 at Elswick (Cumberland 2-3, Royal Oak 4-5) | 6-17 |
| 274G | One 4.5in HA/LA | | Admiralty | Nov 1934 | Mid-1936 | 13647 | | | | Prototype? | |
| 275G | Three twin 6in MK XXI | Aurora | Admiralty | Dec 1934 | Mid-1937 | 164770 | | | 54923 | | 47-49 |
| 276G | Four triple 6in Mk XXII | Glasgow | Admiralty | Dec 1934 | Mid-1937 | 278521 | | | 69630 | Includes 110 tons of gunshield armour £21300. Excludes guns c£2750 each, directors £25000. Four each Birmingham & Sheffield at Elswick | 14-17? |
| 277G | One spare triple 6in Mk XXII | | Admiralty | Dec 1934 | Oct 1938 | 51679 | | | | To Portsmouth. Probably not complete mounting | 18? |
| 278G | Forty-three single 4.7in CP XVIII | 'H' and 'I' class destroyers | Admiralty | Dec 1934 | Mar 1937 | 126492 | | | 2942 | Hostile 1-4, Hunter 5-8, Hero 9-12, Hotspur 13-16, Hyperion 17-20, Hardy 21-25, Hereward 26-7, 36-7, Hasty 28-31, Havoc 32-35, Icarus 38-41, Imogen 42-3 | 1-43 |
| 279G | One twin 4.5in HA BD Mk II | Iron Duke | Admiralty | Jan 1935 | Sep 1936 | 12000 | | | | Prototype | 1 |
| 280G | Eight 4 barrel 2pdr Mk VII | | Admiralty | Jan 1935 | Oct 1936 | 41917 | | | 5240 | Cumberland 2-3, Suffolk 4-5, Southampton 6-7, Newcastle 8-9 | 2-9 |
| 281G | Four single 4in HA Mk XX | Cruisers | Admiralty | Jan 1935 | Dec 1936 | 24671 | | | 6168 | Suffolk 1-2, Adelaide 3, Shoeburyness 4 | 1-4 |
| 282G | Twenty-six twin 4in HA Mk XIX | | Admiralty | May 1935 | Dec 1937 | 153751 | | | 5914 | Plus eight at Elswick. Penelope 18-19, Newcastle 20-23 (Elswick), Southampton 24-27, Warspite 28-31, Sheffield 47, 32-34 (Elswick), Malaya 35 (Elswick), Glasgow 36-39, Aurora 40-42, Birmingham 43-46, Neptune 48-49, Orion 50-51 | 18-51 |
| 284G | Ten twin 4.5in HA/LA UD Mk III | Ark Royal | Admiralty | Jun 1935 | Sep 1938 | 182338 | | | 18234 | Ark Royal 1-8, Tyne 9-10 | 1-10 |
| 285G | One 5.1in HA | | Admiralty | 1935 | Mar 1936 | | | | | Experimental. Planned as secondary armament for battleships. Trial in Kempenfelt not successful. Twin 5.25in developed | |

| Contract Number | Mountings | Ship | Customer | Ordered | Completed | Price excluding guns £ | Cost | Profit | Average price per mounting | Notes | Registered Numbers |
|---|---|---|---|---|---|---|---|---|---|---|---|
| 286G | One twin 4.7in CP.XIX | Hereward | Admiralty | Jul 1935 | End 1936 | 25908 | | | 25908 | Prototype. Trials at Woolwich | |
| 287G | Three single 4.7in UD | João de Lisboa | Portugal | Aug 1935 | Mid-1936 | 7964 | | | 2655 | Two for sloop. One spare? | |
| 289G | Fifteen 4 barrel 0.5in Mk III | 'H' class destroyers | Admiralty | Aug 1935 | Mid-1936 | 17552 | | | 1170 | Hero 1-2, Hostile 3-4, Hunter 5-6, Hyperion 7-8, Hardy 9-10, Hotspur 11-12, Hasty 13-14, Fleetwood 15 | 1-15 |
| 290G | Two single 15in CD Mk II | Singapore | War Office | Oct 1935 | End 1937? | 141825 | | 10% on cost | 70913 | For Buona Vista battery. Shields 114 tons @ £212/T. V-A Barrow 65%, Elswick 35% | A7630-31 |
| 291G | Two single 3in CP V | 'S' class Submarines | Admiralty | Aug 1935 | Sep 1936 | 3568 | | | 1784 | Spearfish, Sunfish | 11-12 |
| 292G | Eight 4 barrel 2pdr Mk VII | | Admiralty | Sep 1935 | Oct 1937 | 36656 | | | 4582 | Sheffield 10-11, Glasgow 12-13, Birmingham 14-15, Zulu 16, Afridi 17 | 10-17 |
| 293G | Three triple 6in | La Argentina | Argentina | Nov 1935 | 1938 | 184322 | | | 61441 | Cruiser | |
| 294G | Six 8 barrel 2pdr Mk VI | | Admiralty | Oct 1935 | Aug 1938 | 54018 | | | 9003 | Ark Royal 31-34, Kent 35-36 | 31-36 |
| 296G | Six 8 barrel 2pdr Mk VI | | Admiralty | Oct 1935 | Jun 1937 | 57473 | | | 9579 | Warspite 25-27, trials 28, Dorsetshire 29-30 | 25-30 |
| 297G | Four 4in AA | La Argentina | Argentina | Nov 1935 | 1938 | 25905 | | | 6476 | | |
| 298G | Six twin 25mm AA | La Argentina | Argentina | Nov 1935 | 1938 | 6449 | | | 1075 | | |
| 299G | Four triple 6in Mk XXII | Liverpool | Admiralty | Nov 1935 | Mid-1938 | 316924 | | | 79231 | Includes work at Elswick and Crayford. Armour weight per mounting 46 tons. Four each Manchester & Gloucester at Elswick (27-34) | 23-26 |
| 300G | Forty single 4.7in CP XVIII | 'I' class destroyers | Admiralty | Nov 1935 | Dec 1937 | 111640 | | | 2791 | Imogen 44-5, Inglefield 46-50, Ilex 51-4, Intrepid 55-8, Ivanhoe 59-62, Imperial 63-6, Impulsive 67-70, Hunter 71-74 (Isis?), spare 76. Other 8 to Brazilian destroyers? Harland & Wolff also built 143 at Scotstoun | 44-83 |
| 301G | Eight twin 4in Mk XIX | | Admiralty | Nov 1935 | Feb 1938 | 48534 | | | 6067 | Dorsetshire 52-3, Leander 54-5, Protector 56, Bittern 57-9 | 52-59 |
| 302G | Six 8 barrel 2pdr Mk VI | | Admiralty | Nov 1935 | Sep 1938 | 56580 | | | 9430 | Norfolk 37, Cornwall 38-9, Resolution 40-1, Berwick 42 | 37-42 |
| 303G | Twenty-one 4 barrel 2pdr Mk VII | | Admiralty | Nov 1935 | Early 1939 | 113138 | | | 5388 | Maidstone 23-4, Gloucester 32-3, Manchester 19-20, Liverpool 21, 26, Cossack 18, Mohawk 22, Sikh 25, Gurkha 27, Maori 28, Nubian 29, Ashanti 30, Somali 31, Eskimo 34, Matabele 35, Tartar 36, Bedouin 37, Mashona 38 | 18-38 |
| 304G | Twelve 4 barrel 0.5in Mk III | 'I' class destroyers | Admiralty | Nov 1935 | Early 1937 | 13045 | | | 1087 | Imogen 48-9, Inglefield 50-1, Icarus 52-3, Isis 54-5, Intrepid 56-7, Imperial 58-9 | 48-59 |
| 305G | Thirty-two twin 4.7in CP XIX | 'Tribal' destroyers | Admiralty | Jan 1936 | Mid-1939 | 360408 | | | 11263 | Zulu 2-5, Gurkha 6-9, Nubian 10-13, Ashanti 14-16, Matabele 17-20, Bedouin 21-24, Punjabi 25-28, Janus 29, Jupiter 30-32. 33 spare to Eskimo 8/40. Elswick: Janus 2, Jervis 3. Price excludes guns at c£2200 each | 2-33 |
| 306G | Four twin 4in HA Mk XIX | Cruisers | Admiralty | Feb 1936 | Mid-1937 | 24677 | | | 6169 | Dorsetshire 64-5, Leander 66-7. Price excludes guns £1700 each | 64-67 |
| 307G | Eight 4 barrel 2pdr Mk VII | | Admiralty | Mar 1936 | Early 1939 | 40751 | | | 5094 | Punjabi 39, Jackal 40, Jervis 41, Cairo 42, Calcutta 43, Jersey 44, Jaguar 45, Javelin 46 | 39-46 |
| 308G | Twenty-eight single 4.7in CP XVII | 7 destroyers | Argentina | Jun 1936 | Mid-1938 | 89037 | | | 3180 | 4 each in Buenos Aires, Entre Rios, Corrientes, Misiones, Santa Cruz, San Juan, San Luis. Also supplied fire control gear. Excludes guns | 34539-34556 |
| 309G | Fourteen 0.5in M mountings | 7 destroyers | Argentina | Jun 1936 | 1938 | | | | | 2 each as 308G | |
| 310G | Twelve twin 4.5in HA BD Mk II | Renown | Admiralty | Apr 1936 | End 1939 | 230362 | | | 19197 | 2-7 transferred to Scotswood, 8-11 (Barrow), 2 spare (12 to Implacable, 13 to Indefatigable in 1943) | 2-13 |
| 311G | Sixty-one 4 barrel 0.5in Mk III | | Admiralty | Apr 1936 | End 1938 | 75367 | | | 1236 | 4 Ark Royal, 3 Effingham, 2 in cruisers, 2 in destroyers, 1 in sloops | 90-150 |
| 312G | Twenty-two 8 barrel 2pdr Mk VI | | Admiralty | May 1936 | End 1939 | 201870 | | | 9176 | Valiant 59-62, Renown 54-5, 57, Rodney 43, Revenge 50-1, Belfast 48-9, Edinburgh 45-6, Australia 52-3, Nelson 58, Furious 47, Coventry 56, AA School Devonport 44, Portsmouth 63-4 | 43-64 |
| 313G | Thirty twin 4in HA Mk XIX | | Admiralty | May 1936 | End 1938 | 181792 | | | 6060 | Orion 68, Norfolk 69-70, Royal Sovereign 71-2, Ajax 73-6, Devonport 77, Manchester 78-9, Barham 80-3, Resolution 84-7, Egret 88-9, Berwick 90-1, Auckland 92-5, Pelican 96-7. | 68-97 |
| 316G | Two quadruple 14in Mk III | Prince of Wales | Admiralty | May 1936 | Aug 1940 | 1334201 | | | 667101 | Includes £97,000 drawings and tools, and gunshield armour for 3 mountings. | |
| 317G | One twin 14in Mk II | Prince of Wales | Admiralty | May 1936 | Sep 1940 | 279817 | | | 279817 | All shipped in coaster Sea Fisher Barrow-Birkenhead | |
| 318G | Thirteen twin 4in HA Mk XIX | | Admiralty | May 1936 | Early 1939 | 85786 | | | 6599 | Guardian 108, Furious 109-10, Amphion 111-4, Revenge 115-8, Wallace 119-20 | 108-120 |
| 319G | One twin 5.25in HA/LA Mk I | Iron Duke | Admiralty | May 1936 | Early 1939 | 84215 | | | 84215 | Prototype. Includes design £35000 | 1? |
| 320G | Forty one twin 4.7in CP XIX | Destroyers | Admiralty | Jun 1936 | | | | | | Transferred to Elswick Feb 1937. Harland & Wolff also built 91 mountings at Scotstoun | |
| 321G | Four triple 6in Mk XXIII | Belfast | Admiralty | Jun 1936 | Early 1939 | 271216 | | | 67804 | Plus four at Elswick for Edinburgh (5-8) | 1-4 |
| 322G | Eighteen 4 barrel 2pdr Mk VII | | Admiralty | Jul 1936 | End 1939 | 91232 | | | 5068 | Jupiter 47, Janus 48, Juno 49, Kelly 50, Kingston 51, Kashmir 52, Kandahar 53, Kelvin 54, Carlisle 55, Flamingo 56, Khartoum 57, Curacao 58, Naiad 61-2, Palmers 63, Smiths Dock 64 for conversions | 47-64 |
| 323G | Nine 4 barrel 2pdr Mk VII | | Admiralty | Aug 1936 | 1940 | | | | | Transferred to Scotswood | 65-73 |
| 324G | Two single 3in CP.V | 'S' class submarines | Admiralty | Aug 1936 | Mid-1938 | 3708 | | | 1854 | Sterlet 13, spare 14 | 13-14 |
| 325G | Seven 8 barrel 2pdr Mk VI | | Admiralty | Aug 1936 | Early 1940 | 65187 | | | 9312 | 1 each Glengyle, Glenroy, plus ships under conversion | 65-71 |
| 326G | Four twin 4.5in HA UD Mk III | Maidstone | Admiralty | Aug 1936 | Early 1939 | 64179 | | | 16045 | Fitted at Malta | 11-14 |
| 327G | Twenty-two 4 barrel 0.5in Mk III | | Admiralty | Sep 1936 | Early 1939 | 27690 | | | 1259 | 2 in Edinburgh, Calcutta, 1 or 2 in Tribals, sloops and river gunboats | 151-172 |
| 328G | Ten twin 5.25in HA/LA Mk II | Dido, Bonaventure | Admiralty | Nov 1936 | Mid-1940 | 509508 | | | 50951 | Originally 51 ordered (2-52). 12 for re-arming Hawkins and Frobisher cancelled. Barrow: only four fitted in Bonaventure (1-4) & Dido (6, 7, 9, 10), Phoebe X 24, Hermione Q 28. Others to Elswick/Scotswood. Naiad (11-15), Euryalus (16-20), Phoebe Y 21, A 22, B 23; Hermione B 27, X 29, Y 30. See 348G | (1-10) |
| 329G | Sixteen twin 4.5in HA BD Mk II | Aircraft carriers | Admiralty | Nov 1936 | Mid-1941 | 303104 | | | 18944 | Illustrious 34-41, Indomitable 42-49. Price excludes guns at c£2400 each | 34-49 |
| 330G | Twenty twin 4.5in HA BD Mk II | Queen Elizabeth, Valiant | Admiralty | Nov 1936 | End 1940 | | | | | Transferred to Scotswood Feb 1937. Valiant 14-23, QE 24-33 | 14-33 |
| 331G | Eight triple 6in Mk XXIII | Kenya, Trinidad | Admiralty | Dec 1936 | Early 1941 | 451521 | | | 56440 | Kenya 21-24, Trinidad 25, 26, Elswick 20, A mounting (28) lost in Shoal Fisher 23.2.41, replaced by Y removed from Belfast repairing at Devonport. Plus twelve at Elswick Fiji 9-12, Nigeria 13-16, Mauritius 17-19 + Barrow X 27. Price excludes guns at c£2700 each. | 21-28 |
| 333G | Two quadruple 14in Mk III | Duke of York Y, Howe A | Admiralty | Jan 1937 | Apr 1941, Apr 1942 | 1232600 | | | 616300 | Originally ordered as three quadruple. Approx price includes 3 sets of gunshield armour but excludes guns c£22000 each. Duke of York A & B Elswick. DoY Y shipped in coaster Sea Fisher Barrow-Clydebank | |
| 334G | One quadruple 14in Mk III | Howe Y | Admiralty | Jan 1937 | Feb 1942 | 446843 | | | 446843 | Originally ordered as one twin for Duke of York. Not final price? Shipped in Sea Fisher Barrow-Govan | 23 |
| 335G | One twin 14in Mk II | Howe B | Admiralty | Jan 1937 | May 1942 | 306314 | | | 306314 | Originally ordered as reserve mounting. Not final price? Erection of 3 mountings per ship approx £150,000 | 25 |
| 336G | Reserve 14in mounting | | Admiralty | Jan 1937 | | | | | | Cancelled | |
| 337G | Thirty twin 4in HA Mk XIX | | Admiralty | Mar 1937 | End 1940 | 168154 | | | 5605 | Belfast 269-74, Vega 275-6, Westminster 277-8, Unknown 279-82, Chatham DY 283-6, Alynbank 287-90, Cattistock 291-2, Cleveland 293-4, Whaddon 295-6, Viceroy 297-8 | 269-298 |

| Contract Number | Mountings | Ship | Customer | Ordered | Completed | Price excluding guns £ | Cost | Profit | Average price per mounting | Notes | Registered Numbers |
|---|---|---|---|---|---|---|---|---|---|---|---|
| 338G | Four twin 4.5in HA UD Mk III | Depot ships | Admiralty | Apr 1937 | Mid-1940 | 67325 | | | 16831 | Tyne 19-20, Hecla 21-22 | 19-22 |
| 339G | Four twin 4.5in HA UD Mk III | Forth | Admiralty | May 1937 | Mid-1939 | 64084 | | | 16021 | | 15-18 |
| 340G | Eight twin 5.25in HA/LA Mk I | 1936 battleships | Admiralty | Nov 1936 | End 1940 | 263831 | | | 32979 | Ordered for Prince of Wales but King George V got 2-5 (+ Elswick 10-13) and PoW 6-9 (+ H&W Scotstoun 18-21). Price excludes guns at c£2600 each | 2-9 |
| 341G | Sixty-three 8 barrel 2pdr Mk VI | | Admiralty | Jun 1937 | End 1941 | 546832 | | | 8680 | 6 each Duke of York 112, 126-130, Formidable 87-92, Victorious 102-7, Indomitable 120-5. 4 each King George V 78-9, 81-2, Prince of Wales 97-100, Illustrious 72-5. 2 Anson 133-4. 1 Roberts 131. Rest to dockyards and shipyards for refits: Queen Elizabeth 93-6, Sussex 85-6, London 80, 101 | 72-134 |
| 342G | Sixty-two 4 barrel 2pdr Mk VII | | Admiralty | Jun 1937 | | | | | | Transferred to Scotswood. Barrow costs £165254 | 74-102+ |
| 343G | Sixteen twin 4.5in HA BD Mk II | Aircraft carriers | Admiralty | Jun 1937 | End 1940 | | | | | Transferred to Scotswood Aug 1938. 8 Formidable, 8 Victorious | 50-65 |
| 344G | Twenty-eight 0.5in Mk III | | Admiralty | Aug 1937 | Mid-1939 | 34723 | | | 1240 | 2 each Belfast, Forth, Greenwich. 1 or 2 in destroyers and sloops | 247-274 |
| 345G | Twenty-eight twin 4.7in Mk XX | 'L' & 'M' class destroyers | Admiralty | Oct 1937 | End 1942 | 831320 | | | 29690 | Lightning 1-3, Lookout 4-6, Laforey 7-9, Matchless 10-12, Meteor 13-15, Musketeer A & B 16-17, Milne A & Y 18-19, Myrmidon/Orkan 20-22, Mahratta A & B 23-24, spares 25-28. Elswick/Scotswood (29-40) 3 each Loyal, Martin, Marne, 1 each Milne, Musketeer, Mahratta | 1-28 |
| 346G | Two twin 4.5in HA BD Mk II | | Admiralty | Oct 1937 | | | | | | Transferred to Scotswood. Possibly one converted to Mk IV for Savage | 66-67 |
| 347G | Five twin 4.5in HA UD Mk III | | Admiralty | Oct 1937 | | | | | | Planned for Repulse. Suspended and material transferred to 353G | 23-27 |
| 348G | Eleven twin 5.25in HA/LA Mk II | Dido class cruisers | Admiralty | Nov 1937 | Mid-1942 | 418047 | | | 38004 | Hermione A 26, Sirius Q 8, X 39, Y 40 (Scotswood B 5, A 43), Dido Q 37 at Brooklyn Navy Yard, Argonaut 46-50 (2 RPC), Spartan A 68. Plus others ordered Elswick, transferred to Scotswood. See 328G. | |
| 349G | Twenty-four single 4.7in CP XVIII | Destroyers | Brazil | Jul 1938 | Early 1940 | 107601 | | | 4483 | Cancelled. Admiralty mountings substituted. 2 ships Barrow, 2 White, 2 Thornycroft. Taken over by Admiralty 1939 as Handy, Hurricane, Havant, Havelock, Hesperus, Highlander | |
| 351G | Eight triple 6in Mk XXIII | Jamaica, Ceylon | Admiralty | Sep 1938 | End 1942 | 480757 | | | 68680 | Ceylon suspended 1940, then mountings re-allocated. Jamaica 33-36. Three planned for Ceylon to Bermuda 37-39 + 4th Y under 369G (40). Seven completed under 351G. See 369G. Excludes directors. Plus Elswick Gambia 29-32, Uganda 41-43. Bermuda mountings shipped in coasters Empire Jack and Empire Jill Barrow-Clydebank. | 33-40 |
| 352G | Twenty-six 4 barrel 0.5in Mk III | | Admiralty | Jul 1938 | Early 1940 | 30354 | | | 1167 | 2 each in Fiji, Naiad, Bonaventure, Glenearn. Others to dockyards | 370-395 |
| 353G | Twelve twin 4.5in HA UD Mk III | | Admiralty | Feb 1938 | End 1941 | 192619 | | | 16052 | Planned for modernised 'D' class cruisers. Hecla 28-29, Adamant 30-33, Charybdis 34-37, Scylla 39, Shoeburyness 38 | 28-39 |
| 354G | Three triple 16in Mk II | Temeraire | Admiralty | Jul 1938 | | | | | | Delivery due 1942. Suspended 13.6.40 later cancelled. £23672 costs. Lion ordered Elswick | |
| 356G | Seventeen 4 barrel 2pdr Mk VII | | Admiralty | Sep 1938 | | | | | | Transferred to Scotswood. £18104 Barrow costs | 136-152 |
| 357G | Thirty 8 barrel 2pdr Mk VI | | Admiralty | Sep 1938 | Early 1943 | 303190 | | | 10106 | 6 Howe 139-44, 4 Anson 135-38, 1 Abercrombie 159. Rest to dockyards and shipyards for refits | 135-164 |
| 358G | Fifty 4 barrel 0.5in Mk III | | Admiralty | Oct 1938 | Mid-1940 | 58913 | | | 1178 | 2 each Fiji, Dido, Phoebe, Locust, Berkeley, Springdale, Mosquito. 1 each Bangor, Bridlington. Rest to dockyards, shipyards and depots | 396-445 |
| 359G | Nine 4 barrel 2pdr Mk VII | | Admiralty | Oct 1938 | | | | | | Transferred to Scotswood. Barrow costs £6490 | 153-161 |
| 360G | Thirty-four 4 barrel 0.5in Mk III | | Admiralty | Feb 1939 | End 1940 | 40003 | | | 1177 | 4 Tyne, Hecla. 2 Larne, Nerissa, Isaac Sweers. 1 in Flowers and minesweepers | 306-339 |
| 361G | Two single 15in CD Mk II | | Turkey | Mar 1939 | | | | | | Suspended 1941, later cancelled. Value £281k | |
| 362G | Ten 8 barrel 2pdr Mk VIA | | Admiralty | Mar 1939 | End 1943 | 137220 | | | 13722 | Remote Power Control RP.50. 5 Indefatigable 167-71, 3 Implacable 172-4, 2 refits | 165-174 |
| 363G | Sixteen single 4.7in Mk XVIII | Four destroyers | Turkey | Mar 1939 | End 1941 | 84771 | | | 5298 | Sultanhisar, Demirhisar (Denny). Barrow ships taken over as Inconstant, Ithuriel | 35521-35536 |
| 364G | Eight 4 barrel 0.5in HA | Four destroyers | Turkey | Mar 1939 | End 1941 | 16031 | | | 2004 | As 363G | |
| 365G | Four single 4in mountings | Four submarines | Turkey | Mar 1939 | End 1941 | 19790 | | | 4948 | Murat Reis, Oruc Reis. British P.614, P.615 | 35505-35508 |
| 366G | Five twin 4.5in HA UD Mk III | | Admiralty | Mar 1939 | 1942 | 79925 | | | 15985 | Spares | 47-51 |
| 367G | Five single 3in CP V | 'S' class submarines | Admiralty | Apr 1939 | Mid-1941 | 9074 | | | 1815 | Safari 16, Sahib 17, Seraph 18, P222 19, store 15 | 15-19 |
| 368G | Six 8 barrel 2pdr Mk VIA | | Admiralty | Apr 1939 | Early 1944 | 80918 | | | 13486 | RP.50 gear. 2 Implacable 175-6. 2 each refits Howe 179-80, Formidable 177-8. | 175-180 |
| 369G | Eight triple 6in Mk XXIII | 1939 Fiji class cruisers | Admiralty | May 1939 | Early 1943 | 236911 | | | 59228 | Five suspended 1940 then re-allocated. Ceylon's 4th from 351G to Bermuda Y (40). Three Ceylon 44-46. Four completed under 369G, others cancelled. Plus Elswick Newfoundland 47-49 | |
| 370G | Three triple 16in Mk II | Thunderer? | Admiralty | May 1939 | | | | | | For 1939 battleship. Suspended June 1940, later cancelled. £5639 costs | |
| 371G | Two single 15in CD Mk II | For South Africa | War Office | Jul 1939 | | | | | | Cancelled Nov 1939. Monitor Erebus planned to be sent instead. £8888 costs | |
| 372G | Ten twin 5.25in Mk II | 1939 Dido class cruisers | Admiralty | 1939 | Mid-1943 | | | | | RPC. Spartan 68, 70-72, Black Prince 58-59, 61-62. 2 replacements with RPC 1944 for Euryalus X & Y 56-57. Spares 60, 69. Scotswood: Royalist 44-45, 53, 65 & Bellona 51-52, 54-55 | |
| 373G | Four 4 barrel 0.5in Mk III | | Australia | Aug 1939 | Mid-1941 | 5119 | | | 1280 | Sent to Garden Island, Sydney | 498-501 |
| 374G | Forty 4 barrel 0.5in Mk III | | Admiralty | Oct 1939 | Early 1941 | 51694 | | | 1292 | 2 each in destroyers, 1 each in sloops and minesweepers. Conversions | 502-541 |
| 375G | Thirty 4 barrel 2pdr Mk VII | | Admiralty | Nov 1939 | | 133265 | 127526 | 5739 | 4442 | Erected at Openshaw (Barrow 75% costs) | 226-255 |
| 376G | Sixteen 4 barrel 0.5in Mk III | | Admiralty | Jan 1940 | Mid-1941 | 19882 | | | 1243 | 2 each in destroyers, 1 each in minesweepers | 568-583 |
| 377G | Ten 8 barrel 2pdr Mk VIA | | Admiralty | Jan 1940 | Mid-1944 | 135362 | | | 13536 | RP.50 gear. Refits: 7 King George V 184-190, 3 Formidable 181-83 | 181-190 |
| 378G | Eight single 3in CP V | Submarines | Admiralty | Apr 1940 | Early 1942 | 13516 | | | 1690 | Unbroken 20, Unison 21, United 22, Unrivalled 23, Unruffled 24, Unshaken 25, Shakespeare 26, Saracen 27 | 20-27 |
| 379G | Thirty-five 4 barrel 0.5in Mk III | | Admiralty | May 1940 | End 1942 | 43507 | | | 1243 | To Gun Mounting Depot, Coventry | 600-634 |
| 380G | Thirty-one twin 4in Mk XIX | | Admiralty | Jul 1940 | Early 1943 | 166899 | | | 5384 | Derwent 691-2, Cattistock 693-4, Jamaica 695-8, Penylan 699-700, Brissenden 701-3, Rockwood 704, 706, Woodpecker 705, 707-8, Eggesford 709-10, Wren 711-3,Whimbrel 714-6, Redpole 717-9, Ariadne 720-1 | 691-721 |
| 381G | Forty 4 barrel 2pdr Mk VII | | Admiralty | Jul 1940 | Early 1943 | 166372 | | | 4159 | Ulster Queen 316-7, Heythrop 318-9, Cleopatra 320, 322, Exmoor 321, Charybdis 332, Gambia 333-4, Scylla 335, 341-2, Sirius 336, Jamaica 339-40, Argonaut 343-4, Antonia 345, Bermuda 346-7, Unicorn 349-52, Uganda 353-4, Newfoundland 323-4, Danae 329-30, Portsmouth DY 325-6, 355, Devonport DY 327-8, 337-8, 348, Cammell Laird 331 | 316-355 |

| Contract Number | Mountings | Ship | Customer | Ordered | Completed | Price excluding guns £ | Cost | Profit | Average price per mounting | Notes | Registered Numbers |
|---|---|---|---|---|---|---|---|---|---|---|---|
| 383G | Conversion of four twin 15in Mk I* to Mk I/N | Vanguard | Admiralty | End 1940 | 1945 | | | | | Mountings ex-Courageous and Glorious converted by Barrow, re-erected at Harland & Wolff, Scotstoun. RP.12. Some work by Elswick | |
| 384G | Fifty-six single 3in CP V | 'S','U' &' V' class submarines | Admiralty | Dec 1940 | End 1943 | 92252 | | | 1647 | Sibyl, Splendid, P47, P48, Sea Nymph, Unseen, Unruly, Sea Dog, Sickle etc | 28-83 |
| 385G | Seventy-two 4 barrel 0.5in Mk III | | Admiralty | Jan 1941 | | | | | | Cancelled 2nd qtr 1943 | 670-741 |
| 386G | Eight twin 5.25in Mk I* | Vanguard | Admiralty | 1941 | Early 1946 | | | | | RP.10. Also one spare mounting | 1-8 |
| 387G | Forty single 4.7in Mk XXII | 'S', 'T', 'U', 'V' class destroyers | Admiralty | Feb 1941 | End 1943 | 275969 | | | 6899 | Troubridge 1-4, Tumult 5-8, Scorpion 9-12, Scourge 13-16, Success 17-20, Teazer 21-24, Venus 25-28, Urchin 29-32, Tenacious 33-34, Urania 35, 39-40, Ulysses 36-38. Plus 35 mountings Scotswood (41-75) | 1-40 |
| 388G | Fourteen single 4in S 1 | 'T' class submarines | Admiralty | Mar 1941 | Mid-1943 | 35084 | | | 2506 | Tutankhamen, Trespasser, Taurus, Tactician, Truculent, Templar, Tally Ho, Tantalus. Others dockyards and store | 74-87 |
| 389G | Fifty twin 3-axis 40mm Mk IV | | Admiralty | Mar 1941 | | | | | | Dutch Hazemeyer design transferred to Openshaw. Barrow costs £43057 | 1-50 |
| 390G | Three triple 6in Mk XXIII | Ontario | Admiralty | Apr 1941 | Mid-1944 | 238222 | | | 79407 | Originally ordered for Tiger at Vickers-Armstrongs (Tyne). When latter redesigned for 6in Mk XXIV (406G), order transferred to Minotaur/Ontario (Harland & Wolff). Plus six mountings Elswick for Swiftsure 50-52 & Superb 56-58. | 53-55 |
| 391G | One triple 6in Mk XXIII | Belfast | Admiralty | May 1941 | Jul 1942 | 47404 | | | | Replacement for Y mounting transferred to Trinidad. Excludes gunhouse armour (which had been removed before transfer as too heavy for Trinidad) | 59 |
| 392G | Ten single 4.7in Mk XXII | Emergency destroyers | Admiralty | Jul 1941 | Early 1944 | 61930 | | | 6193 | Urania 76-7, Shark 78-81, Wizard 82-3, Wakeful 84-5. Plus 15 mountings Scotswood (86-100) | 76-85 |
| 394G | Ten single 4.7in Mk XXII | 'W' class destroyers | Admiralty | Aug 1941 | Early 1944 | 62830 | | | 6283 | Wizard 141-2, Wager 143-6, Wrangler 147-50. Plus 20 mountings Scotswood (151-170) | 141-150 |
| 395G | Twenty single 3in CP V | 'S', 'U' & 'V' class submarines | Admiralty | Nov 1941 | Mid-1944 | 33022 | | | 1651 | Sturdy, Pipinos, Strongbow, Vigorous, Vineyard etc | 84-103 |
| 396G | Twenty-two twin 4in Mk XIX | | Admiralty | Jan 1942 | End 1944 | 89802 | | | 4082 | RP.51. Completion of mountings from Harland & Wolff? All for refits or stores. Plus 50 mountings Scotswood. | 468-489 |
| 397G | Three triple 8in Mk V | Cruisers | Admiralty | Nov 1941 | | | | | | Suspended Jul 1942. Costs £5497 | |
| 398G | Twenty-one single 4in S 1 | 'S' & 'T' class submarines | Admiralty | Nov 1941 | Mid 1944 | 44422 | | | 2115 | Tantivy, Tradewind, Tudor, Telemachus, Trenchant, Thule, Zwaardvisch, Terrapin, Thorough, Tireless, Tiptoe, Token, Totem, Trump, Taciturn, Tapir, Seascout, Selene. Spares | 88-108 |
| 399G | Six triple 6in Mk XXIV | 1941 cruisers | Admiralty | 2nd qtr 1941 | 1949 | | | | | Originally ordered as Mk XXIII for Defence (Scotts) and Bellerophon (John Brown - Tiger). End 1942 changed to 6in Mk XXIV 60° elev RP.10. 90% complete by 1946. Stored at Rosyth. Ships redesigned for two twin 6in Mk 26 | |
| 400G | Ten single 4.5in Mk V | Destroyers | Admiralty | Aug 1941 | Early 1944 | 68670 | | | 6867 | Savage 1-2, Caprice 3-4, 9-10, Myngs 5-8 | 1-10 |
| 401G | Twenty-two 4 barrel 2pdr Mk VII | | Admiralty | | 1944 | | | | | RP.50 | |
| 402G | Sixteen twin 4.5in Mk IV | 8 'Battle' class destroyers | Admiralty | 2nd qtr 1942 | 1945 | | | | | RP.10. One flotilla | |
| 403G | Sixteen twin 4.5in Mk II | 2 aircraft carriers | Admiralty | 2nd qtr 1942 | 1950? | | | | | Probably for Eagle and Ark Royal. RP.10. 3 to Scotswood | |
| 404G | Sixteen single 4in S 1 | 'S' & 'T' class submarines | Admiralty | Jun 1942 | End 1944 | 37574 | | | 2348 | Scythian, Scotsman, Sleuth, Sidon, Seraph, Sea Devil, Spur, Scorcher, Turpin, Tijgerhaai, Truncheon, Thermopylae, Amphion. Spares | 109-124 |
| 405G | Nine single 3in CP V | 'V' class submarines | Admiralty | Jun 1942 | End 1944 | 18518 | | | 2058 | Plus 8 cancelled (113-120). Volatile, Vengeful, Votary, Virulent, Vortex, Vagabond. Spares | 104-112 |
| 406G | Six triple 6in Mk XXIV | 1942 cruisers | Admiralty | 2nd qtr 1942 | | | | | | 3 for Hawke (Portsmouth) and 3 spares. Cancelled Apr 1945 when 40% complete. Plus six at Elswick for Tiger (V-A Tyne - cancelled 1946) and Blake (Fairfield - completed with two twin Mk 26). | |
| 408G | Thirty single 4.5in Mk V | 'C' class destroyers | Admiralty | Nov 1942 | Mid-1945 | 378136 | | | 12605 | RP.50. 4 each Chevron 72-5, Cheviot 76-9, Comet 80-1, 84-5, Crescent 82-3, 86-7, Croziers/Trondheim 88-9, 92-3, Chequers 90-1, 94-5, Crusader 96-9. Coventry store 100. Plus 30 mountings Scotswood (101-130). Harland & Wolff also built 50 at Scotstoun | 71-100 |
| 409G | Twenty-six single 4in S1 | 'S', 'T' & 'A' class submarines | Admiralty | Apr 1943 | Mid-1946 | 101760 | | | 3914 | Original order for 40. Seneschal, Sanguine, Saga, Sentinel, Tudor, Tiara, Talent, Teredo, Tabard, Astute, Aeneas, Alaric, Alliance, Aurochs, Ambush, Auriga, Affray, Andrew. Others to depots | 125-150 |
| 410G | Twenty-four 8 barrel 2pdr Mk VIA | | Admiralty | Jul 1943 | End 1945 | 254862 | | | 10619 | RP.50. 7 Duke of York 196-202. 6 Devonport (for Valiant?) 208-213. 4 Sussex 191-4. 1 Anson 195. Others to dockyards or stores as spares | 191-214 |
| 411G | Twenty one twin 4.5in Mk IV | 'Battle' class destroyers | Admiralty | 1943 | | | | | | RP.10. 12 cancelled | |
| 412G | Fifteen twin 4.5in Mk VI | Destroyers | Admiralty | Aug 1943 | 1953 | 970863 | | 128702 | 97086 | For Daring class destroyers. Five cancelled | |
| 413G | Thirty-six single 4in S1 | Submarines | Admiralty | Aug 1943 | | | | | | Cancelled Dec 1943. 'T' and 'A' class? | 165-200 |
| 414G | Two prototype 6 barrel 40mm Mk VI | | Admiralty | Aug 1943 | Aug 1945 | | | | | Portsmouth S1, Vanguard S2 | S1-2 |
| 415G | Five single 4.5in Mk V | Later 'Battle' class destroyers | Admiralty | Aug 1943 | Mid-1946 | 135022 | | | 27004 | RP.50. Original order 18. Four transferred to Scotswood (186-189). Nine cancelled. Dunkirk 182, Barrosa 183, Matapan 184, Jutland 185, spare 181. | 181-185 |
| 416G | Fourteen twin 4.5in Mk VII | Malta class carriers | Admiralty | 4th qtr 1943 | | | | | | Cancelled 1946. Also fourteen Elswick | |
| 417G | Twenty two 6 barrel 40mm Mk VI | | Admiralty | 4th qtr 1943 | | | | | | RP.50. Vanguard (1946) and fleet carriers | |
| 418G | Thirteen twin 4.5in Mk VI | Destroyers | Admiralty | 1st qtr 1944 | | | | | | Cancelled 1946 | |
| 419G | Conversion gear for 91 8 barrel 2pdr Mk VI | | Admiralty | 1st qtr 1944 | | | | | | RP.11 gear. 30 suspended | |
| 420G | Ten twin 4.5in Mk II | | Admiralty | 1944 | | | | | | No details | |
| 421G | Twenty 4 barrel 2pdr Mk VII | | Admiralty | Mar 1944 | | 432645 | | | 21632 | RP.50. Transferred from Openshaw | |
| 422G | Conversion gear for 50 4 barrel 2pdr Mk VII | | Admiralty | Mar 1944 | | 136432 | | | | RP.50 gear for Openshaw. | |
| 423G | Five single 3.5in | Submarines | Admiralty | 2nd qtr 1944 | | | | | | New design. Not built? | |
| 424G | Twenty-four single 4in S2 | 'A' class submarines | Admiralty | May 1944 | Mid-1949 | 148342 | | | 6181 | Alaric, Alderney, Artemis, Aurochs, Alliance, Artful, Ambush, Anchorite, Andrew, Acheron, Alcide. (Some replaced 4in S1). Others to depots | 1-24 |
| 425G | Ten twin 4.5in Mk VI | Destroyers | Admiralty | 1945? | | | | | | Daring class? Also Elswick orders | |
| 426G | One protoype twin 6in Mk 26 | Cumberland | Admiralty | 3rd qtr 1947 | 1956 | 1610811 | | 122093 | | Prototype for trials. | 1 |
| 427G | One twin 3in HA Mk 6 | Cumberland | Admiralty | 2nd qtr 1948 | 1955 | 1126285 | | 90170 | | Prototype for trials. | |
| 433G | Three twin 6in Mk 26 and spares | Tiger + Blake A | Admiralty | 3rd qtr 1949 | 1959 | 2336681 | | 183616 | 778894 | RP.50. Tiger A (3), Y (2), Blake A (4). Plus three RP.15 from Elswick for Blake Y (5) and Lion A (7), Y (6) | 2-4 |
| 434G | Five twin 3in Mk 6 and spares | Tiger class | Admiralty | 4th qtr 1949 | 1958 | 2045973 | | 118908 | 409195 | Presume with five from Elswick includes Tiger Q3 (3), Q1 (12), Q2 (13), Lion Q2 (9), Q3 (10), Q1 (15) and Blake (19-21) | |
| 435G? | Six twin 4.5in Mk IV? | Destroyers | Venezuela | 1950 | 1954 | | | | | Assumed contract for Nueva Esparta and Zulia. 1956 3 more for Aragua | |
| 438-440G | Modernisation of 5.25in mountings | | Admiralty | c1950 | c1955 | 700000 | | | | Conversion to RPC incl Royalist. Approx price | |
| 441G | Seven twin 4.5in Mk 6 RP.41 | Frigates | Admiralty | 1951 | 1957 | | | | | Type 12 | |
| 442G | Four twin 3in Mk 6 | Destroyer escorts | Canada | c1952 | 1960 | 1077400 | | 77808 | 269350 | Assumed for Restigouche class forward mounting. Also some Elswick | |
| 447G | Seven twin 3in Mk 6 and spares | Destroyer escorts | Canada | 1954? | 1962 | 2333253 | | 201480 | 333322 | Assumed for Restigouche class forward mounting. | |
| 450G | Six twin 4.5in Mk 6* Mod 1 | Frigates | Admiralty | 1954? | 1962 | 774133 | | 91036 | 129022 | Type 12, 41, 61 plus spares | |
| 457G | Seven? single 4.5in Mk 8 | Destroyers and frigates | Min of Defence | c1970 | | | | | | Some for overseas navies | |

APPENDIX TWO

# Elswick Heavy Ordnance Output

ARMSTRONG'S ELSWICK WORKS produced a huge range of products both military and civilian. No useful contracts ledgers have been found, but after the Vickers-Armstrongs merger separate product groups were introduced: *eg* TM Transferable mountings, HM Hydraulic mountings, GL Large guns, GS Small guns, BM Breech mechanisms, etc. The annual works accounts running to about 200 pages a year showed under each of those product groups one line for each contract under Work in Progress and two to ten lines under Profit & Loss. There were thousands of entries in total, including not only major ordnance items but spare parts, modifications, repairs etc. It is thus possible to extract major ordnance contracts for guns and mountings of 4.5in and over from the mid-1920s to the 1950s, although excluded are smaller items such as shells and fuzes and most War Office contracts with their huge number of anti-aircraft guns. Work was also done on components for mountings being built at Barrow and Scotswood, *eg* VSG pumps. Some medium calibre mounting contracts were transferred to Scotswood from Elswick but are not detailed in the Scotswood accounts. The table reveals the torrent of orders resulting from re-armament in 1936–37. The contracts, particularly during WW2, were subject to transfers, delays and cancellations, while values were subject to later adjustments, but are the best figures that can now be determined.

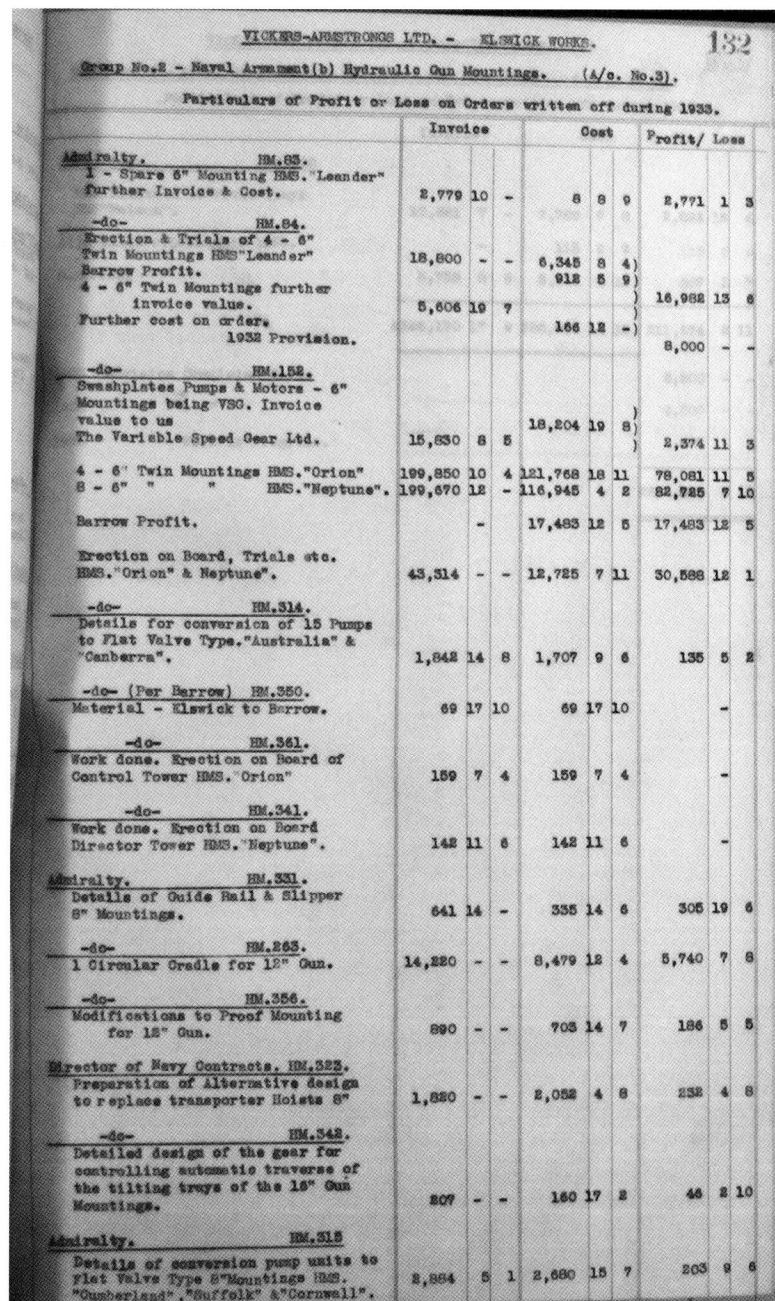

VICKERS-ARMSTRONGS LTD. - ELSWICK WORKS. 132

Group No.2 - Naval Armament(b) Hydraulic Gun Mountings. (A/c. No.3).

Particulars of Profit or Loss on Orders written off during 1933.

| | Invoice | | | Cost | | | Profit/ Loss | | |
|---|---|---|---|---|---|---|---|---|---|
| **Admiralty. HM.83.** 1 - Spare 6" Mounting HMS."Leander" further Invoice & Cost. | 2,779 | 10 | - | 8 | 8 | 9 | 2,771 | 1 | 3 |
| **-do- HM.84.** Erection & Trials of 4 - 6" Twin Mountings HMS"Leander" | 18,800 | - | - | 6,345 | 8 | 4) | | | |
| Barrow Profit. | | | | 912 | 5 | 9) | | | |
| 4 - 6" Twin Mountings further invoice value. | 5,606 | 19 | 7 | | | ) | 16,982 | 13 | 6 |
| Further cost on order. | | | | 166 | 12 | -) | | | |
| 1932 Provision. | | | | | | | 8,000 | - | - |
| **-do- HM.152.** Swashplates Pumps & Motors - 6" Mountings being VSG. Invoice value to us | | | | 18,204 | 19 | 8) | | | |
| The Variable Speed Gear Ltd. | 15,830 | 8 | 5 | | | ) | 2,374 | 11 | 3 |
| 4 - 6" Twin Mountings HMS."Orion" | 199,850 | 10 | 4 | 121,768 | 18 | 11 | 78,081 | 11 | 5 |
| 8 - 6" " " HMS."Neptune". | 199,670 | 12 | - | 116,945 | 4 | 2 | 82,725 | 7 | 10 |
| Barrow Profit. | | - | | 17,483 | 12 | 5 | 17,483 | 12 | 5 |
| Erection on Board, Trials etc. HMS."Orion" & Neptune". | 43,314 | - | - | 12,725 | 7 | 11 | 30,588 | 12 | 1 |
| **-do- HM.314.** Details for conversion of 15 Pumps to Flat Valve Type."Australia" & "Canberra". | 1,842 | 14 | 8 | 1,707 | 9 | 6 | 135 | 5 | 2 |
| **-do- (Per Barrow) HM.350.** Material - Elswick to Barrow. | 69 | 17 | 10 | 69 | 17 | 10 | | - | |
| **-do- HM.361.** Work done. Erection on Board of Control Tower HMS."Orion" | 159 | 7 | 4 | 159 | 7 | 4 | | - | |
| **-do- HM.341.** Work done. Erection on Board Director Tower HMS."Neptune". | 142 | 11 | 6 | 142 | 11 | 6 | | - | |
| **Admiralty. HM.331.** Details of Guide Rail & Slipper 8" Mountings. | 641 | 14 | - | 335 | 14 | 6 | 305 | 19 | 6 |
| **-do- HM.263.** 1 Circular Cradle for 12" Gun. | 14,220 | - | - | 8,479 | 12 | 4 | 5,740 | 7 | 8 |
| **-do- HM.356.** Modifications to Proof Mounting for 12" Gun. | 890 | - | - | 703 | 14 | 7 | 186 | 5 | 5 |
| **Director of Navy Contracts. HM.323.** Preparation of Alternative design to replace transporter Hoists 8" | 1,820 | - | - | 2,052 | 4 | 8 | 232 | 4 | 8 |
| **-do- HM.342.** Detailed design of the gear for controlling automatic traverse of the tilting trays of the 16" Gun Mountings. | 807 | - | - | 160 | 17 | 2 | 46 | 2 | 10 |
| **Admiralty. HM.318** Details of conversion pump units to Flat Valve Type 8"Mountings HMS. "Cumberland","Suffolk" &"Cornwall". | 2,884 | 5 | 1 | 2,680 | 15 | 7 | 203 | 9 | 6 |

A page from the Elswick accounts for 1933 showing profit and loss on some Hydraulic Gun Mountings. For Contract HM.152 *Orion* and *Neptune*'s eight twin 6in the total invoice (sales) value was £458,665, costs £287,128, profit £171,537. *(DS.VA/3/5/6)*

*Elswick Heavy Ordnance Contracts*

| *Contract Number* | *Item* | *For* | *Ordered* | *Completed* | *Sales value £* | *Notes* |
|---|---|---|---|---|---|---|
| | Six triple 16in Mk I mtgs | Nelson & Rodney | Dec 1922 | 1926 | 3.266M | 6in mtgs from Barrow |
| D.1001 | One 16in BL I gun body | Nelson & Rodney | 1923 | 1925 | 40300 | + BM £4268. Gun S.10 |
| D.1025 | Four twin 8in mtgs Mk I | Berwick | Jun 1924 | 1927 | 160000 each | Approx price. Probably includes Mk I design costs |
| D.1025 | Four twin 8in mtgs Mk I | Suffolk | Jun 1924 | 1927 | 160000 each | + 2 spare mtgs |
| D.1025 | Four twin 8in mtgs Mk I | Cornwall | Jun 1924 | 1927 | | Openshaw |
| D.1027 | Three 8in BL VIII gun bodies | Cruisers | Jul 1924 | 1926 | 5300 each | + BM £1120 |
| D.1056 | Four twin 8in mtgs Mk I | Australia | May 1925 | 4Q 1927 | | 2 from Elswick, 2 from Openshaw + 1 spare mtg |
| D.1057 | Six 8in BL VIII* gun bodies | Cruisers | Jun 1925 | 1927 | 5300 each | |
| D.1069 | Four twin 8in mtgs Mk I* | London | 4Q 1925 | 4Q 1928 | | Openshaw |
| D.1069 | Four twin 8in mtgs Mk I* | Sussex | 1Q 1926 | 3Q 1928 | 589112 | + 2 spare mtgs. Profit £26814 |
| D.5206 | Four 8in BL VIII gun bodies | Cruisers | 1926 | 1928 | 6010 each | |
| HM.3 | Four twin 8in mtgs Mk II | Dorsetshire | 4Q 1926 | 3Q 1930 | 450239 | + 2 spare mtgs |
| HM.4 | Three twin 8in mtgs Mk II | York | 1Q 1927 | 4Q 1930 | 338559 | + 1 spare mtg |
| 304003 | Four 15in BL guns | Spain CD | 1927 | 1929 | 18000 each | Made at Sheffield |
| 304005 | Twenty-eight 4.7in gun bodies | Admiralty | 1927 | 1929 | 1294 each | Contract profit £11010 |
| D.5262 | Nine 8in BL VIII gun bodies | Cruisers | 1927 | 1929 | 7008 each | + BM £925 |
| HM.5 | Three twin 8in mtgs Mk II* | Exeter | 3Q 1928 | 1931 | 329103 | Shields from Openshaw. 1 spare mtg |
| D.1219 | One single 15in CD mtg | War Office | Oct 1928 | 1Q 1932 | 208828 | To store at Woolwich where 2 more made |
| GL.1 | Two 8in BL VIII | Spain | 1929 | 1930 | 11265 | For Canarias mountings at Barrow |
| HM.17 | Four twin 8in mtgs Mk II | Surrey | Feb 1929 | | | Cancelled Feb 1930 |
| GS.1 | Twenty 4.7in QF IX gun bodies | Destroyers | 1929 | 1930 | 1378 each | + BM £438 |
| HM.83 | One twin 6in mtg | Admiralty | 1Q 1930 | | 51415 | Prototype Mk XXI? |
| HM.84 | Four twin 6in Mk XXI mtgs | Leander | 2Q 1930 | 4Q 1932 | 260102 | |
| GL.3 | Repair of two 14in Mk I guns | Almirante Latorre | 1930 | | 11680 | Chile ex-RN Canada. Refitting at Devonport |
| HM.152 | Eight twin 6in Mk XXI mtgs | Orion, Neptune | Dec 1930 | 3Q 1933 | 458665 | £57333 each. Contract profit £171537 |
| GL.7 & 9 | Retubing four 15in BL I | Admiralty | 1Q 1931 | 1932 | 5433 each | To store at Woolwich |
| GM.4 & 5 | Twenty-two 6in BL XXIII gun bodies | Cruisers | 2Q 1931 | 1932 | 2099 each | + BM £860 |
| HM.205 | Removal of six twin 13.5in mtgs | Iron Duke 2, Tiger 4 | 2Q 1931 | 1Q 1932 | 19350 | To store at Rosyth |
| GS.18 | Eleven 4.7in QF IX gun bodies | Destroyers | 1931 | 1932 | 1322 each | Total profit £1948 |
| GM.30 | Twenty 4.7in QF IX* gun bodies | Destroyers | 1932 | 1933 | 1303 each | |
| HM.276 | Four twin 6in Mk XXI mtgs | Amphion/Perth | May 1932 | 4Q 1934 | 236380 | Ship completed with new gearing Jul 1936 |
| HM.277 & 284 | Two twin 6in Mk XXI spare mtgs | Admiralty | 1932 | 1934 | 49900 each | |
| GS.44 | Thirteen 6in BL XXIII gun bodies | Cruisers | 2Q 1932 | 1934 | 2083 each | + BM £860 |
| GS.45 | Twenty-one 4.7in QF IX* gun bodies | Destroyers | 3Q 1932 | 1934 | 1305 each | + BM £438 |
| HM.332 | Four twin 6in Mk XXI mtgs | Phaeton/Sydney | Dec 1932 | 2Q 1935 | 232028 | |
| HM.332 | Two twin 6in Mk XXI mtgs | Apollo/Hobart | Dec 1932 | Mid 1935 | 116014 | + 2 at Barrow 252G |
| GL.10 | Retubing one 16in BL I gun S.11 | Admiralty | 4Q 1932 | 1934 | 6480 | Profit £1691 |
| GL.12, 13 | Retubing three 15in BL I | Admiralty | 1933 | 1935 | 5550 each | |
| GS.59 | Eleven 6in BL XXIII gun bodies | Cruisers | 1933 | 1935 | 1895 each | |
| GS.60 | Twenty 4.7in QF IX* gun bodies | Destroyers | 1933 | 1934 | 1300 each | |
| HM.383 | Four triple 6in Mk XXII mtgs | Newcastle | Dec 1933 | 4Q 1936 | 325414 | Includes design |
| HM.404 | One triple 6in Mk XXII spare mtg | Admiralty | 1934 | 1936 | 57021 | |
| GS.75 | Fifteen 6in BL XXIII gun bodies | Cruisers | 1934 | 1936 | 2059 each | + BM £735. Contract profit £7393 |
| GS.76 | Twenty one 4.7in QF IX* gun bodies | Destroyers | 1934 | 1935 | 1317 each | + BM £480 |
| HM.460 & 471 | Conversion of four twin 15in Mk I mtgs to Mk I/N | Warspite | 4Q 1934 | 4Q 1936 | 182594 | Modernised at Portsmouth |
| HM.495 | Four triple 6in Mk XXII mtgs | Sheffield | Dec 1934 | 1Q 1937 | 256776 | |
| HM.496 | Four triple 6in Mk XXII mtgs | Birmingham | Dec 1934 | 2Q 1937 | 256776 | |
| HM.514 | Convert Warspite 15in to pneumatic run-out | | 1935 | | 12920 | |
| GS.91 | Nineteen 6in BL XXIII gun bodies | Cruisers | 2Q 1935 | 1937 | 2021 each | + BM £735 |
| GS.93 | Eighteen 4.7in QF IX** gun bodies | Destroyers | 2Q 1935 | 1936 | 1317 each | |
| GL.15 | Retubing one 15in BL I | Admiralty | 2Q 1935 | 1937 | 5140 | |
| GL.16 | Relining 15in to 14in SE.59 | Admiralty | 1935 | 1937 | 5831 | |
| HM.596 | Four triple 6in Mk XXII mtgs | Manchester | Nov 1935 | 4Q 1937 | 302368 | |
| HM.597 | Four triple 6in Mk XXII mtgs | Gloucester | Nov 1935 | 4Q 1938 | 288372 | |
| GL.17 | One 14in BL gun body SE.60 | Admiralty | 4Q 1935 | 1938 | 18500 | |
| GS.115 | Twenty 4.7in QF IX** gun bodies | Destroyers | 1936 | 1937 | 1482 each | Reg Nos S.2296-2315 |
| GS.121 | Thirty 4.7in QF XII gun bodies | 'Tribal' class destroyers | 1Q 1936 | 1938 | 1671 each | Contract profit £2759 |
| GS.122 | Nineteen 6in BL XXIII gun bodies | Cruisers | 1Q 1936 | 1938 | 2020 each | + BM £695. Reg Nos S.4789-4807 |
| GL.18 | Fourteen 14in BL VII gun bodies | KGV class battleships | 2Q 1936 | 1939 | 19050 each | 7 left, 7 right. + BM £3200 |
| GL.19 | Retubing two 15in BL I | Admiralty | 2Q 1936 | 1938 | 5085 each | |
| GS.131 | Thirteen 6in BL XXIII gun bodies | Cruisers | 3Q 1936 | 1938 | 2000 each | |
| GS.132 | Thirty-eight 4.7in QF XII gun bodies | 'Tribal' class destroyers | 3Q 1936 | 1938 | 1696 each | + BM £500 |
| TM.72 | Twenty twin 4.7in CP XIX mtgs | 'Tribal' class destroyers | 3Q 1936 | 1939 | 12590 each | |
| HM.609 | Two quad 14in Mk III mtgs | King George V | May 1936 | May 1940 | | £1.69M all KGV mtgs. Design, mock ups and tools £445k |
| HM.642 | One twin 14in Mk II mtg | King George V | May 1936 | 2Q 1940 | | |
| HM.647 | One spare 14in mtg | King George V | May 1936 | 1941 | 40000 | Approx price. Elevating portion only |

| *Contract Number* | *Item* | *For* | *Ordered* | *Completed* | *Sales value £* | *Notes* |
|---|---|---|---|---|---|---|
| HM.650 | Four triple 6in Mk XXIII mtgs | Edinburgh | Jun 1936 | 1Q 1939 | 303945 | Profit c£36000 |
| HM.658 & 679 | Conversion of three twin 15in mtgs to Mk I/N | Renown | 3Q 1936 | 1939 | 125010 | Modernised at Portsmouth |
| GS.140 | Thirty-one 4.5in QF I gun bodies | Battleships & carriers | 4Q 1936 | 1938 | 1935 each | + BM £562. Contract profit £9524 |
| HM.710 | Four triple 6in Mk XXIII mtgs | Fiji | Dec 1936 | 4Q 1939 | 263760 | |
| HM.710 | Four triple 6in Mk XXIII mtgs | Nigeria | Dec 1936 | 2Q 1940 | 263760 | |
| HM.710 | Four triple 6in Mk XXIII mtgs | Mauritius | Dec 1936 | 3Q 1940 | 263760 | |
| HM.716 & 717 | Conversion of four twin 15in mtgs to Mk I/N | Valiant | 4Q 1936 | 4Q 1939 | 153524 | Modernised at Devonport |
| HM.718 & 719 | Conversion of four twin 15in mtgs to Mk I/N | Queen Elizabeth | 4Q 1936 | 4Q 1940 | 153524 | Modernised at Portsmouth |
| | Twenty twin 4.5in BD II mtgs | Valiant, Queen Elizabeth | 4Q 1936 | 1939 | | Transferred to Scotswood |
| | Sixteen twin 4.5in BD II mtgs | Formidable, Victorious | 1937 | 1940 | | Transferred to Scotswood 1938 |
| TM.114 | Twenty-one twin 4.7in CP XIX mtgs | 'J','K','N' class destroyers | 1Q 1937 | 1940 | 12590 each | |
| TM.118 | Ten twin 5.25in Mk II mtgs | Dido class cruisers | Nov 1936 | 1940 | 50147 each | Transferred to Scotswood |
| TM.121 | Eight twin 5.25in Mk I mtgs | King George V | Nov 1936 | 3Q 1940 | 37096 each | 4 transferred to Scotswood |
| GS.146 | Seventy 5.25in QF I gun bodies | Battleships & cruisers | 1Q 1937 | 1940 | 2312 each | + BM £527. Contract profit c£13000 |
| GS.150 | Fifty 4.5in QF I loose barrels | Admiralty | 1Q 1937 | 1939 | 512 each | Contract profit £3885 |
| GS.151 | Fifty 5.25in QF I loose barrels | Admiralty | 1Q 1937 | 1940 | 723 each | |
| GL.20 | Eighteen 14in BL VII gun bodies | KGV class battleships | 1Q 1937 | 1941 | 18777 each | + BM £2300 |
| GS.152 | Forty 4.5in QF III gun bodies | Battleships & carriers | 1Q 1937 | 1939 | 1800 each | + BM £480 |
| GS.153 | Fifty 4.7in QF XII* gun bodies | 'J','K','N' class destroyers | 1Q 1937 | 1939 | 1691 each | + BM £555 |
| HM.721 | One quad 14in Mk III mtg | Duke of York A | Jan 1937 | Jun 1941 | | Two originally ordered for Anson A, Y |
| HM.722 | One twin 14in Mk II mtg | Duke of York B | Jan 1937 | May 1941 | | Originally ordered for Anson B |
| HM.725 | Two quad 14in Mk III mtgs | Anson A, Y | Jan 1937 | 1Q 1942 | | One originally ordered for Duke of York A. Barrow Y |
| HM.726 | One twin 14in Mk II mtg | Anson B | Jan 1937 | 1Q 1942 | | Originally ordered for Duke of York B |
| GS 163 | Thirty-nine 6in BL XXIII gun bodies | Cruisers | 1937 | 1939 | 2140 each | + BM £682 |
| TM.513 | Twelve twin 4.7in Mk XX mtgs | 'L' & 'M' class destroyers | Oct 1937 | 1941 | 34500 each | Original order 24 |
| GS.175 | Thirty-six 4.7in QF XI gun bodies | Destroyers | 1938 | 1940 | 1900 each | + BM £503 |
| GS.196 | Thirty-six 4.5in QF III gun bodies | Battleships & carriers | 3Q 1938 | 1940 | 1478 each | + BM £457 |
| GL.25 | Eighteen 16in BL II & III gun bodies | Lion class battleships | 3Q 1938 | 1944? | | Only 4 or 5 completed |
| HM.925 | Four triple 6in Mk XXIII mtgs | Gambia | Sep 1938 | 4Q 1941 | 274044 | |
| HM.925 | Three triple 6in Mk XXIII mtgs | Uganda | Sep 1938 | 3Q 1942 | 205553 | Original order 4 mtgs |
| HM.929 | Three triple 16in Mk II mtgs | Lion | Jul 1938 | | | Cancelled 1945 after £395k spent |
| GS.206 | Eighteen 6in BL XXIII gun bodies | Cruisers | 1938 | 1940 | 2481 each | + BM £936 |
| GS.232 | Fifty 4.5in QF III loose barrels | Admiralty | 1939 | 1941 | 450 each | |
| GS.233 | Twenty-six 4.5in QF III gun bodies | Admiralty | 1939 | 1941 | 1558 each | |
| GS.235 | Twenty 5.25in QF I gun bodies | Admiralty | 2Q 1939 | 1941 | 1856 each | |
| HM.5070 | Convert three twin 13.5in mtgs to CD | Turkey | 2Q 1939 | 2Q 1941 | | Ex-Iron Duke and Tiger. 2 completed then stored |
| HM.1034 | Three triple 6in Mk XXIII mtgs | Newfoundland | May 1939 | 4Q 1942 | 230000 | Original order 4 mtgs |
| GS.244 | Sixty 4.7in QF XII** gun bodies | Destroyers | 3Q 1939 | 1941 | 1479 each | + BM £468 |
| GS.246 | Twenty-five 4.7in QF XI gun bodies | 'L' & 'M' class destroyers | 3Q 1939 | 1941 | 1660 each | + BM £480 |
| GS.249 | Thirty 5.25in QF I loose barrels | Admiralty | 1939 | 1941 | 625 each | |
| GL.27 | Relining 14in S.68 | Admiralty | 1939 | 1940 | 4115 | |
| | Twenty two 6in BL XXIII gun bodies | Cruisers | 3Q 1939 | 1942 | | 14 right, 8 left guns |
| GS.267 | Seventy seven 4.7in QF IX** gun bodies | Destroyers | 1940 | 1941 | 1335 each | Contract profit £5473 |
| HM.1146 & 1176 | Refurbish twin 15in mtg ex-Marshal Soult | Roberts | 1Q 1940 | 2Q 1941 | 80233 | Profit £2777 |
| GS.285 | Sixty-four 4.7in QF IX** gun bodies | Destroyers | 1940 | 1942 | 1142 each | |
| GL.38 | Eight 8in BL VIII* gun bodies | Cruisers | 1940 | 1942 | 5554 each | |
| GS.299 | Forty 4.5in QF loose barrels | Admiralty | 1940 | 1942 | 454 each | |
| GS.300 | Twelve 4.7in QF IX loose barrels | Admiralty | 1940 | 1941 | 542 each | |
| HM.1259 | Conversion work on four 15in Mk I* | Vanguard | End 1940 | 1945 | | Some work also Barrow. Re-erected H&W Scotstoun |
| GL.50 | Four 14in BL VII gun bodies | Battleships | 1941 | 1943 | 25000 each | |
| GL.52 | Twelve 8in BL VIII* gun bodies | Cruisers | 1941 | 1943 | 5360 each | + BM £990 |
| GS.332 | Eighty 5.25in QF I loose barrels | | 1941 | 1943 | 563 each | Most for War Office? |
| GS.334 | Forty-six 4.7in loose barrels | Admiralty | 1941 | 1943 | 471 each | |
| HM.1314 & 1325 | Convert twin 15in to Mk I/N | Abercrombie | 2Q 1941 | 1Q 1943 | 212856 | Planned for Furious. Profit £8908 |
| HM.1320 | Three triple 6in Mk XXIII mtgs | Swiftsure | Apr 1941 | 4Q 1943 | 79000 each | Planned for Newfoundland |
| HM.1320 | Three triple 6in Mk XXIII mtgs | Superb | 4Q 1941 | 4Q 1944 | 81000 each | Approx price |
| HM.5091 | Modify two mountings for 15in CD | War Office | 1941 | 1942 | 152228 | For Dover |
| HM.5092 | Six 8in CD mtgs | War Office | 1941 | 1943 | 208509 | For Dover |
| GL.58 & 67 | Reline three 13.5in guns to 8in HV | War Office | 1942 | 1943 | 2812 each | For Dover |
| GS.364 | Eighty-four 4.7in QF gun bodies | Destroyers | 1942 | 1944 | 1280 each | |
| GS.365 | Sixty-five 4.7in loose barrels | Admiralty | 1942 | 1944 | 366 each | |
| GS.396 | Seventy-three 4.5in QF IV gun bodies | 'C' class destroyers | 1942 | 1944 | 1304 each | |
| HM.1465 | Three triple 6in Mk XXIV mtgs | Tiger/Bellerophon | 2Q 1942 | c1948 | | Planned as Mk XXIII. Completed. To store at Rosyth |

| *Contract Number* | *Item* | *For* | *Ordered* | *Completed* | *Sales value £* | *Notes* |
|---|---|---|---|---|---|---|
| HM.1466 | Three triple 6in Mk XXIV mtgs | Blake | 2Q 1942 | c1948 | | Only 1 or 2 completed. 2 twin Mk 26 fitted 1960 |
| GS.408 | One hundred and eight 4.5in QF IV gun bodies | 'C' class destroyers | 1943 | 1945 | 974 each | |
| GS.410 | Thirty-six 6in BL XXIII gun bodies | Admiralty | 1943 | 1944 | 1800 each | |
| GS.450 | Sixty-six 4.5in QF III gun bodies | Battle class destroyers | 1943 | 1945 | 1510 each | |
| GS.451 | Ten 4.5in QF V gun bodies | Daring class destroyers | 1943 | 1947 | 1950 each | Original order 36 |
| GS.455 | Sixteen 4.5in QF IV gun bodies | 'C' class destroyers | 1943 | 1946 | 1467 each | |
| TM.953 | Fourteen twin 4.5in Mk VI mtgs | Daring class destroyers | Aug 1943 | 1948-51 | | |
| TM.989 | Fourteen twin 4.5in Mk VII mtgs | Malta class carriers | 3Q 1943 | | 107641 | Cancelled 1946. 14 Barrow |
| GS.481 | Fifty-two 4.5in QF V loose barrels | Admiralty | 1943 | 1947 | 438 each | |
| GS.482 | Sixty 5.25in loose barrels | | 1943 | 1945 | 717 each | Most for War Office? |
| GS.485 | Forty 4.5in QF III gun bodies | 'Battle' class destroyers | 1944 | 1946 | 1704 each | |
| TM.1183 | Sixteen twin 4.5in Mk VI mtgs | Daring class destroyers | 1944 | | | All cancelled? |
| GS.504 | Two 6in 50 cal guns | Admiralty | 1944 | 1948 | 30450 | Prototype N.5. Plus two spare barrels |
| GS.505 | Thirty-four 4.5in QF V gun bodies | Daring class destroyers | 3Q 1944 | 1947 | 1335 each | |
| GS.506 | Twenty-six 4.5in loose barrels | Admiralty | 1945 | 1948 | 436 each | |
| TM.1359 | Two twin 4.5in Mk 6 mtgs | Decoy | c1946 | 1952 | | |
| HM.2323 | Five twin 3in Mk 6 mtgs | Tiger class cruisers | 1949 | 1960 | 2.064M | Plus five from Barrow |
| HM.2328 | Three twin 6in Mk 26 mtgs | Lion A, Y + Blake Y | 1949 | 1959 | 866200 each | Barrow Tiger A, Y and Blake A |
| TM.1495 | Six twin 4.5in Mk 6 mtgs | Frigates | 1951 | 1954-57 | 98606 each | |
| HM.2360 | Four twin 3in Mk 6 mtgs | Canada | c1952 | c1962 | 269100 each | Assumed for Restigouche class |
| TM.1742 | Eight single 4in mtgs | Chile | c1955 | 1961 | 192232 each | For Almirante Williams and Riveros. Loss £497k |
| TM.1774 | Four twin 4.5in Mk 6 mtgs | Frigates | 1955 | 1960 | 112118 each | |
| TM.1826 | Four twin 4.5in Mk 6 mtgs | Devonshire, Whitby, Falmouth | 1956 | 1960 | 108648 each | |
| | Fourteen 4.5in QF V guns + 18 loose barrels | Admiralty | 1957 | | 102808 | |
| TM.2019 | Six twin 4.5in Mk 6 mtgs | Destroyers and frigates | c1959 | 1962 | c120000 each | |
| TM.2119 | Five twin 4.5in Mk 6 mtgs | Frigates | c1960 | 1963 | | |
| | Twelve twin 4.5in Mk 6 mtgs | Frigates | c1963 | 1965-68 | | Last Elswick gun mounting contracts. 6 for Netherlands frigates |

A twin 4.7in Mark XX being moved from an Elswick shop to rail wagon before transfer to shipyard. *(HMS* Excellent*)*

APPENDIX THREE

# Conqueror Gun Trials

*(See Frontispiece and back jacket)*

Elswick Works
NEWCASTLE-UPON-TYNE
19 June 1912

Gentlemen

H.M.S. "Conqueror" – Gun Trials

We went out for trial at 8.20 a.m. on Monday, the 17th instant. There was a very large attendance.

Mr Winston Churchill [First Lord of the Admiralty]

Admiral [A G H W] Moore, Controller of the Navy.

Captain [F C T] Tudor, D.N.O.

Engineer Adml. [E F] Ellis, R.N.

Admiral [R H] Bacon [Managing director of Coventry Ordnance Works]

Mr C G Walker [C.O.W., formerly of Elswick]

Mr McCartney from Coventry &c &c.

I have been shown every courtesy and attention. Admiral Bacon and Mr Walker took me round and explained all the machinery to me.

At 10.40 A turret was manned.

The programme was to fire four rounds from each gun; the following was the result:-

LEFT GUN:-

- Reduced charge. All well.
- Loading cage would not come fully up. Great efforts were made to re-adjust it, but these failing the gun was loaded with the hand tray and hydraulic rammer.
- Ditto. Full elevation.

RIGHT GUN:-

- Reduced charge. Gun only recoiled 10 inches. The reason for this was not discovered. Probably the run in and out byepass [*sic*] valve would not lift properly, as the gun could be run in and out by working the control valves without any difficulty.

It was decided not to fire any more rounds from either gun, as the fourth round from both was to be simultaneous and packed back.

We left A turret at 12 noon and went to B turret, but failed to get ready to fire for 20 minutes, when one reduced from right gun was fired then it was decided to go to dinner. … There was much trouble in clearing the loading cage of the right gun from its interlocking gear; when at last this was done and the cage was raised, the shot [inert shell] fell out from the rear of the cage down into the working chamber. …

The left gun cage was brought up and the shot rammed home, but they could not get the cordite charges to drop in front of the rammer. So eventually after 1½ hours had been spent in the turret waiting for range [to clear?] and endeavouring to load the guns, orders were given to go to Q turret, but not to load until the weather cleared. …

Generally speaking, the gun machinery is very much more complicated, heavier, and more costly, than anything we have ever done. I was told frankly that £150,000 had been lost to the firm over the order, but probably there was a little exaggeration in this. [A not implausible figure as further costs would be incurred correcting the faults and COW reported losses of £187,243 in 1911 and £145,407 in 1912, requiring more funds from the parent companies][1] Anyhow, I am sure I am correct in saying that such machinery could not be produced at Elswick under 25% more cost than what we usually supply.

[1] Warren. *op cit*, p149

The weakest point was the gun loading hoist. It weighed over 3 tons!! was extremely complicated and inaccessible. It is impossible to imagine that such a piece of mechanism will ever be allowed to go into the service. …

The central cage arrangement is clumsy and complicated but it worked all right. There are independent cordite and shot hoists, so here again they have one more press than we do. The operations which the cordite charges pass are really quite interesting. They roll from the magazine hopper into the central hoist, then out into the waiting tray, then rammed into other waiting trays, then rolled into a cage which works up and down in the gun loading cage, then hoisted in the latter, then hoisted again in the internal cage, then rolled in front of the rammer and lastly rammed into the gun. Is it possible to imagine a more complicated process? …

There are two good things in the installation:-

- A grab which enables you to tilt the shell.
- A very neat hand pump.

…

The roof is given a very considerable curve, so that the walls are low, but the drawback is a dreadfully complicated contour to the top of the side and front armour. It must have been a most costly matter to cut the rabbet on the top of the armour.

The turret trains well, but I did not see the creep tried. The engine is, however, much more noisy than our swash plate engine.

The elevating seems excellent, but the cylinders appear to have been very difficult to cast; 29 castings were made to obtain 10 sound ones. [one for each gun] …

<u>Tuesday 18th June</u>

The ship left her anchorage at 9 a.m.

At 10.52 a.m. B turret was manned and both guns were loaded, the right gun with a full charge and the left gun with a reduced charge. The damage done by the falling of a projectile in the rear of the right gun had been repaired, and the left cage had been adjusted so as to work more correctly. There was, however, difficulty in making the projectile tilt correctly in the upper position of the gun loading cage. I noticed that if the breech of the gun were raised or lowered there was a distinct relative motion of the gun and the gun loading cage. This occurred in every gun loading installation, and the smallest motion of the gun loading cage produced a tilt on the shell. The result was that accurate adjustment was almost impossible and if obtained in one instance would be unlikely to continue; as a matter of fact it never was obtained during the whole of the trial. With the short shot the method of loading which was adopted with some success was to use a stick and push the shot forward 1 foot, or more, through the rear opening in the cage. The centre of gravity of the shot then got in front of the point of oscillation of the tilting cage and that generally brought the cage into proper line with the gun, so that the rammer could go forward.

When the right gun was fired the non return valve casing in connection with the run in and out cylinder burst. There was a great loss of water, and pressure was therefore taken from the left gun at the moment when the cage was up and the rammer partly in the cage. The cage promptly dropped because the stop from the rammer was not in place as it should have been. …

Having fired one full and one reduced charge from B turret we proceeded to Q turret. The results in Q turret were much the same. One non return valve burst, the length of the recoil being far too little in the mounting where the burst took place. On the other mounting pressures were taken. The length of recoil with that gun was also short. …

X turret was manned immediately after dinner. It was worked with the auxiliary gear and hand ramming. From the order "Go" until the left gun, which was the first of the two, was ready for firing occupied 4 minutes 23 seconds. The official time which was given was 3 minutes 24 seconds, but I am certain they were a minute wrong as I took the time with a stop watch. The second round was done in 2 minutes 45 seconds. The non return valve burst in the left gun after the second round. Otherwise everything worked well.

Y turret was worked by hydraulic means. The same difficulties occurred, that is to say the shell did not tilt properly in the cages and both non return valves burst after the full charge was fired. … The full recoil of the gun I was told is 36 inches. With full charges the recoils in various guns varied from 22 inches to 35 inches. … Generally speaking the machinery worked fairly well, except the gun loading cages which, as before remarked, were most complicated, unnecessarily heavy, and must have been very costly indeed. I do not think it would be possible to make them without their being constantly in need of adjustment.

From conversations which I overheard I think the Admiralty will insist upon their being entirely remodelled. The rammers worked quite as quickly if not more quickly than ours. …

A feature of the mounting was the construction of the loading arm which was a very strongly built box girder of square section. The tail of the rammer went inside this girder. It would, however, be very difficult to examine it if ever examination was required. The engine for working the rammer and for working the breech was a three cylinder engine. I was informed it was giving a good deal of trouble in adjustment and therefore resembled ours in this respect. … The man who worked the rammer was provided with a seat having a coal scuttle back. He stepped off the loading arm as soon as the cage was lowered. I was informed that if he remained there the recoiling gun would strike his knee. The quickest motion of ramming which I was able to time was 8 seconds including ramming the shot and charges. In the MONARCH the same operator completed these operations in 7 seconds. … The hand gear for working the breech was an exact copy of ours, so the drawings must have been furnished by the Admiralty for them to work to.

The structure was rather curiously arranged. To support the shield there were two columns, the front one which was about the centre of the turret, was an exact copy of the column used in MONARCH, the rear one was of the most complicated system and must have been very costly to produce; it was a solid forging cut out in the same way as the forward one except that the pillars were skewed. To pass by this column, so as to get to the sighting position, was only possible for a thin man. Mr Winston Churchill did it with the greatest difficulty [!]. It was necessary to work one's body in and out of the pillars. … I was told that the Coventry Ordnance Company expected to show that there was far less vibration when the guns fired than there was in the MONARCH class. I can only say that the vibration in my mind was much worse, but this may have been due to the short recoils. …

In the shell rooms there was a bogie designed on the same principle as our bogie but with a considerable difference in detail. The rack attached to the ship's deck is of very large diameter, coming right outside the bogie, and the bogie must weigh quite double what ours does, although it is considerably lower. … The interlocking gear was on the principle adopted in the SWIFTSURE that is to say, unless the man in the shell room moved his lever the control lever in the working chamber was locked but this was coupled up with the usual complicated indicator. The gear for locking the doors was strong and reliable and better than what we generally use. I was told that segments of roller paths had been supplied as spare but it would be impossible to place them without lifting the turret out. …

The trial impressed me very much by the difficulties a new firm must have in endeavouring to construct hydraulic mountings, and there was abundant evidence to prove that they could not have been so successful unless they had been assisted by some efficient Engineer like Engineer Rear-Admiral Ellis. Efforts were made everywhere to make the design differ from our designs, but the principles on which they worked were very thinly veiled.

The firing was completed by 3 o'clock and the ship returned to Spithead. Another trial is to take place when the ship commissions, when two rounds will be fired from each gun with full charge. There is also to be a rapidity trial from one turret.

Tyne & Wear Archives DS.VA/6/34/1

*Conqueror* showing her five twin Coventry Ordnance Works 13.5in turrets. *(World Ship Society)*

## APPENDIX FOUR

# 13.5-inch guns for Turkish Coast Defences

BATTLECRUISER *TIGER* was required to be broken up by 1932 under the terms of the Washington Treaty, but the Admiralty decided to retain her four twin 13.5in mountings for possible future use. Early in 1932, they were removed at Rosyth Dockyard before she was towed to nearby Inverkeithing for scrapping. *Iron Duke* was required to be partially demilitarised at the same time so her side belt armour and B and Y mountings were removed before she became a gunnery training ship. All six mountings were stored at Rosyth, probably close to the 250-ton crane which would have lifted *Tiger*'s out. The guns themselves were probably removed to the gun storage yard which each dockyard had.

In May 1938 the British Government granted Turkey a £6 million credit for buying armaments in Britain. Four destroyers and four submarines were ordered plus their associated armament. Turkey also wished to modernise its coast defences at the Dardanelles, and in June 1938 inquired about obtaining 15in guns. The Admiralty was initially not keen, possibly because they did not have enough 15in guns in reserve to sell (the Army was also using the same 15in for coast defences), and perhaps also because manufacturing new mountings would divert resources from its own programme. Originally the guns had cost £17,270 each. It offered instead three of the 13.5in mountings in store at Rosyth. Vickers-Armstrong would be able to convert them quite readily, and there were enough 13.5in guns and ammunition in store, as no front-line battleships now mounted them. Indeed, there were no fewer than 52 13.5in Mark V still on the Admiralty books in 1939, out of 206 originally made. Only six were afloat in *Iron Duke,* while eighteen were supposedly being retained for mounting in monitors. Three were eventually sent on old railway mountings to Dover in 1941.

Discussions took place between the Admiralty and the Treasury as to how much to charge for this ordnance.[1] Eventually it was agreed that three mountings and ten guns would be supplied at 50% of their original cost or £175,955. The guns would be nearly new with a projected life of over 400 equivalent full charges. APC and CPC shells in stock could also supplied (APC costing £142 each) but new propellant charges would be needed. The Turks would have to pay for all of these

[1] Turkey: Purchase of Guns and Ammunition from the Admiralty. T161/1396

*Iron Duke*'s B and Y twin 13.5in mountings were removed at Devonport in 1932. Y's gunshield seen here may well have been part of the sale of three 13.5in mountings to Turkey for coast defences in 1939. *(Tyne & Wear Archives)*

plus the mounting conversion costs and shipment to Turkey out of the £6M credit.

The contract to convert the twin mountings from *Tiger* and *Iron Duke* was placed with Vickers-Armstrong by the Turkish Ministry of National Defence dated 24 March 1939. It was signed by F H Yapp, a V-A director, and Tevfik Rustu Aras, the Turkish Ambassador in London. The scope of the contract included:[2]

- Removal of the three mountings and ten guns from Rosyth to Elswick (£19,550)
- Supplying necessary parts for the conversion (£128,000)
- Dismantling the mountings and fitting new or renovated parts and re-assembling (£64,000)
- Supplying spares and accessories (£17,600)
- Delivering to jetty at Elswick free on board ship (£2700)

This provisional price totalling £231,850 was to be paid in ten instalments: 10% on contract signing, then for each of the three mountings 15% when 50% of parts had been re-erected, 10% when shop trial completed and 5% when loaded on to ship. These instalments would be paid when the work done had been certified by the Admiralty, drawing down the monies from the British Treasury credit. This sum was in addition to the initial purchase from the Admiralty of the 10 guns (£63,500) and three mountings (£112,500) and 3000 rounds of APC (worth £213,780, possibly both 13.5in and 15in). Adding in something for 1800 propellant charges would bring the total to about £750,000 to which would have to be added shipment and construction costs of the three gun emplacements in Turkey, say £1 million in total. Delivery of the first mounting was to be in fifteen months, the last in twenty-four, *ie* March 1941. The delivery times could be extended in the event of strikes, fire, accidents, Act of God or war. There was an eighteen-month guarantee for faulty parts or workmanship. Fifty sets of handbooks and 300 sets of drawings were also to be supplied, all in Turkish.

The guns were the regular RN Mark V, 45-calibre, 76 tons, firing a 1400lb APC at 2491 ft/sec with a 297lb charge. As the maximum elevation remained at 20°, the range was only 23,740 yards, though siting them on high ground would extend the range. The mounting itself was adapted to train through 300° at 2° per sec. Each weighed 553 tons, which suggests that the gunhouse armour remained the same as when built; it would only have added to the cost to change it.

A second contract was signed the same day for two single 15in Mark II coast defence mountings to be built new at Barrow, similar to those mounted at Singapore. The contract to deliver free on board at Barrow amounted to £281,350, which was substantially more than two Singapore mountings. Delivery of the first was to be by December 1941 and the second by April 1942. Only two guns would be supplied, already purchased from the Admiralty at £19,342 each, plus 400 rounds of ammunition. They were existing Mark I guns, *ie* 42-calibre, 100 tons, firing a 1938lb 6 crh APC at 2400 ft/sec, giving a range of 36,900 yards at the maximum elevation of 45°.

The Turkish Ministry signed another contract with Vickers-Armstrongs on 31 January 1941 for the transport and erection of the 13.5in and 15in mountings.[3] Consulting engineers Sir Alexander Gibb & Partners had been engaged for the design of the emplacements in Turkey and the provision of landing places, wharves and railway routes to the emplacements, and ordering the necessary equipment and materials. V-A were also responsible for making a 120-ton gantry crane and for erecting the mountings and carrying out proof firings. Erection was to be completed in August 1941 for the first 13.5in, October for the second and December for the third. The first 15in was to be erected by March 1942 and the second by July, all somewhat ambitious dates. The proof firings were to be of four rounds each from 0° to maximum elevation. Holloway Brothers (London) Ltd would be responsible for the actual construction of emplacements, employing largely Turkish labour but under their supervision. Three sites had been selected for the 13.5in: Çimenlik (in Cannakale) on the Asian side, Alçitepe (known as Achi Baba during the 1915 Gallipoli campaign) a high hill on the European side and Dedebayir, on the south side of the Sea of Marmara.

Most of the 13.5in components had been delivered to Elswick from Rosyth by the third quarter of 1939, with the first mounting re-erected in a pit by December.[4] The twin mountings were being refurbished but the hoist mechanisms probably had to be adapted to loading from magazines at the sides rather than below. The first two mountings were fitted with four guns from *Tiger* and were ready for shipment in the first quarter of 1941 with work about to start on the third. But shipment was now a problem. A ship was needed with wide hatches and heavy lift derricks, as there was no heavy cranage at the Dardanelles. There were only three suitable ships, Christen Smith Bel ships with 100-ton derricks, which could not be risked in the contested waters of the eastern Mediterranean. With delivery thus impossible, the Admiralty suspended both the 13.5in and 15in contracts in May 1941, by which time Elswick had spent £175,000.[5] Also suspended was the fire control gear and refurbished rangefinders from Crayford and the supply of ammunition.

A considerable amount of work had been done on the 15in mountings at Barrow before that contract too was suspended: 85% of the cradles and slides, 75% of the turntable, 70% of the training rack and gear, 60% of the loading gear and 45% of the gun shields.[6] These components were stored at Barrow to await a decision on their future. Meantime Barrow had completed two of the Turkish destroyers in 1941 (the other two at Denny) but they were taken over by the Admiralty to become *Ithuriel* and *Inconstant.* Barrow also completed the four Turkish submarines in 1941, but only two were delivered, the Admiralty taking over two as *P.614* and *P.615.*

There was little space at the congested Elswick site to store the two completed 13.5in mountings at the end of 1941. The working chambers were moved to V-A's Naval Yard (probably under the 250-ton

[2] 13.5" and 15" Coast Defence Equipments – Turkey. ADM 1/12529

[3] *Ibid*

[4] Cambridge University Library. Manuscripts. Vickers Archive 194

[5] Tyne & Wear Archives. VA.DS/3/5/15

[6] Gun Emplacements for Coast Defence Arms (Dardanelles). ADM 1/12146

crane), Scotswood took the roller ring segments, Palmers Hebburn the flashtight doors, girders, shell rails and bins, while Richardsons Westgarth at Sunderland took the hydraulic equipment, training rack, compressors, loading cages, grabs and bogies plus two diesel engines for powering the equipments. The actual 13.5in guns remained at Elswick. In 1942 three of them were relined with extra long 8in liners to become 90-calibre hypervelocity guns capable of firing 256lb shells up to 105,000 yards with 4500ft/sec muzzle velocity.[7] They were originally N$^{os}$ 590, 647 and 648 from *Iron Duke* but renumbered 172, 170 and 171. One went to Shoeburyness for trials. The Great Western Railway Swindon workshops were given a contract in April 1941 to produce a suitable coast defence mounting for the gun, capable of elevating up to 50°. But they struggled with this novel work, so Elswick had to take much of it over, with pit trials there in December 1942.[8] One gun was installed on a railway mounting near Dover early in 1943, manned by the RMA Siege Regiment. It had such a high rate of bore wear that after only 25 efc it was sent back to Elswick for relining and replaced by the third gun in May 1944. That too was condemned in 1945 after 29 efc. Hogg gives the mounting as adapted from Mark II* N$^{o}$ 27; if actually Mark II** this would have been one of *Iron Duke*'s, so quite likely adapted from the third mounting planned for Turkey at Elswick.

Although work had continued on two of the Dardanelles emplacements, the Turks were beginning to have second thoughts about continuing with the mountings whose delivery might now be delayed until after the war. In December 1942 the Turkish Naval Attaché wrote to H V Markham, Secretary to the Admiralty, proposing to cancel the contracts. That set off several rounds of discussion in official circles, DNO, M[ilitary] Branch, Director of Contracts, Director of Armament Supply, the Foreign Office, the Treasury and the British Ambassador in Turkey about the situation. Suggestions were made including asking if the Turks could take delivery at Port Said or sending them six 6in CD mountings instead. It would be less risky to deliver the 13.5in mountings to Port Said where there was a 150-ton floating crane to discharge the components. But the Turks had no suitable ship to take them on to Turkey, while the rail system was quite inadequate for transporting such large items. The War Office did not want them, and the components were taking up valuable factory storage space. While scrapping would yield over 2000 tons of valuable steel, there was a question whether the material now belonged to the Turks, it having been paid for out of the British credit. According to the mounting contract payment terms, only about half would have been paid to V-A by then, but all ten of the 13.5in guns had been paid for. The Admiralty was advised that legally there was no need to reimburse the Turks. Only some of auxiliary equipment was found to be re-usable, with two of the diesels being used for new floating grain elevators. For the 15in mountings again only about half the instalments had been paid (so Barrow were probably out of pocket) but none for the 15in guns.

Discussions continued throughout 1943 without any firm response from Turkey. By April 1944 it was clear that they no longer wanted the heavy mountings, while from the British side there was now less need to keep the Turks sweet with the Germans being driven out of the Aegean area. The files peter out early in 1945. The assumption must be that no more payments were made or received and it is not known if the Turks ever paid off all the outstanding credit post war, or if most was written off. Elswick were probably out of pocket for their 13.5in work, as their 1947 accounts still recorded £176,615 spent but only £136,479 received.[9] It seems most likely that all the 13.5in and 15in material was scrapped about 1945–47, as well as the 40 plus 13.5in guns.

[7] Hogg, Ian V. *British and American Artillery of World War 2*. Arms & Armour Press, 1978; and ADM 1/17228

[8] Tyne & Wear Archives, DS.VA6/86/27

[9] Tyne & Wear Archives, DS.VA3/5/20

# Sources and Bibliography

OVERALL MANY HUNDREDS OF SOURCES have been consulted, the key ones noted below. Other references are cited in the text or notes. Particularly useful have been the many pages of my own notes, started when researching monitors in the 1960s, and never thrown out. These were not always referenced and indeed some original documents have since been destroyed. There are a large number of books that refer to heavy naval guns, for example those on classes of ship or individual ships, but few address gun mountings in any depth, despite their key role in procuring ships. The most valuable have been:


Buxton, Ian. *Big Gun Monitors: Design, Construction and Operations 1914–1945*. Seaforth Publishing, 2008.

Buxton, Ian and Johnston, Ian. *Battleship Duke of York: An Anatomy from Building to Breaking*. Seaforth Publishing, 2021.

Campbell, John. *Naval Weapons of World War Two*. Conway Maritime Press, 1985.

Friedman, Norman. *Naval Weapons of World War One*. Seaforth Publishing, 2011.

Hodges, Peter. *The Big Gun: Battleship Main Armament 1860–1945*. Conway Maritime Press, 1981.

Johnston, Ian and Buxton, Ian. *The Battleship Builders: Constructing and Arming British Capital Ships*. Seaforth Publishing, 2013.


Other books and articles plus references in endnotes:


Dawson, A T. *The Engineering of Ordnance*. Junior Institution of Engineers, 1909.

Dodson, Aidan. *The Windfall Battleships*. Seaforth Publishing, 2023

*Engineering*, Weekly journal. Articles as cited, especially on Vickers.

Friedman, Norman. *The British Battleship 1906–1946*. Seaforth Publishing, 2015.

Friedman, Norman. *Naval Firepower: Battleship Guns and Gunnery in the Dreadnought Era*. Seaforth Publishing, 2008.

Friedman, Norman. *British Cruisers: Two World Wars and After*. Seaforth Publishing, 2010.

Marland, Peter. 'Postwar Weapons in the Royal Navy'. *Warship*, 2015

*The Naval Annual* (Brassey's, various years), J Griffin & Co.

Raven, Alan & Roberts, John. *British Battleships of World War Two*. Arms & Armour Press, 1976

Richardson, Alex. 'Vickers, Sons & Maxim Ltd: Their Works and Manufactures'. *Engineering*, 1902

Trebilcock, Clive. *The Vickers Brothers*. Europa Publications, 1977.

*Warship*. Quarterly 1977–89, then Annual. The years 1981–85 include a series of articles on RN guns.


The most useful archival sources have been:

The National Archives, ADM, CAB, SUPP, WO, MUN, T(reasury) references, especially gun mounting handbooks

Tyne & Wear Archives, particularly Armstrong and Vickers records

Cambridge University Library, Manuscripts Department. Vickers archive

National Maritime Museum, Plans and Photos, Woolwich. Ships Covers and plans

BAE Systems, Barrow shipyard archives. Financial records. Access by special permission

Scottish Business Archives, Glasgow University. Shipbuilder records

National Records of Scotland. Clydebank photographs

Sheffield City Archives. Vickers records

Imperial War Museum, London. Gunnery publications

Admiralty Library, Portsmouth. Gunnery publications

National Museum of the Royal Navy Library and HMS *Excellent*, Portsmouth. Gunnery publications

Portsmouth Royal Dockyard Historical Trust. Gunnery publications

Marine Technology Special Collection, Newcastle University. Shipbuilding and shipbreaking archive.


**ACKNOWLEDGEMENTS**

For their ever-ready willingness to help:

Ian Johnston, for Ships' Cover files and improving the drawings

Roy Metcalfe, for researching the National Archives and taking images

Peter Marland, for advice on post WW2 gun mountings

John Roberts, for gun mounting drawings

Andrew Choong, for finding those elusive NMM documents

Richard Holme, for finding photographs

Jenny Wraight, Admiralty Library

Pete Cannon, for Australian cruisers

John Jeremy, for Australian gun mountings

and of course the staff of the Archives and Libraries listed under Sources.

# Index